Making New Out of the Old: Recent Biological Advances on Mesozoic Marine Reptiles

Guest Editor

Nathalie Bardet

MDPI

Basel • Beijing • Wuhan • Barcelona • Belgrade • Novi Sad • Cluj • Manchester

Guest Editor
Nathalie Bardet
CR2P – Centre de Recherche
en Paléontologie - Paris
Muséum National d'Histoire
Naturelle
Paris
France

Editorial Office
MDPI AG
Grosspeteranlage 5
4052 Basel, Switzerland

This is a reprint of the Special Issue, published open access by the journal *Diversity* (ISSN 1424-2818), freely accessible at: https://www.mdpi.com/journal/diversity/special_issues/238GA5HNT9.

For citation purposes, cite each article independently as indicated on the article page online and as indicated below:

Lastname, A.A.; Lastname, B.B. Article Title. *Journal Name* **Year**, *Volume Number*, Page Range.

ISBN 978-3-7258-4771-6 (Hbk)
ISBN 978-3-7258-4772-3 (PDF)
https://doi.org/10.3390/books978-3-7258-4772-3

Cover image courtesy of Princeton University Art Museum, Princeton, NJ, USA
"Early Jurassic Marine Reptiles" by Benjamin Waterhouse Hawkins

Contents

About the Editor . . . . **vii**

Preface . . . . **ix**

José Patricio O'Gorman
How Elongated? The Pattern of Elongation of Cervical Centra of *Elasmosaurus platyurus* with Comments on Cervical Elongation Patterns among Plesiosauromorphs
Reprinted from: *Diversity* **2024**, *16*, 106, https://doi.org/10.3390/d16020106 . . . . **1**

Nikolay Zverkov, Maxim Arkhangelsky, Denis Gulyaev, Alexey Ippolitov and Alexey Shmakov
Callovian Marine Reptiles of European Russia
Reprinted from: *Diversity* **2024**, *16*, 290, https://doi.org/10.3390/d16050290 . . . . **20**

María Eurídice Páramo-Fonseca, José Alejandro Narváez-Rincón, Cristian David Benavides-Cabra and Christian Felipe Yanez-Leaño
Exceptional In Situ Preservation of Chondrocranial Elements in a Coniacian Mosasaurid from Colombia
Reprinted from: *Diversity* **2024**, *16*, 285, https://doi.org/10.3390/d16050285 . . . . **69**

Marta S. Fernández, Lisandro Campos, Agustina Manzo and Evangelos Vlachos
Bone Connectivity and the Evolution of Ichthyosaur Fins
Reprinted from: *Diversity* **2024**, *16*, 349, https://doi.org/10.3390/d16060349 . . . . **81**

Ida Bonnevier Wallstedt, Peter Sjövall, Ben Thuy, Randolph G. De La Garza, Mats E. Eriksson and Johan Lindgren
Skin Anatomy, Bone Histology and Taphonomy of a Toarcian (Lower Jurassic) Ichthyosaur (Reptilia: Ichthyopterygia) from Luxembourg, with Implications for Paleobiology
Reprinted from: *Diversity* **2024**, *16*, 492, https://doi.org/10.3390/d16080492 . . . . **93**

Rémi Allemand, Michael J. Polcyn, Alexandra Houssaye, Peggy Vincent, Camilo López-Aguirre and Nathalie Bardet
First Virtual Reconstruction of a Mosasaurid Brain Endocast: Description and Comparison of the Endocast of *Tethysaurus nopcsai* with Those of Extant Squamates
Reprinted from: *Diversity* **2024**, *16*, 548, https://doi.org/10.3390/d16090548 . . . . **109**

Nathalie Bardet, Valentin Fischer, Nour-Eddine Jalil, Fatima Khaldoune, Oussama Khadiri Yazami, Xabier Pereda-Suberbiola and Nicholas Longrich
Mosasaurids Bare the Teeth: An Extraordinary Ecological Disparity in the Phosphates of Morocco Just Prior to the K/Pg Crisis †
Reprinted from: *Diversity* **2025**, *17*, 114, https://doi.org/10.3390/d17020114 . . . . **125**

Femke M. Holwerda, Mark T. Mitchell, Madelon van de Kerk and Anne S. Schulp
Mosasaur Feeding Ecology from the Campanian Bearpaw Formation, Alberta, Canada: A Preliminary Multi-Proxy Approach
Reprinted from: *Diversity* **2025**, *17*, 205, https://doi.org/10.3390/d17030205 . . . . **170**

Michael J. Polcyn, John A. Robbins, Anne S. Schulp, Johan Lindgren and Louis L. Jacobs
The Evolution of Mosasaurid Foraging Behavior Through the Lens of Stable Carbon Isotopes
Reprinted from: *Diversity* **2025**, *17*, 291, https://doi.org/10.3390/d17040291 . . . . **184**

About the Editor

Nathalie Bardet

Nathalie Bardet is a vertebrate paleontologist, CNRS Research Director (Prof. Dr.), working at the *Centre de Recherche en Paléontologie de Paris* (CR2P) in the *Muséum National d'Histoire Naturelle* (MNHN) of Paris, France. She is a specialist in Mesozoic marine reptiles (mosasaurs, ichthyosaurs, plesiosaurs) and her fieldwork is mainly concentrated in the Late Cretaceous Phosphate deposits in North Africa and the Middle East, where she has worked for almost 30 years. Her main research interest is the comparison of marine reptile faunas from the Late Cretaceous of the northern (Europe) and southern (North Africa and Middle East) margins of the Mediterranean Tethys, which has allowed her to highlight different faunal paleoprovinces possibly linked to paleolatitudes, and also to address issues related to mass extinctions, especially those of the K/Pg (66 My). She also works on marine reptile faunas from the Jurassic of Europe. She mainly uses comparative anatomy for her studies, and her skills of interest include systematics, phylogeny, paleobiogeography, paleobiology, paleoecology, and history of sciences. She has published more than 150 scientific papers and erected more than 20 new genera/species of marine reptiles, mainly mosasaurs from the Phosphates of Morocco. She has been president of the French Paleontological Association (APF) for 4 years, has supervised 4 PhD thesis students, and has trained Master's students for more than a decade of. She also enjoys science history, namely aspects linked to the first discoveries of marine reptile remains at the turn of the 18th and 19th centuries, as well as science divulgation, having been involved in videos and movies, exhibitions, interviews in radios and TV, and writing about 40 popular science papers and books, the main being *Ocean Life in the Time of Dinosaurs* (Princeton University Press, 2023), with the original French version (*La Mer au temps des Dinosaures*, Belin, 2021) having been awarded by the French Marine Academy.

Preface

In January 2025, as this Special Issue was nearly completed, our colleague María Páramo-Fonseca (1963–2025), an internationally recognized Colombian specialist in marine reptiles, passed away. A brilliant paleontologist and geologist, Maria was a pioneer in her country, having described most of the Mesozoic marine vertebrates from Colombia. She also trained numerous young researchers who, we hope, will carry on her scientific and social work of promoting Colombian paleontological heritage.

As a colleague and, above all, as a friend and "scientific sister" of more than 30 years, I am very sad to hear of her early loss. However, I am happy that one of her final contributions can form part of this Special Issue, which is dedicated to this exceptional woman.

Nathalie Bardet
Guest Editor

diversity

MDPI

Article

How Elongated? The Pattern of Elongation of Cervical Centra of *Elasmosaurus platyurus* with Comments on Cervical Elongation Patterns among Plesiosauromorphs

José Patricio O'Gorman [1,2]

[1] División Paleontología Vertebrados, Museo de La Plata, Universidad Nacional de La Plata, Paseo del Bosque s/n., La Plata B1900FWA, Argentina; joseogorman@fcnym.unlp.edu.ar

[2] CONICET—Consejo Nacional de Investigaciones Científicas y Técnicas, Buenos Aires C1425FQB, Argentina

Abstract: Elasmosaurids comprise some of the most extreme morphotypes of plesiosaurs. Thus, the study of their neck and vertebrae elongation patterns plays a crucial role in understanding the anatomy of elasmosaurids. In this study, the taphonomic distortion of the holotype of *Elasmosaurus platyurus* and its effects on the vertebral length index (VLI) values are evaluated, and a new index to describe the neck is proposed (MAVLI = mean value of the vertebral elongation index of the anterior two-thirds of neck vertebrae). The results provide a strong foundation for a new scheme of neck elongation patterns that divide the diversity of the neck elongation of plesiosauriomorphs into three categories: not-elongate (MAVLI < 95 and Max VLI < 100), elongate (125 > MAVLI > 95 and 100 < Max VLI < 135), and extremely elongated (MAVLI > 125 and Max VLI > 135).

Keywords: Plesiosauria; Elasmosauridae; *Elasmosaurus platyurus*

1. Introduction

Elasmosaurid comprises some of the most extreme plesiosauromorph morphotypes [1,2] with *Elasmosaurus platyurus* being recognized as the vertebrate with the highest cervical account (72 cervical vertebrae [2,3]). Although more than 40 cervical vertebrae are shared by all members of the family, a high variability in both cervical count and the relative elongation of cervical centra are recorded among elasmosaurids [3–6].

Among the records of plesiosaurians, the cervical centra are the most frequently preserved elements due to their high number per individual and preservational characteristics (i.e., box-like compact bones [7,8]). Therefore, cervical vertebrae have played an important role in the taxonomy of elasmosaurids since the 19th century (see [8] for a detailed review). However, after Welles [8] and its comprehensive discussion, the use of isolated cervical centra as diagnostic elements at the genus level has been dismissed. Despite this, the study of variation in vertebral elongation and the patterns of elongation of complete cervical regions among elasmosaurids continued during the second half of the 20th century and the first years of the 21st century. Brown, [9], created the Vertebral Length Index (VLI), while [1] used the term "plesiosauromorph" to describe the plesiosaur morphotype characterized by a long neck and a small skull. Later, O'Keefe and Hiller [4]) studied the patterns of the VLI of cervical regions among elasmosaurids. Their general conclusions followed those of Welles [7] about the low diagnostic value of isolated vertebrae at the genus or species level and drew attention to the high variability in neck elongation, commenting that "variability reigns" [4] (p. 216). However, they made the first steps in the classification of neck elongation patterns, indicating that *Elasmosaurus platyurus* and *Styxosaurus* sp. (their "elongate" group) strongly differed from other elasmosaurids, showing a pattern of a high number of cervical vertebrae and vertebral elongation values far higher than in other elasmosaurids. Despite some difficulties related to the extreme variability observed, O'Keefe and Hiller attempted to create criteria to differentiate between these two groups based on complete and incomplete necks. Additionally, those authors stated that the "elongate" group was restricted to the Western Interior Seaway (WIS) [4]. The absence of the recognition of a third (with short vertebrae) group is related to the systematic affinities of aristonecines considered by the authors. Later, the recognition of the elasmosaurid affinities of aristonectines and the description of *Nakonanectes bradti* increased the known range of variation in cervical elongation patterns among elasmosaurids by adding elasmosaurids with relatively shorter cervical centra [6,10,11]. Otero et al. [12] proposed a modification in neck elongation terminology, using the term "extremely elongated form" to nominate the member of the "elongate" group of O'Keefe and Hiller [4] and adding the category "plesiomorphic elasmosaurids" (Otero et al. [12] (p. 254), also called "intermediate elasmosaurids" in the same contribution (Otero et al. [12] (p. 256), for the more typical elasmosaurid neck. They also used the term "aristonectine" to group the neck elongation pattern observed among elasmosaurids that are members of the Aristonectine subfamily (*Aristonectes parvidens*, *Aristonectes quiriquinensis*, *Kaiwhekea katiki*, and *Wunyelfia maulensis)* for the first time. Otero [5] added the following names for the vertebral morphotypes characteristic of the latter three groups: "can-shaped" cervical vertebrae, "*Cimoliasaurus*-grade" cervical vertebrae, and "aristonectine-type" cervical vertebrae.

This short introduction can be summarized as follow: (1) the neck elongation pattern (in terms of the number of cervical vertebrae and elongation of vertebral centra) has played a central role in the study of elasmosaurid diversity, and (2) different terms were proposed to describe this diversity (Tables 1 and 2). This long story all started with the description of *Elasmosaurus platyurus*, which coincidentally is also one of the extreme forms in terms of cervical elongation [4].

Table 1. Comparison of terminology regarding the neck elongation pattern of elasmosaurids.

	Range VLI	Average VLI	Other Features	Vertebra Type
		O'Keefe and Hiller, 2006		
Non-elongate group	30	~100		*Cimoliasaurus*-grade
Elongate Group	60–100	125–138	Some cervical VLI 150 to 200/erratic variation	Can-shape (some vertebrae)
		Otero et al., 2015		
Plesiomorphic/ Intermediate elasmosaurids		99	Cervical count~60	*Cimoliasaurus*-grade
Extremely long-necked forms		135	Cervical count~72–76	Can-shape
Aristonectine		80	Cervical count~43	Aristonectine

Table 2. Approximate equivalence between [4,12] categories of neck elongation pattern of elasmosaurids.

O'Keefe and Hiller, 2006 [4]	Otero et al., 2015 [12]
Non-elongate group	Plesiomorphic elasmosaurids/intermediate elasmosaurids
Elongate group	Extreme long-necked elasmosaurids
-	Aristonectinae

The type genus of the family Elasmosauridae, *Elasmosaurus*, comprises only one species, *Elasmosaurus platyurus* (ANSP 10081, holotype). This species is a key member of the "elongate" group among elasmosaurids, characterized by its extremely long neck, as evidenced by the high number of cervical vertebrae and the values of the vertebral elongation index (VLI). A detailed examination of the cervical vertebrae count has been conducted by [2], revealing 72 cervical vertebrae. However, less attention has been given to the taphonomic distortion of the holotype specimen (ANSP 10081) and its impact on the VLI values of the cervical vertebrae. It is clear that the taphonomic distortion of ANSP 10081, the holotype of *E. platyurus*, was evident since its original description by Cope. He noted that "the whole skeleton has been under considerable pressure so that most of the ribs have been pressed flat out the vertebrae; the long parapophyses (cervical ribs) of the cervicals have most of them been fractured at their bases and compressed, those of opposite sides thus approaching more nearly in the form of the chevron bones than otherwise would have done" [13] (p. 48). Cope also mentioned that "the proximal cervicals are obliquely flattened by the pressure" [13] (p. 48) and that "many of the preserved ribs have been pressed upon the vertebrae and crushed" (Cope, 1869 [13] (p. 49)). However, a few years later Cope wrote that "the other cervicals have the bodies naturally flat, with the articular surface much less so than the median portions" [14] (p. 80) without any mention of taphonomic distortion.

Later, Welles [15] (p. 185) commented on ANSP 10081 (indicated as ANSP 18001 by [7,8]), stating that it is the only known elasmosaurid with compressed cervicals and suspecting that the compression is due to crushing. However, a decade later, in Welles [7]: Table 1 the measurements L, H, and B of the vertebral centra of ANSP 10081 were provided, but the strong deformation of the cervical centra was not explicitly addressed, and only a few values were given in italics, indicating estimated values. This suggests a possible change in Welles's perception of the extent of the deformation of the specimen. Sachs later mentioned that "most vertebrae are laterally compressed, with this compression being stronger in the middle cervical region than in the anterior or posterior section." [16] (p. 97). Finally, O'Keefe and Hiller acknowledged that previous researchers relied on the assumption that they included only measurements that appeared close to the original values. However, they also noted that it would be "naive" to assume that there is no preservation error in the data, although they believed the generated error was minimal [4].

In summary, the recognition of the taphonomic distortion of the ANSP 10081 has been variable among researchers. However, while several researchers have acknowledged the taphonomic distortion of the holotype, no attempt has been made to quantify or analyze the influence of this factor on the VLI values.

The main goal of this contribution is to analyze the taphonomic distortion of the ANSP 10081, re-evaluate the neck elongation pattern of *Elasmosaurus platyurus*, simplify the definition and terminology of neck elongation patterns, and extend the conclusions to other plesiosauromorphs to describe their evolutionary history.

1.1. Anatomical Abbreviations

BHI, breadth to length index; BI: breadth index; cr, cervical rib; HI, height index; MAVLI: mean of the VLI values of the cervical vertebrae of the anteriormost two-thirds of the neck; nc, neural canal; VLI, Vertebral Length Index.

1.2. Institutional Abbreviations

AMNH, American Museum Natural History, New York, USA; CM, Conway Museum, Christ Church, New Zealand; DMNS, Denver Museum of Nature and Science, Denver, Colorado, USA; MGUAN, Museu de Geologia da Universidade Agostinho Neto, Luanda, Angola; ANSP, The Academy of Natural Sciences of Drexel University, Philadelphia, USA; MLP, Museo de La Plata, Buenos Aires Province, Argentina; NHMUK, Natural History Museum, London, United Kingdom.

2. Materials and Methods

2.1. Methods

2.1.1. Indices

Linear measurements were taken using a mechanical caliper, which allowed for an accuracy of 0.1 mm. The indices considered are those proposed by Welles (1952), which take into account the ratio between the centrum length (L) and height (H) ($HI = 100 \times H/L$), and the ratio between centrum breadth (B) and length (BI = 100 × B/L). Both breadth and height were measured on the posterior articular face. The Vertebral Length Index (VLI = 100 L/(0.5 × (H + B))) proposed by Brown was also used [9]. Additionally, a new index is defined: MAVLI = mean of the VLI values of the anteriormost two-thirds of the neck.

2.1.2. Retrodeformation

The process of restoring the original form of a fossil body is called retrodeformation. General methods for retrodeformation require a final shape to reach, the restitution of bilateral symmetry, or an external source that indicates the direction and magnitude of strains [17]. Complete retrodeformation is unachievable as there is no certainty about the original shape, the lateral compression maintains an approximate bilateral symmetry (Figure 1), and there is no external indicator of the main strain direction. However, a complete retrodeformation is not the objective attempted here. Instead, it is intended to determine the degree to which taphonomic distortion affects the VLI.

The former problem was dealt with by following an approach divided into three steps. First, the material is described, and the main direction of distortion is determined (based on the reduction of the neural canal and the direction of cervical ribs, the main direction is lateral compression, mainly affecting the B value; see description, Supplementary Materials 1). It is clear that each area of the vertebrae was affected in different grades, however, as the objective here is punctual (retrodeformed the B values) the accounting for anisotropy in distortion is unnecessary for our current purposes.

The second step is to estimate the value of B (cervical centrum width) prior to the deformation. To test the tool used, the series of the B values of cervical vertebrae in three mostly undistorted or slightly distorted specimens of elasmosaurids were analyzed: the elasmosaurids with high vertebral elongation, Elasmosaurinae *Styxosaurus* sp. (AMNH 5835, from [18], Supplementary Materials 2 Table S1), and the more conservative weddellonectians *Tuarangisaurus* sp. (CM Zfr 115, from [4], Supplementary Materials 2 Table S2) and *Vegasaurus molyi* (MLP 93-I-5-1, from [19], Supplementary Materials 2 Table S3). The objective was to analyze the significance of the regression line between B value and cervical position and its statistical significance (referred to as "complete analysis" hereafter). An additional regression line was calculated using only the B values of the first and last ten cervical centra (referred to as "reduced analysis" hereafter). As both regression lines are statistically significant but extremely similar, it is assumed that the reduced regression line could be used as an accurate proxy model of the complete regression lines, generating a useful tool to estimate B values when there is a gap in the middle cervical series of elasmosaurids but the number of missing values (missing or distorted vertebrae) is known.

Following this strategy, the original sequence of B values of ANSP 10081 is obtained using a regression line of the first and last ten cervical centra of ANSP 10081, as they are less deformed (first ten cervicals, ST2) or nearly undeformed (last ten cervical, Supplementary Materials 1 Table S1). Lastly, using the estimated B values of ANSP 10081, an estimated series of VLI is obtained (Supplementary Materials 1 Table S1). All of the analyses were carried out with R software package stats and nortest [20,21]. The complete script is given as supplementary data in Supplementary Materials 3. All of the analyses were performed using R software [21], complete script and data used is given as Supplementary Materials 3.

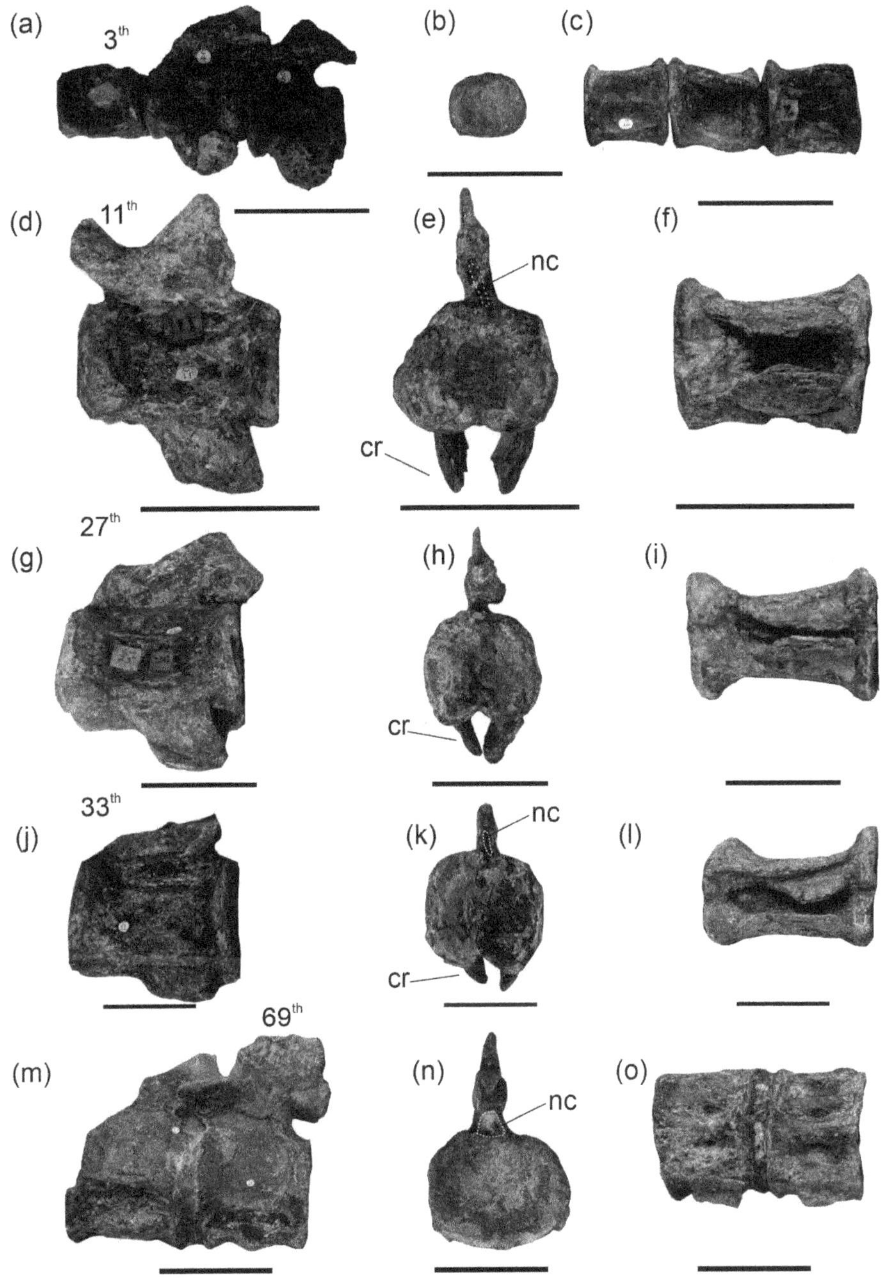

Figure 1. Cervical vertebrae of *Elasmosaurus platyurus* (ANSP 10081). (**a**) The 3–5th cervical vertebrae in left lateral view, (**b**) 3rd cervical vertebra in posterior view, and (**c**) 3–5th cervical centra in ventral view. (**d–f**) The 11th cervical centra in (**d**) left lateral, (**e**) posterior, and (**f**) ventral views. (**g–i**) The 27th cervical centra in (**g**) left lateral, (**h**) posterior, and (**i**) ventral views. (**j–l**) The 33rd cervical vertebra in (**j**) left lateral, (**k**) posterior, and (**l**) ventral views. (**m–o**) The 68–69th cervical vertebrae in left lateral view. (**n**) The 69th cervical vertebra in posterior view. (**o**) The 68–69th cervical vertebrae in ventral view. Scale bar = 50 mm. cr, cervical rib, nc, neural canal.

3. Results

3.1. Systematic Paleontology

Sauropterygia Owen, 1860 [22]
Plesiosauria de Blainville, 1835 [23]
Elasmosauridae Cope, 1869 [13]
Elasmosaurus platyurus Cope, 1868 [24]

Materials: ANSP 10081, skeleton, comprising both premaxillae, part of the maxilla, two maxillary fragments with teeth, anterior part of dentaries, three jaw fragments, and indeterminable cranial fragments, 72 cervical vertebrae, including the atlas–axis complex, three pectoral, six dorsal, four sacral, and 18 caudal vertebrae, and rib fragments. Additionally, originally preserved pectoral and pelvic girdles are now missing.

Locality and Horizon. Near McAllaster, Logan Country, Kansas. Sharon Springs Shale Member, Pierre Shale (lower Campanian).

3.2. Numeration and Order of Vertebrae

The vertebrae are labeled using paper labels, inked in black and labelled with white circles (Figures 1 and S1). None of these are completely coincident. Here, the latter labels (in white circles) are followed, except for vertebrae 9 and 10, which are interpreted as inverted. Additionally, half of a centrum relocated by [2] is intercalated as vertebrae 14 (between the 14th and 15th labeled in white circles).

3.3. Taphonomic Distortion

The cervical region of ANSP 10081 is laterally compressed, but this compression is variable along the neck and affects the vertebral centra, cervical ribs, and neural canal. The atlas–axis complex shows a strong lateral compression, even showing an asymmetrical distorted neural canal (Figure 1e,k). The post-axis cervical vertebrae also show evidence of distortion as the cervical ribs are unnaturally displaced to the sagittal plane (vertebrae 7–10, Figure 1d–f). In ventral view, the third cervical vertebra shows the typical morphology, but in the third post, the cervical shows a deep concavity formed partially by the displacement of cervical ribs. The same is observed in vertebrae 11–18 (Figure 1f), with cervical ribs misplaced to the sagittal plane, almost touching in the midline, and the distortion of articular faces. The same vertebrae show that the neural canal is also strongly compressed. The same pattern is observed in vertebrae 19–27, although the depression generated in the ventral surface is more marked (Figure 1g–i). In the vertebrae 28–33 (Figure 1j–l), the lateral compression becomes more marked, and cervical ribs are also displaced toward the sagittal plane. The vertebrae 34–43 show an increased ventral concavity generated by the displacement of the cervical ribs. Vertebrae 48–54 are not well prepared. Therefore, it is impossible to evaluate them. Vertebrae 55–59 show a less depressed ventral surface, but the ventral foramina are elongated. The vertebrae 60–64 show a less deformed vertebra, and the ventral surface becomes flat to convex. Vertebrae 65–72 show the least distortion of the series with a wide neural canal and visible and elliptical ventral foramina (Figure 1m–o).

3.4. Measures of Cervical Centra

The values of L, H, and B are given in Supplementary Materials 1 Table S1 and Figure 2. In the cranio–caudal direction, the length (L) values increase approximately from the 45th to the 55th cervical vertebrae and then tend to decrease until the posteriormost cervicals. The height (H) values show a growing tendency in the cranio–caudal direction along the neck, with few exceptions. Finally, the width (B) values show a clear growing cranio–caudal sequence with a long plateau between the 35th and 50th cervical vertebrae and middle cervicals.

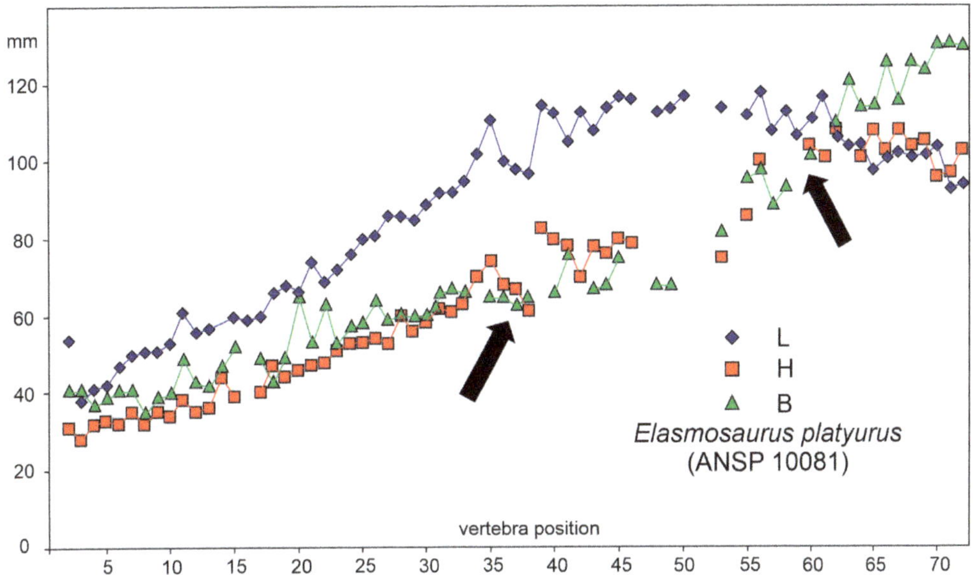

Figure 2. Values of length (L), height (H), and width (B) of cervical centra of the holotype of *Elasmosaurus platyurus* (ANSP 10081). Black arrow indicate the vertebral range with the main taphonomical distortion.

3.5. Estimation of Retrodeformed B Values in Elasmosaurus platyurus

The results of regression lines of the complete B value sequences and the first ten B and last ten B values against the cervical position are given in Table 3. All (complete and reduced data) show a high r^2 value and are statistically significant ($\alpha < 0.05$). Additionally, the comparison between each part indicates that, in each case, the regression lines show no significant differences. Therefore, the B value of the cervical centra of *Elasmosaurus platyurus* (ANSP 10081) is estimated using the regression lines of reduced data (Table 4).

Table 3. Regression line of vertebral position vs. B values for first ten and last ten cervical vertebrae: ac = linear coefficient of complete regression; ar = linear coefficient of reduced regression.

Regression	r^2	Regression Line	*p*-Value	ac/ar
Styxosaurus sp. (complete)	0.9919676	y = 1.795x + 27.41	$<2.10^{-16}$	0.988
Styxosaurus sp. (reduced)	0.9935	y = 1.816x + 29.625	$<2.10^{-16}$	
Tuarangisaurus sp. (complete)	0.9848	y = 1.08x + 36.8	$<2.10^{-16}$	1.018
Tuarangisaurus sp. (reduced)	0.9933	y = 1.06x +35.30	$<2.10^{-16}$	
Vegasaurus molyi (complete)	0.979	y = 1.411x + 27.534	$<2.10^{-16}$	1.016
Vegasaurus molyi (reduced)	0.9941	y = 1.389x + 25.837	$<6.71.10^{-16}$	

Table 4. Regression model obtained for B values of cervical centra of ANSP 10081 vs. vertebral position.

Regression	r^2	Regression Line	p-Value
Elasmosaurus platyurus (reduced)	0.9828	y = 1.46x + 34.61	$<2.20^{-16}$

The ten anteriormost cervicals of (ANSP 10081) show lateral compression (although less than the middle cervical vertebrae). Therefore, although arbitrary, 10% is added to the original B values in order to diminish its effect in the complete regression prior to the calculation of the regression line. The regression line of vertebral position vs. B values obtained for the first ten (corrected by adding 10% as they show a visible lateral compression) and the posteriormost ten (mostly uncompressed) cervical vertebrae of *Elasmosaurus platyurus* is given in Table 4.

4. Discussion

4.1. Elongation Pattern of *Elasmosaurus platyurus*

The first point to discuss is the differences in values between different versions of the data sets of the VLI values of ANSP 10081 (Table 5) and graphs (Figure 3). The differences between the data of Welles [8] and Sachs [16] are explicable as differences in measures (including the articular face, which is considered an observational error) and some changes in vertebral order. On the other hand, the differences between the latter two VLI sequences and the VLI estimated here are quite high and require other explanations. The main differences are explainable by the strong lateral compression suffered by the cervical vertebrae, almost closing the neural canal, displacing the ribs and reducing the B values. Additionally, the description reinforces this assumption regarding a significant distortion, and indicates that the most affected zone is the middle cervical zone (also commented on by Sachs, 2005 [16]), where the L values reach their maximum. The combination of the latter two elements (high L values and B maximum reduction by deformation) generates the highest VLI values of the neck of *Elasmosaurus platyurus*. These values correspond with the middle cervical region of *Elasmosaurus platyurus*.

Table 5. Comparison of VLI mean value, VLI standard deviation, MAVLI, and Max VLI.

Data Source	VLI Mean	VLI sta. Dev.	MAVLI	Max VLI
Welles, 1952, O'Keefe and Hiller, 2006 [4,8].	137.72	27.15	150	174
Sachs, 2005 [16].	136.07	22.73	143	175
New measures, this paper	129.72	23.22	141	160
Estimated measures, this paper	118.29	17.65	127	145

The estimation of the undeformed B and VLI values (Supplementary Materials 1 Table S1, Figure 3) indicates that *Elasmosaurus platyurus*, the iconic species of the elongated group, seems to show a VLI lower than previously considered. Therefore, the definitions of the neck elongation patterns should be modified. The division into fixed categories is not possible at all. However, the diversity seems to fall into three main groups following the ideas of [4,5], and some criteria appear to be valid. Particularly important is to remark on the mean value of VLI in the anterior two-thirds of the neck (MAVLI, hereafter). The new definitions of categories proposed here are given below (Table 6).

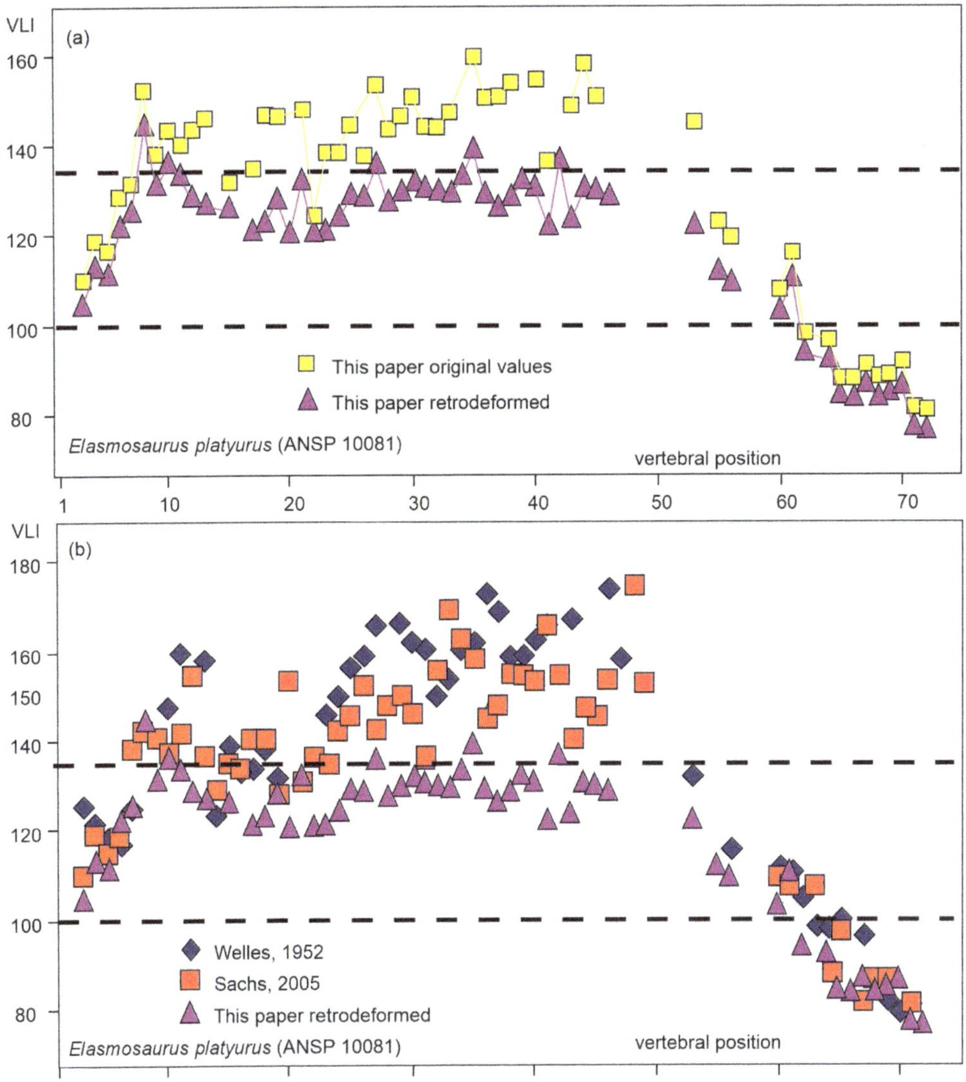

Figure 3. Values of VLI (vertebral length index = 100*L/((H + B)/2) of cervical centra of the holotype of (ANSP 10081) *Elasmosaurus platyurus*. (**a**) comparison of original VLI values and thse obtained after correction of taphonomic distortion; (**b**), Comparison of VLI values after [8,16] and this paper after retrodeformation.

Table 6. Definitions of categories of neck elongation patterns proposed in this contribution.

	MAVLI	Max VLI	HI, BI Middle Cervical
Non-elongated	MAVLI < 95	≤100	HI~100; BI > 140
Elongated	125 > MAVLI > 95	100 < max < 135	HI~70–80; BI < 140
Extremely elongated	MAVLI > 125	>135	Some vertebrae HI < 65

4.2. *Classification of Elasmosaurid Neck Elongation Pattern*

Each elasmosaurid is classified based on its neck pattern using the new criteria. In cases where no complete neck is available, the possible category is discussed (Tables 6 and 7). Among the basal elasmosaurids, *Jucha squalea* only preserves two groups of non-continuous cervical vertebrae, of which one of them (UPM NV 15, [25]) has VLI values well above 100 but lower than 125. Therefore, *J. squalea* seems to exhibit an elongated neck pattern. The holotype of *Callawayasaurus colombiensis* (UCMP 38349, Welles [8], Figure 4a (this paper) shows VLI values higher than 100 but never reaching 110, thus fitting the elongated pattern. There is no available list of vertebral measurements for *Zarafasaura oceanis*. However, based on photos of the specimen (WDC CMC-01, Supplementary Material 1, Figure S8), the cervical proportions indicate a non-elongated pattern.

Table 7. Plesiosauromoph taxa, MAVLI, MAX VLI, and Mean HI and BI values. For citation, see Table 8. * indicate that is based on incomplete cervical regions, ~ indicate approximate values.

Taxa	MAVLI	Max VLI	Mean HI/BI of Middle Cervicals
Neoplesiosauria			
Rhaeticosaurus mertensi	-	-	-
Plesiosauroidea			
Eoplesiosaurus antiquior	-	-	-
Plesiosauridae			
Plesiosaurus dolichodeirus	91	102	HI = 103/BI = 79
Microcleididae			
Microcleidus tournemirensis	129	143	HI = 61/BI = 86
Microcleidus homalospondilus	-	-	-
Seeleyosaurus guillelmiimperatoris	100	111	HI = 101/BI = 80
Cryptoclididae			
Ophthalmothule cryostea	109	118	HI = 77/BI = 93
Spitrasaurus wensaasi	105	129	HI = 67/BI = 87
Spitrasaurus larseni	-	105?	-
Muraenosaurus leedsi	103	109?	HI = 99/BI = 83
Tricleidus seeleyi	92	98	HI = 100/BI = 80
Cryptoclidus eurymerus	~85	~88	-
Colymbosaurus megadeirus	~85	~94	-
Djupedalia engeri	85	88	HI = 70/BI = 95
Abyssosaurus nataliae	87	98	HI = 98/BI = 130
Picrocleidus beloclis	102?	106?	HI = 80/BI = 89
Tatenectes laramiensis	-	-	-
Kimmerosaurus langhami	-	-	-
Colymbosaurus svalbardensis	-	-	-
Leptocleididae			
Brancasaurus brancai	94	110	HI = 96/BI = 95
Elasmosauridae			
Callawayasaurus colombiensis	~96.3	107	HI = 82/BI = 109
Jucha squalea	-	124	-
Zarafasaura oceanis	-	-	-
Euelasmosaurida			
Thalassomedon haningtoni	108	135	HI = 78/BI = 92
Libonectes morgani	108	128	HI = 67/BI = 101
Cardiocorax mukulu	-	-	-

Table 7. *Cont.*

Taxa	MAVLI	Max VLI	Mean HI/BI of Middle Cervicals
Elasmosaurinae			
Hydrotherosaurus alexandrae	105	118	HI = 91/BI = 102
Nakonanectes bradti *	91	100	HI = 88/BI = 134
Elasmosaurus platyurus (estimated)	127	145	HI = 65/HI = 87
Styxosaurus sp. (AMNH 5835)	135	147	HI = 63/BI = 89
Albertonectes vanderveldei *	139	141	HI = 66/BI = 79
Terminonatator ponteixensis	129	151	HI = 65/BI = 96
Weddellonectia			
Kawanectes lafquenianum	-	-	-
Vegasaurus molyi	97	108	HI = 86/BI = 120
Morenosaurus stocki	-	-	-
Futabasaurus suzukii	-	-	-
Aristonectinae			
Wunyelfia maulensis	-	-	-
Kaiwhekea katiki	-	-	-
Aristonectes quiriquinensis *	-	~87	-
Aristonectes parvidens *	~79	~88	-

Table 8. Plesiosauromorph taxa, neck elongation pattern, and data source.

Taxa	Neck Elongation	Data Source
Neoplesiosauria		
Rhaeticosaurus mertensi	Non-elongated	[7]
Plesiosauroidea		
Eoplesiosaurus antiquior	Non-elongated?	[26]
Plesiosauridae		
Plesiosaurus dolichodeirus	Non-elongated	[27]
Microcleididae		
Microcleidus tournemirensis	Extremely elongated	[28]
Microcleidus homalospondilus	Elongated?	[29]: table V
Westphalisaurus simonsensii	Elongated?	[13]
Seeleyosaurus guillelmiimperatoris	Elongated?	[30]
Cryptoclididae		
Ophthalmothule cryostea	Elongated	[31]
Spitrasaurus wensaasi	Elongated	[32]
Tricleidus seeleyi	Non-elongated	[33]
Cryptoclidus eurymerus	Non-elongated	[9]
Colymbosaurus megadeirus	Non-elongated	[34]
Djupedalia engeri	Non-elongated	[35]
Spitrasaurus larseni	Non-elongated	[32]
Abyssosaurus nataliae	Non-elongated	[36]
Picrocleidus beloclis	Non-elongated	[33]
Tatenectes laramiensis	Non-elongated	[37]
Kimmerosaurus langhami	Non-elongated	[38]
Colymbosaurus sp.	Non-elongated	[39]
Leptocleididae		
Brancasaurus brancai	Non-elongated	[4]
Elasmosauridae		
Callawayasaurus colombiensis	Elongated	[7]
Jucha squalea	Elongated	[25]
Zarafasaura oceanis	Non-elongated	[40]

Table 8. *Cont.*

Taxa	Neck Elongation	Data Source
Euelasmosaurida		
Thalassomedon hanintoni	Elongated	[15]
Libonectes morgani	Elongated	[18]
Cardiocorax mukulu	Non-elongated	[41]
Elasmosaurinae		
Hydrotherosaurus alexandrae	Elongated	[15]
Nakonanectes bradti	Non-elongated	[6]
Elasmosaurus platyurus	Extremely elongated	[8]
Styxosaurus sp.	Extremely elongated	[8]
Albertonectes vanderveldei	Extremely elongated	[3]
Terminonatator ponteixensis	Extremely elongated	[42,43]
Weddellonectia		
Kawanectes lafquenianum	Elongated	[44]
Vegasaurus molyi	Elongated	[19]
Chubutinectes carmeloi	Elongated?	[45]
Aristonectinae		
Wunyelfia maulensis	Non-elongated	[46]
Kaiwhekea katiki	Non-elongated	[47]
Aristonectes quiriquinensis	Non-elongated	[48]
Aristonectes parvidens	Non-elongated	[49]

Among euelasmosaurians, *Thalassomedon haningtoni* (DMNS 1588 [18], Figure 4b (this paper) has a cervical vertebra (numbered 35) with relatively high VLI values (134) but lower than 135. Combined with the observation of probable lateral compression of DMNS 1588 (reduction of B values), as well as MAVLI being 108, it indicates that *Thalassomedon haningtoni* exhibits an elongated neck pattern. The VLI pattern of *Libonectes morgani* (SMUSMP 69120, [18]: Table 1) shows VLI values lower than 135 and MAVLI of 108, fitting well with the elongated group. The holotype of *Cardiocorax mukulu* (MGUAN PA103) only preserves a few cervical vertebrae, with one of them appearing to be anterior to middle and showing a VLI value of approximately 98. Additionally, the specimen MGUAN PA278, also referred to as *C. mukulu* [41], shows VLI values of the first seven post-axial cervical vertebrae between 76 and 88 and middle cervical vertebrae with VLI values of 83 and 100. Therefore, *C. mukulu* seems to present a strong case for a non-elongated neck, but as no complete neck is available, further evidence is needed to confirm this assumption.

Among the elasmosaurines, the VLI pattern of *Hydrotherosaurus alexandrae* (UCMP 33912, [15], Figure 4e) fits with the elongated group, as it exhibits middle cervical VLI values higher than 100 but lower than 135, and an MAVLI of 105. The only specimen of *Nakonanectes bradti* preserves only the anterior half of the neck, and its VLI sequence does not reach 100 (MOR 3072, Serratos et al., 2017 [6]), indicating a non-elongated neck. There are no complete sets of vertebral values available for *Albertonectes vanderveldei*; however, the available (Max VLI 141 and MAVLI~139) data indicate a pattern similar to that inferred for *Styxosaurus* (Figure 4f) and *Elasmosaurus platyurus*. Therefore, it is considered to exhibit an extremely elongated pattern. The case for *Terminonatator ponteixensis* is quite easy to determine as, although not complete, the neck is well preserved Sato (2002, 2003 [42,43]). This gives an approximate MAVLI of 129 and an MAX VLI of 151, which indicate an extremely elongated neck pattern, as concluded by (O'Keefe and Hiller, 2006 [4]).

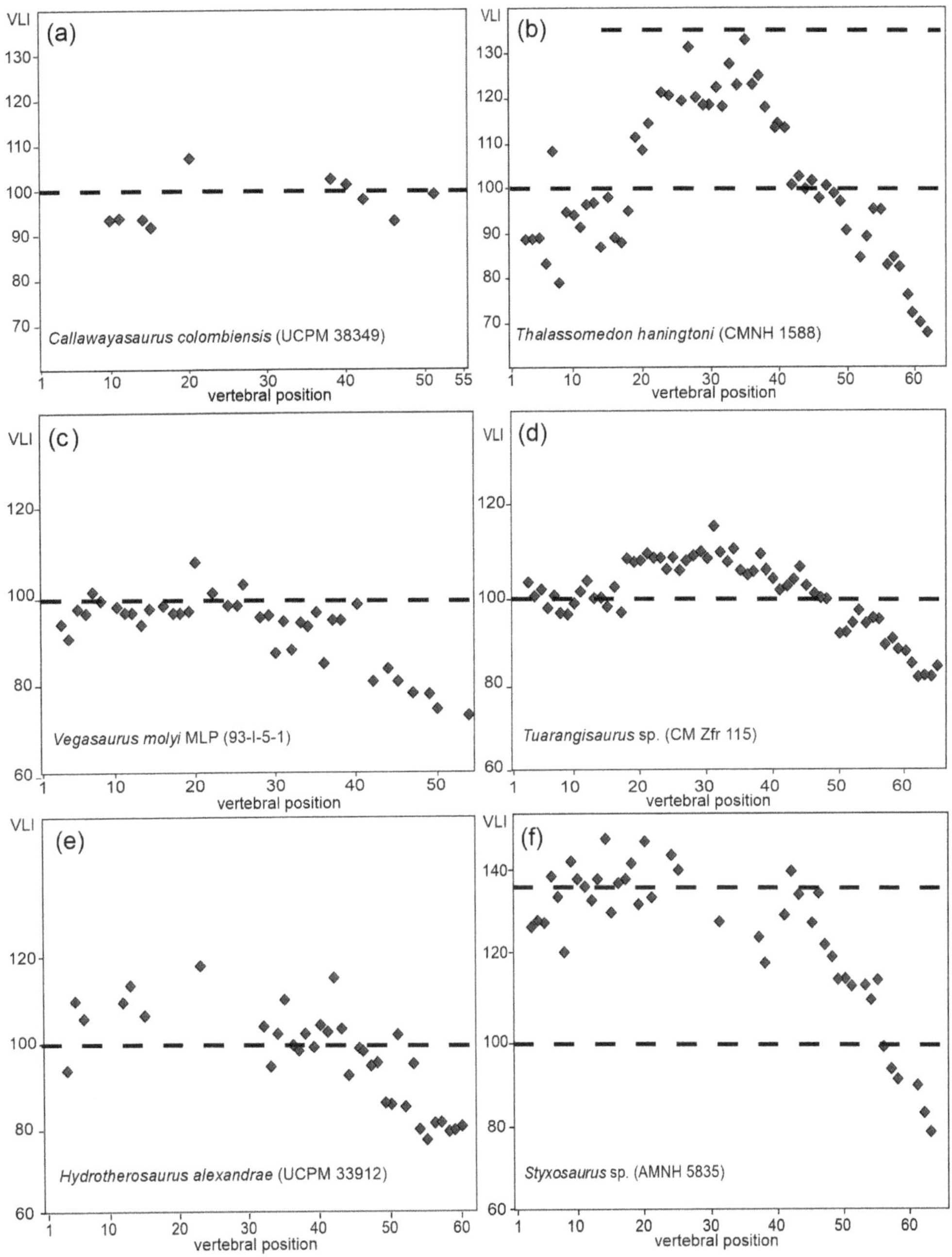

Figure 4. VLI of cervical region of elasmosaurids. (**a**) *Callawayasaurus colombiensis* (UCPM 38349), (**b**) *Thalassomedon haningtoni* (DMNS 1588), (**c**) *Vegasaurus molyi* (MLP 93-I-5-1), (**d**) *Tuarangisaurus* sp. (MC Zfr 115), (**e**) *Hydrotherosaurus alexandrae* (UCPM 33912), and (**f**) *Styxosaurus* sp. (AMNH 5835). Data taken from [4,8,15,19]. Diamond indicate individual vertebra, dotted line indicate vertebral length index = 100.

Among weddellonectians, the holotype of *Morenosaurus stocki* only preserves approximately the posterior half of the neck. The VLI pattern does not reach 95 (Welles, 1943 [15]) in any vertebra. Considering that the posterior cervical region usually shows a VLI lower than the middle cervicals in elasmosaurids, it is impossible to determine the elongation pattern as it falls on the boundary between elongated and non-elongated. A similar case occurs with the holotype of *Futabasaurus suzukii*, where the neck of the holotype (NSM PV15025) only preserves 15 cervical vertebrae representing the posterior half of the neck, with relatively low VLI values ranging from 69 to 82 (Sato et al., 2006 [50]), probably indicating a non-elongated neck. However, this assumption is weak as no complete neck is available. A similar case is observed in *Aphrosaurus furlongi*, where only the 18 posteriormost cervical centra are preserved. The neck of *Vegasaurus molyi* ([19], Figure 4c) exhibits middle cervicals with VLI values around 100, falling within the lower limit of the elongated neck pattern. No complete cervical region of *Kawanectes lafquenianum* is available; however, the specimens at hand comprise middle cervical vertebrae with VLI values around 100 [44], indicating that the neck probably fits well with the elongated pattern. A similar situation happened in *Chubutinectes carmeloi*, and therefore, this species probably shows an elongated neck pattern [45]

The holotype of *A. parvidens* (MLP 40-XI-14-6) only preserves the nineteen anterior and middle cervical, with VLI values lower than 88. Therefore, it exhibits a non-elongated pattern [49]. The holotype of *Aristonectes quiriquinensis* (SGO.PV.957, [48]) preserves anterior, middle, and posterior cervical vertebrae. The maximum VLI value recorded is 95, indicating that *A. quiriquinensis* shows a non-elongated pattern. There are no complete sets of vertebral measurements available for *Kaiwhekea katiki*; however, Cruickshank and Fordyce [47] indicate that the VLI of the anteriormost cervicals of the holotype varies between 61 and 75, fitting well with the non-elongated pattern. The VLI of the anteriormost five post-axial vertebrae of *Wunyelfia maulensis* ranges from 82.9 to 89.4 [46], indicating a non-elongated cervical pattern.

4.3. Plesiosauromorph Neck Elongation Pattern

The modification of the classification of neck elongation patterns and its application to elasmosaurids opens questions about its application to other plesiosauromorph plesiosaurs. The results of this are given in Tables 7 and 8. Although not a clear "plesiosauromorph," the oldest plesiosaur, *Rhaeticosaurus mertensi*, is considered here as non-elongated [51]. Among Plesiosauroidea, the Hettangian *Eoplesiosaurus antiquior* is recorded as non-elongated with doubts, as the B values are unavailable (Benson et al., 2012 [26]). Also, the iconic Sinemurian *Plesiosaurus dolichodeirus* shows a non-elongated pattern (Table 1 [27] and Figure 5a (this paper)).

Among microcleidids, *Seeleyosaurus guillelmiimperatoris* fits well with the elongated neck definition even though only a small number of cervicals show VLI values above 100 ([30]: Table 1, Figure 5c (this paper)). The same is inferred from *Microcleidus homalospondilus* based on Owen's description [29]. The Toarcian *Microcleidus tournemirensis* is a plesiosauromorph with an extremely elongated neck (MAVLI 129; Max VLI = 143, data from Bardet et al. [28], Figure 5b (this paper)), and therefore is the first appearance of elongated patterns in plesiosaurian history.

The cryptoclidids *Ophthalmothule cryostea*, *Spitrasaurus larseni*, *Spitrasaurus wensaasi*, *Picrocleidus*, *Tricleidus seeleyi*, *Abyssosaurus nataliae* and *Muraenosaurus leedsi* (Figures 5d–h and 6a,b) are recorded as elongated plesiosauromorphs [32,33,36], while the other analyzed cryptoclidids, *Cryptoclidus eurymerus* ([9]: Figure 13), are not elongated. The case of *Kimmerosaurus langhami* is complex as only the first seven cervical vertebrae are available and show VLI values higher than 75 (Brown et al., 1986 [38]), indicating a probable non-elongated pattern. However, this inference is weak due to the incompleteness of the neck. Finally, among Leptocleididae, *Brancasaurus brancai* shows a non-elongated pattern [52].

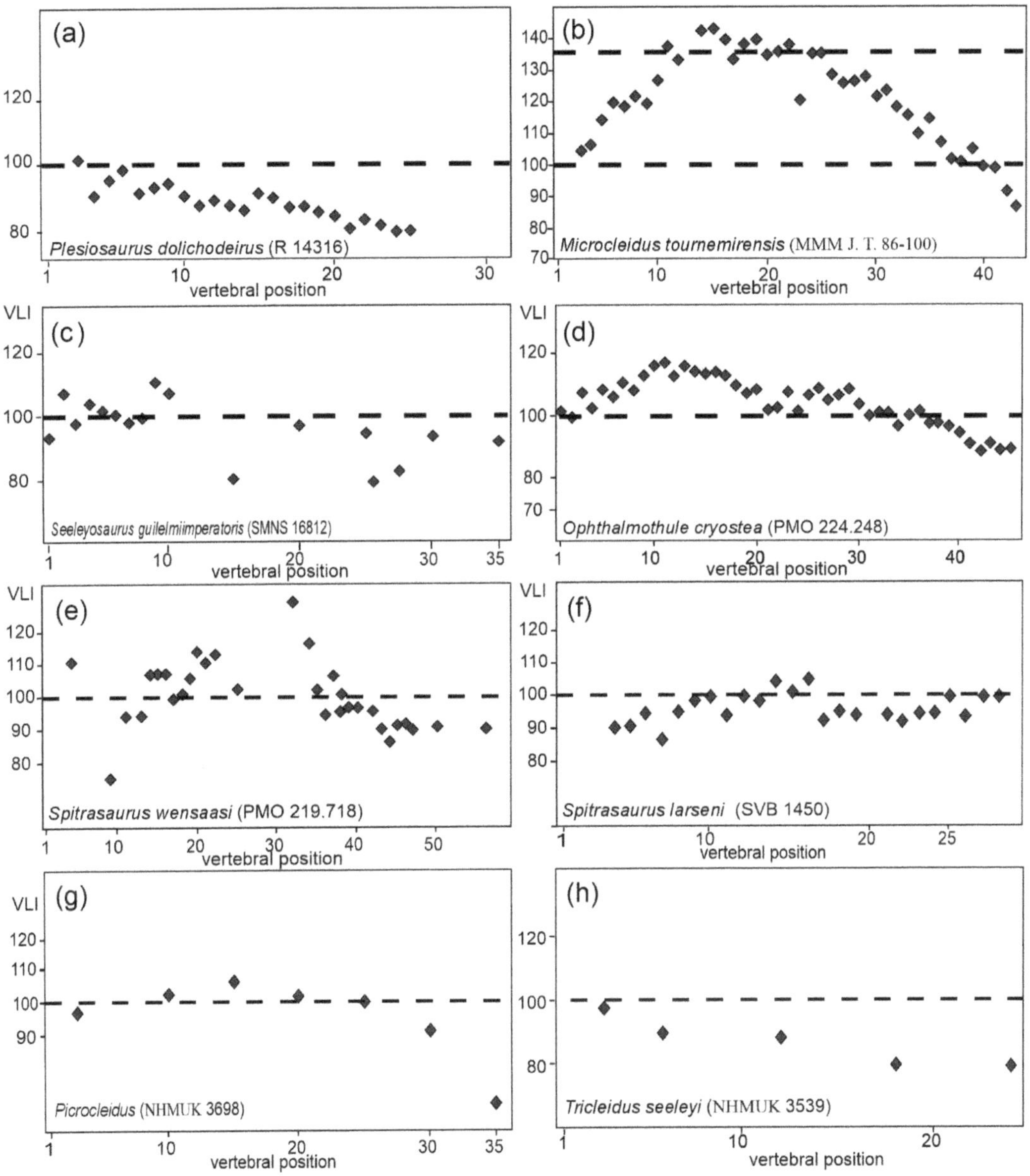

Figure 5. VLI of cervical region of plesiosauromorphs. (**a**) *Plesiosaurus dolichodeirus*; (**b**) *Microcleidus tournemirensis*; (**c**), *Seeleyosaurus guilelemiimperatosis*; (**d**) *Ophthalmothule cryostea*; (**e**) *Spitrasaurus wensaasi*; (**f**) *Spitrasaurus larseni*; (**g**) *Picrocleidus*; and (**h**) *Tricleidus seeleyi*. Data taken from [27,28,30–33]. Diamond indicate individual vertebra, dotted line indicate vertebral length index = 100.

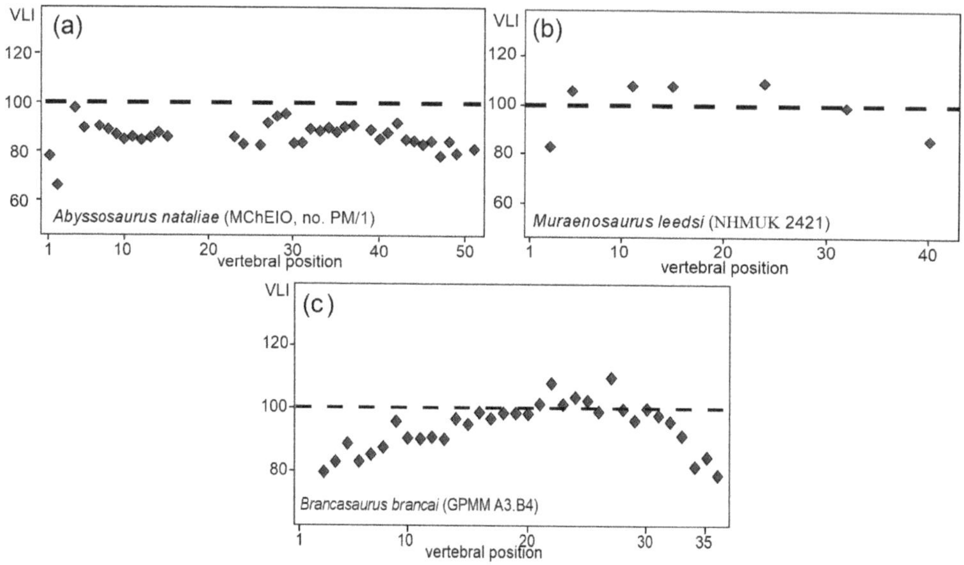

Figure 6. VLI of cervical region of plesiosauromorphs: (**a**) *Abysossaurus nataliae;* (**b**) *Muraenosaurus leedsi;* and (**c**) *Brancasaurus brancai* [4,33,36]. Diamond indicate individual vertebra, dotted line indicate vertebral length index = 100.

In summary, the redefinition of neck elongation patterns allows for a more comprehensive history of plesiosauromorphs. It indicates that the oldest appearance of elongated necks (MAVLI > 100) most likely occurred during the Pliensbachian-Toarcian while the appearance of the extremely elongated pattern (MAVLI > 125, represented by *Microcleidus tournemirensis*) happened during the Toarcian as was previously briefly commented by Benson et al. [26]. This shows that a complete diversity of neck elongation patterns was present during the Early Jurassic and was mirrored during the Late Cretaceous, albeit in a more extreme expression, mainly within elasmosaurids.

5. Conclusions

Partial retrodeformation of the cervical centra of the holotype of *E. platyurus* was performed, leading to the conclusion that the natural sequence of VLI values is lower than previously recorded.

A modification to the classification of neck elongation patterns in plesiosauromorphs is proposed, introducing the following categories: extremely elongated (MAVLI > 125; Max VLI > 135; some cervicals HI < 65), elongated (125 > MAVLI > 100; 100 < Max VLI < 135; middle cervical HI~70-80; BI < 140), and non-elongated (MAVLI < 100; Max VLI < 100; middle cervical HI~100; BI > 140).

The elongated pattern appears during the Pliensbachian-Toarcian, and the extremely elongated necks are recorded for the first time in the Miclocleidid *Microcleidus tournemirensis* (Toarcian). This indicates that the complete diversity of neck elongation was present during the Early Jurassic and reemerged during the Late Cretaceous.

Supplementary Materials: The following supporting information can be downloaded at: https://www.mdpi.com/article/10.3390/d16020106/s1, Supplementary Material 1: Table S1. Values of L, H, B and HI, BI, BHI and VLI. Cervical position (TC, this contribution, S, [2,16]. Table S2 L, H and B est. and resulting HI es, BI est, VLI est; Figure S1: ANSP 10081 (holotype of *Elasmosaurus platyurus*). A, detailed of different labels. B–E, atlas-axis complex in B, right lateral; C, left lateral; D, posterior and dorsal. Scale bar = 20 mm. Figure S2: ANSP 10081 (holotype of *Elasmosaurus platyurus*). A, B, cervical vertebrae 1–10th in A, left lateral and B, ventral view (above) and some articular faces (below). C, D, cervical vertebrae 11–18th in C, left lateral and D, ventral view (above) and some articular faces (below). Scale bar = 50 mm. Figure S3: A, B, cervical vertebrae 19–26th in A, left lateral and B, ventral view (above) and some articular faces (below). C, D, cervical vertebrae 27–32th in C, left lateral and D, ventral view (above) and some articular faces (below). Scale bar = 50 mm. Figure S4. A, B, cervical vertebrae 33–38th in A, left lateral and B, ventral view (above) and some articular faces (below). C, D, cervical vertebrae 39–43th in C, left lateral and D, ventral view (above) and some articular faces (below). Scale bar = 50 mm. Figure S5. A, cervical vertebrae 44–47th in A, left lateral view and articular faces (below). Scale bar = 50 mm. Figure S6. A, cervical vertebrae 48–50th; B, cervical vertebrae 51–54th; C, D, cervical vertebrae 55–59th in C, left lateral and D, ventral views. Scale bar = 50 mm. Figure S7. A, B, cervical vertebrae 60–64th in A, lft lateral and B, ventral views. C, D, cervical vertebrae 65–72th in C, left lateral and D, ventral views. A, B, Scale bar = 50 mm; C, D, Scale bar = 100 mm. Figure S8. WDC CMC-01 *Zarafasaura oceanis*. Not in scale; Supplementary Material 2: Table S1. *Styxosaurus* sp. (AMNH 5835).Values of L (length), H (higth), B (width), B reduced, B estimated, VLI (vertebral length index) and VLI estimated. Taken from [18]. Table S2. *Tuarangisaurus* sp. (Zfr115). Values of L (length), H (higth), B (width), B reduced, B estimated, VLI (vertebral length index) and VLI estimated. Data taken from [4]. Table S3. *Vegasaurus molyi* (MLP 93-I-5-1). Values of L (length), H (higth), B (width), B reduced, B estimated, VLI (vertebral length index) and VLI estimated. Data taken [19]; Supplementary Material 3: Script used for analysis.

Funding: This research was funding by Agencia Nacional de Promoción científica y Tecnológica grant number (PICT 2015-0678, PICT-2018-02443, PICT-2021-GRF-TII-00236).

Institutional Review Board Statement: Not applicable.

Data Availability Statement: The authors confirm that the data supporting the findings of this study are available within the article and its Supplementary Materials.

Acknowledgments: The author thanks Ned Gilmore (Academy of Natural Sciences of Drexel University) for allowing the study of the specimen under his care. I also thank D. Lomax and B. Wahl (Wyoming Dinosaur Center) for provide photographic material of the specimen WDC CMC-01. The author also thanks the two anonymous reviewers.

Conflicts of Interest: The author declares no conflicts of interest.

References

1. O'Keefe, F.R. The evolution of plesiosaur and pliosaur morphotypes in the Plesiosauria (Reptilia: Sauropterygia). *Paleobiology* **2002**, *28*, 101–112. [CrossRef]
2. Sachs, S.; Kear, B.P.; Everhart, M.J. Revised vertebral count in the "longest-necked vertebrate" *Elasmosaurus platyurus* Cope 1868, and clarification of the cervical-dorsal transition in Plesiosauria. *PLoS ONE* **2013**, *8*, e70877. [CrossRef]
3. Kubo, T.; Mitchell, M.T.; Henderson, D.M. *Albertonectes vanderveldei*, a new elasmosaur (Reptilia, Sauropterygia) from the Upper Cretaceous of Alberta. *J. Vertebr. Paleontol.* **2012**, *32*, 557–572. [CrossRef]
4. O'Keefe, F.R.; Hiller, N. Morphologic and ontogenetic patterns in elasmosaur neck length, with comments on the taxonomic utility of neck length variables. *Paludicola* **2006**, *5*, 206–229.
5. Otero, R.A. Taxonomic reassessment of *Hydralmosaurus* as *Styxosaurus*: New insights on the elasmosaurid neck evolution throughout the Cretaceous. *PeerJ* **2016**, *4*, e1777. [CrossRef] [PubMed]
6. Serratos, D.J.; Druckenmiller, P.; Benson, R.B. A new elasmosaurid (Sauropterygia, Plesiosauria) from the Bearpaw Shale (Late Cretaceous, Maastrichtian) of Montana demonstrates multiple evolutionary reductions of neck length within Elasmosauridae. *J. Vertebr. Paleontol.* **2017**, *37*, e1278608. [CrossRef]
7. Welles, S.P. A New species of elasmosaur from the Aptian of Colombia and a review of the Cretaceous plesiosaurs. *Univ. California. Publ. Geol. Sci.* **1962**, *44*, 1–96.
8. Welles, S.P. A review of the North American Cretaceous elasmosaurs. University of California. *Publ. Geol. Sci.* **1952**, *29*, 47–144.
9. Brown, D.S. The English upper Jurassic Pleisauroidea (Reptilia), and a review of the phylogeny and classification of the Plesiosauria. *Bull. Br. Mus. (Nat. Hist.) Geol.* **1981**, *35*, 253–347.

10. Gasparini, Z.; Martin, J.E.; Fernández, M. The elasmosaurid plesiosaur *Aristonectes* Cabrera from the latest Cretaceous of South America and Antarctica. *J. Vertebr. Paleontol.* **2003**, *23*, 104–115. [CrossRef]
11. Otero, R.A.; Soto-Acuña, S.; Rubilar-Rogers, D. A postcranial skeleton of an elasmosaurid plesiosaur from the Maastrichtian of central Chile, with comments on the affinities of Late Cretaceous plesiosauroids from the Weddellian Biogeographic Province. *Cretac. Res.* **2012**, *37*, 89–99. [CrossRef]
12. Otero, R.A.; Soto-Acuña, S.; Salazar, C.; Oyarzún, J.L. New elasmosaurids (Sauropterygia, Plesiosauria) from the Late Cretaceous of the Magallanes Basin, Chilean Patagonia: Evidence of a faunal turnover during the Maastrichtian along the Weddellian Biogeographic Province. *Andean Geol.* **2015**, *42*, 237–267.
13. Cope, E.D. Synopsis of the extinct Batrachia, Reptilia and Aves of North America. *Trans. Am. Philos. Soc.* **1869**, *14*, e252.
14. Cope, E.D. The Vertebrata of the Cretaceous formations of the West. *Rep. United States Geol. Surv. Territ.* **1875**, *2*, 303.
15. Welles, S.P. Elasmosaurid plesiosaurs with description of new material from California and Colorado. *Mem. Univ. Calif.* **1943**, *13*, 125–254.
16. Sachs, S. Redescription of *Elasmosaurus platyurus* Cope 1868 (Plesiosauria: Elasmosauridae) from the Upper Cretaceous (lower Campanian) of Kansas, USA. *Paludicola* **2005**, *5*, 92–106.
17. Cooper, R.A. Interpretation of tectonically deformed fossils. *Nett Zealand J. Geol. Arid. Geophys.* **1990**, *33*, 321–332. [CrossRef]
18. Welles, S.P. A new elasmosaur from the Eagle Ford Shale of Texas. Part I. Systematic description. *Fondren Sci. Ser.* **1949**, *1*, 1.
19. O'Gorman, J.P.; Salgado, L.; Olivero, E.B.; Marenssi, S.A. *Vegasaurus molyi*, gen. et sp. nov.(Plesiosauria, Elasmosauridae), from the Cape Lamb Member (lower Maastrichtian) of the Snow Hill Island Formation, Vega Island, Antarctica, and remarks on Wedellian Elasmosauridae. *J. Vertebr. Paleontol.* **2015**, *35*, e931285. [CrossRef]
20. Gross, J.; Gross, M.J. Package 'nortest'—Tests for Normality. R Package Version. 2009. Available online: https://cran.r-project.org/web/packages/nortest/nortest.pdf (accessed on 20 January 2019).
21. R Core Team. R: Alanguage and Environment for Statistical Computing. 2013. Available online: http://www.r-project.org (accessed on 5 June 2022).
22. Owen, R. On the orders of fossil and recent Reptilia, and their distribution in time. *Rep. Br. Assoc. Adv. Sci.* **1860**, *29*, 153–166.
23. Blainville, H.M.D. Description de quelques espèces de reptiles de la Californie, précédée de l'analyse d'un système général d'Erpétologie et d'Amphibiologie. *Nouv. Ann. Muséum (Natl.) D'histoire Nat. Paris* **1835**, *3*, 233–296.
24. Cope, E.D. Remarks on a new enaliosaurian, *Elasmosaurus platyurus. Proc. Acad. Nat. Sci. Phila.* **1868**, *20*, 92–93.
25. Fischer, V.; Zverkov, N.G.; Arkhangelsky, M.S.; Stenshin, I.M.; Blagovetshensky, I.V.; Uspensky, G.N. A new elasmosaurid plesiosaurian from the Early Cretaceous of Russia marks an early attempt at neck elongation. *Zool. J. Linn. Soc.* **2021**, *1921*, 1167–1194. [CrossRef]
26. Benson, R.B.; Evans, M.; Druckenmiller, P.S. High diversity, low disparity and small body size in plesiosaurs (Reptilia, Sauropterygia) from the Triassic–Jurassic boundary. *PLoS ONE* **2012**, *7*, e31838. [CrossRef] [PubMed]
27. Schwermann, L.; Sander, P.M. Osteologie und Phylogenie von *Westphaliasaurus simonsensii*: Ein neuer Ples osauride (Sauropterygia) aus dem Unteren Jura (Pliensbachium) von Sommersell (Kreis Höxter) Nordrhein-Westfalen, Deutschland. *Geol. Paläont. Westf.* **2011**, *79*, 1–60.
28. Bardet, N.; Godefroit, P.; Sciau, J. A new elasmosaurid plesiosaur from the Lower Jurassic of southern France. *Palaeontology* **1999**, *42*, 927–952. [CrossRef]
29. Owen, R. Monograph of the fossil reptilia of the Liassic formations. Part first. Sauropterygia. *Monogr. Palaeontogr. Soc.* **1865**, *17*, 1–40. [CrossRef]
30. Großmann, F. The taxonomic and phylogenetic position of the Plesiosauroidea from the Lower Jurassic Posidonia Shale of South-West Germany. *Palaeontology* **2007**, *50*, 545–564. [CrossRef]
31. Roberts, A.J.; Druckenmiller, P.S.; Cordonnier, B.; Delsett, L.L.; Hurum, J.H. A new plesiosaurian from the Jurassic–Cretaceous transitional interval of the Slottsmøya Member (Volgian), with insights into the cranial anatomy of cryptoclidids using computed tomography. *PeerJ* **2020**, *8*, e8652. [CrossRef]
32. Knutsen, E.M.; Druckenmiller, P.S.; Hurum, J.H. Two new species of long-necked plesiosaurians (Reptilia: Sauropterygia) from the Upper Jurassic (Middle Volgian) Agardhfjellet Formation of central Spitsbergen. *Nor. J. Geol./Nor. Geol. Foren.* **2012**, *92*, 187–212.
33. Andrews, C.W. *A Description Catalogue of the Marine Reptiles of the Oxford Clay. Part I*; British Museum: London, UK, 1910; pp. 1–205.
34. Benson, R.B.J.; Bowdler, T. Anatomy of *Colymbosaurus megadeirus* (Reptilia, Plesiosauria) from the Kimmeridge Clay Formation of the UK, and high diversity among Late Jurassic plesiosauroids. *J. Vertebr. Paleontol.* **2014**, *34*, 1053–1071. [CrossRef]
35. Knutsen, E.M.; Druckenmiller, P.S.; Jørn, H.H. A new plesiosauroid (Reptilia: Sauropterygia) from the Agardhfjellet Formation (middle Volgian) of central Spitsbergen, Norway. *Nor. J. Geol.* **2012**, *92*, 213–234.
36. Berezin, A.Y. A new plesiosaur of the family Aristonectidae from the Early Cretaceous of the center of the Russian platform. *Paleontol. J.* **2011**, *45*, 648–660. [CrossRef]
37. O'Keefe, F.R.; Street, H.P. Osteology of the cryptocleidoid plesiosaur *Tatenectes laramiensis*, with comments on the taxonomic status of the Cimoliasauridae. *J. Vertebr. Paleontol.* **2009**, *29*, 48–57. [CrossRef]
38. Brown, D.S.; Milner, A.C.; Taylor, M.A. New material of the plesiosaur *Kimmerosaurus langhami* Brown from the Kimmeridge Clay of Dorset. *Bull. Br. Mus. (Nat. Hist.)* **1986**, *40*, 225–234.

39. Roberts, A.J.; Druckenmiller, P.S.; Delsett, L.L.; Hurum, J.H. Osteology and relationships of *Colymbosaurus* Seeley, 1874, based on new material of *C. svalbardensis* from the Slottsmøya Member, Agardhfjellet Formation of central Spitsbergen. *J. Vertebr. Paleontol.* **2017**, *37*, e1278381. [CrossRef]
40. Lomax, D.R.; Wahl, W.R. A new specimen of the elasmosaurid plesiosaur *Zarafasaura oceanis* from the Upper Cretaceous (Maastrichtian) of Morocco. *Paludicola* **2013**, *9*, 97–109.
41. Marx, M.P.; Mateus, O.; Polcyn, M.J.; Schulp, A.S.; Gonçalves, A.O.; Jacobs, L.L. The cranial anatomy and relationships of *Cardiocorax mukulu* (Plesiosauria: Elasmosauridae) from Bentiaba, Angola. *PLoS ONE* **2021**, *16*, e0255773. [CrossRef]
42. Sato, T. Description of Plesiosaurs (Reptilia: Sauropterygia) from the Bearpaw Formation (Campanian–Maastrichtian) and a Phylo-genetic Analysis of the Elasmosauridae. Ph.D. Dissertation, University of Calgary, Calgary, AB, Canada, 2002; p. 391.
43. Sato, T. *Terminonatator ponteixensis*, a new elasmosaur (Reptilia; Sauropterygia) from the Upper Cretaceous of Saskatchewan. *J. Vertebr. Paleontol.* **2003**, *23*, 89–103. [CrossRef]
44. O'Gorman, J.P. A small body sized non-aristonectine elasmosaurid (Sauropterygia, Plesiosauria) from the Late Cretaceous of Patagonia with comments on the relationships of the Patagonian and Antarctic elasmosaurids. *Ameghiniana* **2016**, *53*, 245–268. [CrossRef]
45. O'Gorman, J.P.; Carignano, A.P.; Calvo-Marcilese, L.; Panera, J.P.P. A new elasmosaurid (Sauropterygia, Plesiosauria) from the upper levels of the La Colonia Formation (upper Maastrichtian), Chubut Province, Argentina. *Cretac. Res.* **2023**, *152*, 105674. [CrossRef]
46. Otero, R.A.; Soto-Acuña, S. *Wunyelfia maulensis* gen. et sp. nov.; a new basal aristonectine (Plesiosauria, Elasmosauridae) from the Upper Cretaceous of central Chile. *Cretac. Res.* **2021**, *118*, 104651. [CrossRef]
47. Cruickshank, A.R.; Fordyce, R.E. A new marine reptile (Sauropterygia) from New Zealand: Further evidence for a Late Cretaceous austral radiation of cryptoclidid plesiosaurs. *Palaeontology* **2002**, *45*, 557–575. [CrossRef]
48. Otero, R.A.; Soto-Acuña, S.; O'Keefe, F.R.; O'Gorman, J.P.; Stinnesbeck, W.; Suárez, M.E.; Rubilar-Rogers, D.; Salazar, C.; Quinzio-Sinn, L.A. *Aristonectes quiriquinensis* sp. nov., a new highly derived elasmosaurid from the upper Maastrichtian of central Chile. *J. Vertebr. Paleontol.* **2014**, *34*, 100–125. [CrossRef]
49. O'Gorman, J.P. New insights on the *Aristonectes parvidens* (Plesiosauria, Elasmosauridae) holotype: News on an old specimen. *Ameghiniana* **2016**, *53*, 397–417. [CrossRef]
50. Sato, T.; Hasegawa, Y.; Manabe, M. A new elasmosaurid plesiosaur from the Upper Cretaceous of Fukushima, Japan. *Palaeontology* **2006**, *49*, 467–484. [CrossRef]
51. Wintrich, T.; Hayashi, S.; Houssaye, A.; Nakajima, Y.; Sander, P.M. A Triassic plesiosaurian skeleton and bone histology inform on evolution of a unique body plan. *Sci. Adv.* **2017**, *3*, e1701144. [CrossRef]
52. Sachs, S.; Kear, B.P. Postcranium of the paradigm elasmosaurid plesiosaurian *Libonectes morgani* (Welles, 1949). *Geol. Mag.* **2015**, *152*, 694–710. [CrossRef]

diversity

MDPI

Article

Callovian Marine Reptiles of European Russia

Nikolay Zverkov [1,*], Maxim Arkhangelsky [2,3], Denis Gulyaev [4], Alexey Ippolitov [1,5,6] and Alexey Shmakov [7]

1 Geological Institute of the Russian Academy of Sciences, Pyzhevsky Lane 7, Moscow 119017, Russia
2 Department of General Geology and Minerals, Saratov State University, Astrakhanskaya Str. 83, Saratov 410012, Russia
3 Department of Oil and Gas, Saratov State Technical University, Politekhnicheskaya Str. 77, Saratov 410054, Russia
4 Commission on Jurassic System of the Interdepartmental Stratigraphical Committee (ISC) of Russia, Chekhova St., 25/7, Yaroslavl 150054, Russia
5 School of Geography, Environment and Earth Sciences, Victoria University of Wellington Te Herenga Waka, 21 Kelburn Parade, Wellington 6012, New Zealand
6 Institute of Geology and Petroleum Technologies, Kazan Federal University, Kremlevskaya St., 18, Kazan 420008, Russia
7 Borissiak Paleontological Institute of the Russian Academy of Sciences, Profsoyuznaya St., 123, Moscow 117997, Russia
* Correspondence: zverkovnik@mail.ru

Abstract: Our knowledge of marine reptiles of the Callovian age (Middle Jurassic) is majorly based on the collections from the Oxford Clay Formation of England, which yielded a diverse marine reptile fauna of plesiosaurians, ichthyosaurians, and thalattosuchians. However, outside of Western Europe, marine reptile remains of this age are poorly known. Here, we survey marine reptiles from the Callovian stage of European Russia. The fossils collected over more than a century from 28 localities are largely represented by isolated bones and teeth, although partial skeletons are also known. In addition to the previously described rhomaleosaurid and metriorhynchids, we identify pliosaurids of the genera *Liopleurodon* and *Simolestes*; cryptoclidid plesiosaurians, including *Cryptoclidus eurymerus*, *Muraenosaurus* sp., and cf. *Tricleidus*, and ophthalmosaurid ichthyosaurians, including the iconic *Ophthalmosaurus icenicus*. These findings expand the ranges of several Callovian marine reptile taxa far to the Eastern Europe, and support the exchange of marine reptile faunas between Western and Eastern European seas in the middle to late Callovian. However, some specimens from the lower Callovian of European Russia show differences from typical representatives of the middle Callovian Oxford Clay fauna, possibly representing the earlier stages of evolution of some of these marine reptiles not yet recorded in Western Europe or elsewhere.

Keywords: Cryptoclididae; Ichthyosauria; Metriorhynchidae; Middle Jurassic; Middle Russian Sea; Plesiosauria; Pliosauridae; Paleobiogeography; Thalattosuchia

1. Introduction

The fossil record of marine reptiles is strongly spatially and temporally biased [1,2]. For the Jurassic period, the main sources of information are several formations in Western Europe, which have been the subject of intensive collecting and thus demonstrate "Lagerstätten effects" [1], along with a persisting global imbalance in the distribution of fossil data [3]. In the Middle Jurassic series, only its uppermost part, the Callovian, is well characterized by marine reptile fossils, and this knowledge is based nearly exclusively on the Oxford Clay Formation of England, which has yielded a diverse marine reptile fauna, including plesiosaurians, ichthyosaurians, and thalattosuchians [4–11]. Other Western European occurrences of Callovian marine reptiles in France and Germany yielded similar but scarcer faunas [12–26], whereas outside Western Europe, Callovian marine reptile occurrences are rarely reported. The records include localities in Argentina, Chile,

Mexico [27–30], Arctic Canada [31,32], and European Russia [33–40]. Specimens reported from these regions are mostly fragmentary and comprise isolated bones, teeth, and, more rarely, bone associations, all of which are identified in open nomenclature and generally referred to European taxa, with the only exception being a partial skeleton of a relict rhomaleosaurid, *Borealonectes russelli,* from Arctic Canada [31,32]. This situation hampers the assessment of the diversity of Callovian marine reptiles outside of Western Europe and demonstrates that our knowledge of Callovian marine reptile faunas is strongly spatially biased. The question arises, how justified can an extrapolation of the knowledge of the classic Oxford Clay fauna to a global scale be; whether marine reptile taxa were widespread in the Callovian, or whether there was some provincialism? In this respect, the growing number of discoveries of marine reptile remains from the Callovian of European Russia can be a valuable contribution to our knowledge. Here, we review all the available remains of plesiosaurians and ichthyosaurians from the Callovian of European Russia, taking into account both historical materials and recent finds. We also report new thalattosuchian finds, which supplement the recently described thalattosuchian specimens from Russia [40].

2. Historical Background

The earliest reports of marine reptile finds, probably from the Callovian Stage of European Russia, were made by Riabinin [41] and Bogolubov [42]. Riabinin described a pliosaurid bone association from the condensed upper Callovian and lowermost Oxfordian deposits of the Unzha River, near Gradulevo Village (currently part of Manturovo Town) in the Kostroma Region [41]. This specimen, now kept in the Mining Museum, St Petersburg (MM 219), was referred to *Peloneustes*; however, it was later re-considered as Pliosauridae indet. [43]. Bogolubov [42] described a new species, *Cryptoclidus simbirskensis*, from Gorodischi, Simbirsk Province (currently Ulyanovsk Region). This taxon was later considered Plesiosauria indet. [43,44]. The bones of "*Cryptoclidus simbirskensis*" were collected ex situ; therefore, their precise stratigraphic position is uncertain. Their state of preservation and color of the matrix occurs in the Callovian–lower Volgian interval of the region, although near Gorodischi the Callovian deposits are unknown. Bogolubov referred the specimen to the Callovian genus *Cryptoclidus,* based on its anteroposteriorly short cervical vertebrae with concave articular surfaces [42]. Therefore, the suggestion that this specimen is of Callovian age was the result of its taxonomic referral, not vice versa. Considering the presence of other cryptoclidids with similar morphology of cervical centra and propodials in stratigraphically younger deposits of the Arctic (i.e., *Djupedalia* [45] and an indeterminate cryptoclidid from Greenland [46]), we tentatively consider *Cryptoclidus simbirskensis* a nomen dubium, interpret its age as uncertain within the Callovian to lower Volgian interval, and exclude it from further consideration in our work.

In 1911, Bogolubov described several other plesiosaurian specimens [33], now kept in the V.I. Vernadsky State Geological Museum, Moscow (SGM). An elongated cervical vertebra (SGM 1358-37) from the Callovian of Alpatyevo, Ryazan Region, was identified as *Muraenosaurus* sp. Two articulated vertebrae from the lower Callovian of Gorky railway station (currently Fruktovaya), Ryazan Region [33] (pl. 1), were identified as "sacral vertebrae of *Pliosaurus* (*Liopleurodon*) sp." However, our examination of this specimen, SGM 1358-11, revealed that it is neither diagnostic nor complete enough to identify it more precisely than as the ventral portions of the pectoral or sacral centra of Plesiosauria indet., although their relatively large size implies an affinity with pliosaurids or rhomaleosaurids. A weathered caudal centrum (SGM 1358-53) from the middle Callovian of the Sysola River (Komi Republic) was referred to as "(?)*Cryptoclidus*" [33], but it is not possible to identify it more precisely than as Plesiosauria indet. An isolated tooth crown from the middle Callovian of Rechitsy, Moscow region, was described as a new species, *Thaumatosaurus calloviensis* [33]. This taxon was later referred to *Simolestes* as a valid species by Riabinin [47], synonymized with *Simolestes vorax* by Tarlo [6], and finally considered as Pliosauridae indet. by [43,44]. However, as argued by Tarlo [6] (p. 175), "*ridges which begin part way up the crown on the external surface*" are a diagnostic feature of *Simolestes*, not yet found in any other pliosaurid genus (although, see "Morphotype 2" in [48]). Therefore, the Callovian tooth described and figured by Bogolubov ([33] pl. 2, figs. 1 and 6) can be referred to *Simolestes*. Unfortunately, this specimen is now lost.

In 1921, Milanovsky reported on the finding of a partial skeleton (SGM 1445-97–120, dorsal and caudal centra, partial pelvis and phalanges) of a large plesiosaur from the lower Callovian of the "Konnyi Barak" ravine, near the village of Verkhnaya Dobrinka, Zhirnovsk district, Volgograd Region [34]. This specimen was more thoroughly described and referred to as Rhomaleosauridae indet. by Benson et al. [36]

A finding of ichthyosaurian vertebra from the Callovian of Zamezhnaya, Komi Republic, was mentioned by Nesov et al. [49]. This specimen (ZIN PH 1/215) is depicted herein for the first time.

In 1999, Arkhangelsky described a partial forelimb (SSU 104a/27) from the (?)upper Callovian of Saratov Region [35]. He proposed a new genus and species *Khudiakovia calloviensis*, which was later considered synonymous with *Ophthalmosaurus icenicus* [50–52], or as a valid species of *Ophthalmosaurus* [53]. This specimen is reassessed herein.

More recent works on the Callovian marine reptiles from Russia include (in addition to the aforementioned contribution of Benson et al. [36]) a short report on a finding of the right maxilla of *Simolestes* from the lower Callovian of the Kostroma Region [37], a report on a finding of a partial ophthalmosaurid skeleton from the lower Callovian of the Republic of Mordovia [38], and an account of Jurassic marine reptiles from Moscow and adjacent territories [39] that provided brief descriptions and figures of pliosaurid and thalattosuchian tooth crowns, as well as several plesiosaurian and ichthyosaurian bones from the middle and upper Callovian of the Moscow and Ryazan regions. Finally, Young et al. [40] recently examined all the available thalattosuchian specimens from Russia and described, among others, several metriorhynchid tooth crowns from the Callovian.

3. Geological and Paleogeographic Settings

3.1. Paleogeographic Position of the Region

The Middle Jurassic epoch was a time of distribution of marine environments over the East European Platform. A transgression of shallow epicontinental seas far into its territory led to the formation of the Middle Russian Sea, which connected the Boreal seas and Tethys Ocean for a short time in the Early Bathonian [54–60]; then, after some regression during the mid-Bathonian, the Boreal–Tethys connection through this sea resumed in the course of a new cycle of transgression since the beginning of the Callovian [60–65]. Additionally, since the earliest Callovian, the Middle Russian Sea became connected to the seas of Central and Western Europe via the Dnieper-Donets basin and the sub-latitudinal Pripyat (Brest) Strait (Figure 1). This is evidenced by the exchange of the earliest Callovian cephalopod faunas between these regions [60,65–68]. This three-branch configuration of the Middle Russian Sea, with wide systems of straits connecting it to the northern, southern, and western seas and oceans, persisted to the early Volgian/Tithonian age of the Late Jurassic [69]. Thus, for circa 15 Myr, European Russia became a crossroad of migration routes between the Arctic (Boreal), Western European (Sub-Boreal and Sub-Mediterranean), and Western Tethys (tropical) sea basins. Besides, in the Callovian of the Middle Russian Sea, especially in its early chrons, an active neo-endemic evolution of cephalopods took place [65,70]. The eustatic sea-level fluctuations superimposed by non-uniform tectonic development of this vast territory resulted in complicated and changing coastal outlines of the Middle Russian Sea during its history and, consequently, complex stratigraphic architecture of the Jurassic System in European Russia [69].

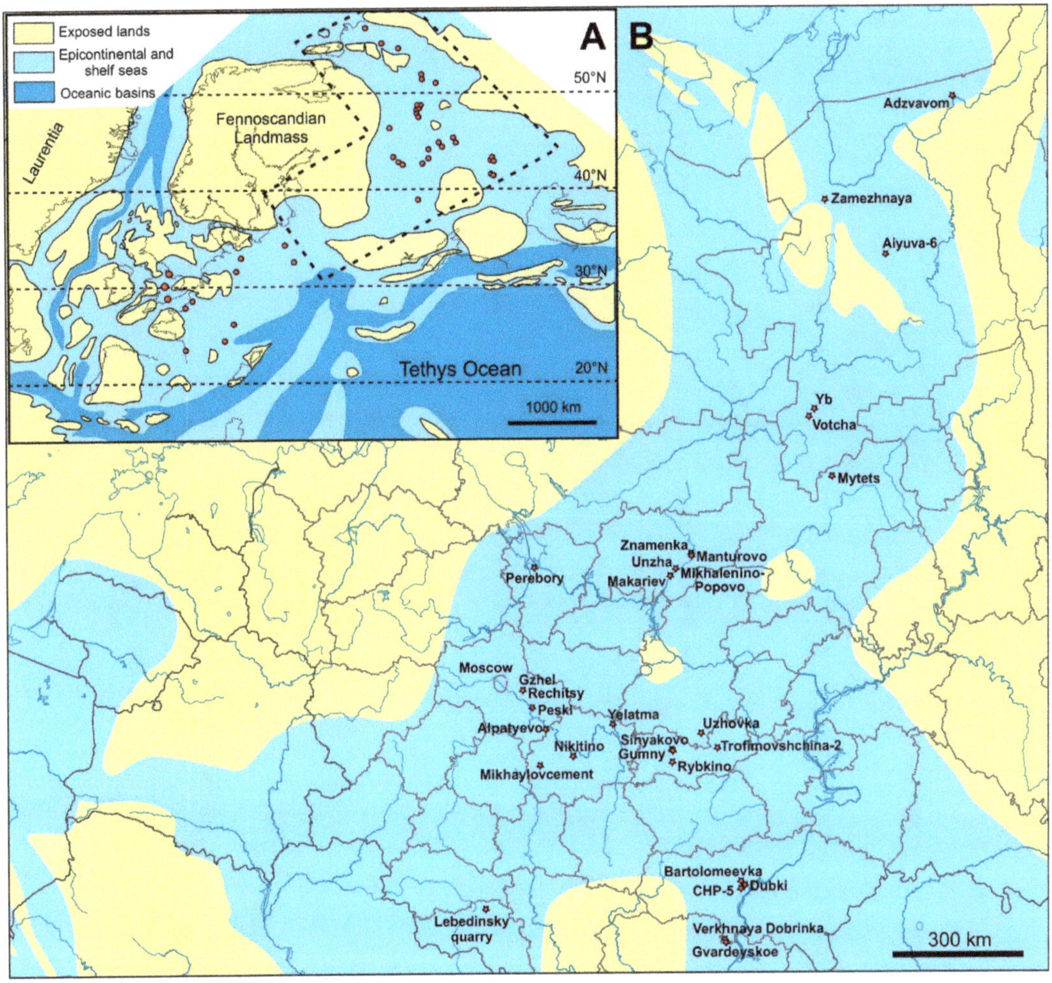

Figure 1. Localities of known marine reptile fossils from the Callovian of European Russia shown by red asterisks (**B**). Blue colorations on the map show outlines of the Middle Russian Sea during the Callovian, based on [71]. The upper left (**A**) shows Callovian paleogeography of Europe (after [72]; modified according to the data of [69,71]), with red dots showing the main localities of Callovian marine reptiles in Europe (i.e., numerous localities with the Oxford Clay Formation in England [4–11,73], localities in Northern and Southern France [12–21,74], Spain [75], Switzerland [76], Northern Germany [22–26], and Poland [77,78]). Dashed contour on (**A**) shows the area depicted in (**B**).

3.2. Localities with Marine Reptiles

At present, no less than 28 localities with Callovian marine reptile remains are known in European Russia (Figure 1). Some of these localities are historical and are not accessible for further excavation nowadays, and the knowledge of their geology is scarce, whereas others are still available or have been discovered and studied in detail in recent decades. We summarize the most important Callovian outcrops in Figures 2–4, based on published literature and personal observations of some of the authors (D.G., A.I., A.S., M.A.).

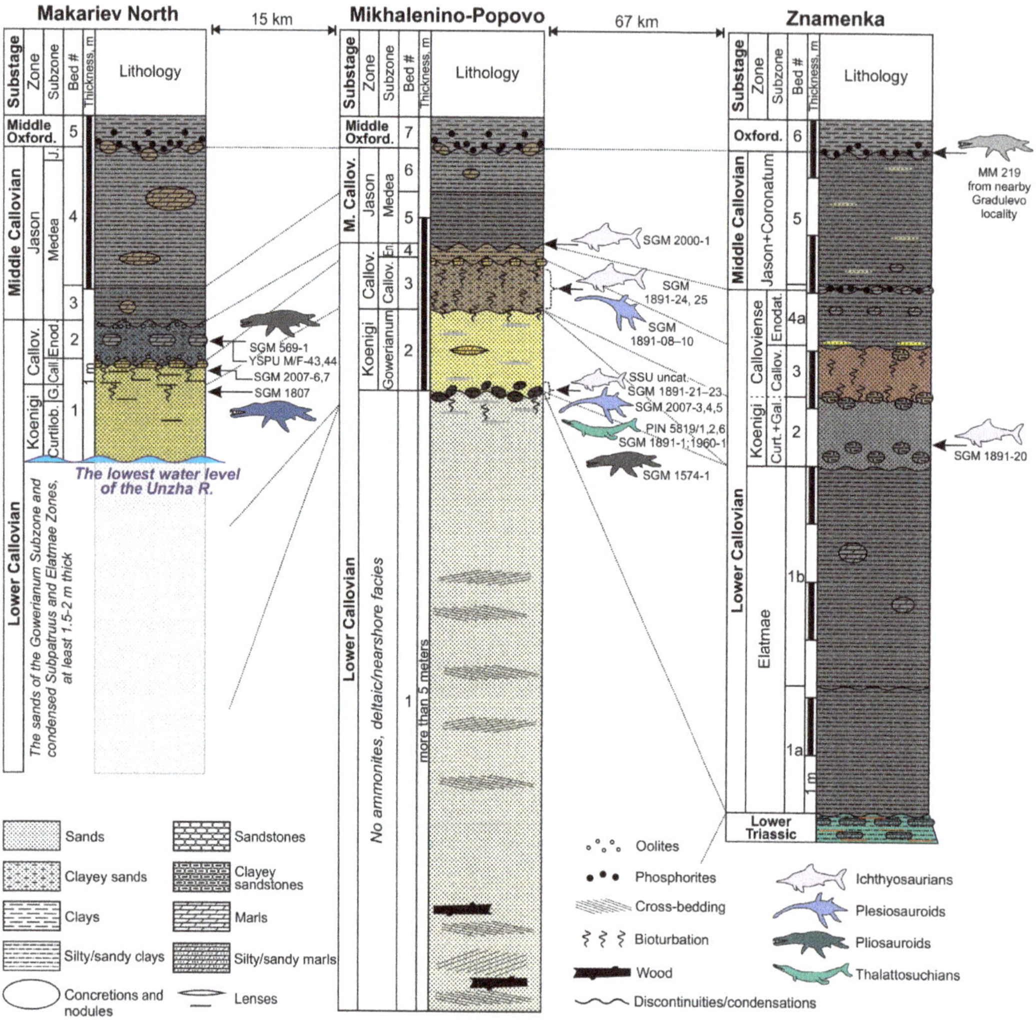

Figure 2. Stratigraphic sections of the Callovian marine reptile localities in the Unzha River basin (Kostroma Region) and distribution of marine reptile remains. Abbreviations of ammonite zones and subzones: Call/Callov, Calloviense; Curt/Curtilob, Curtilobum; En/Enod/Enodat, Enodatum; G/Gal, Galilaeii; J, Jason.

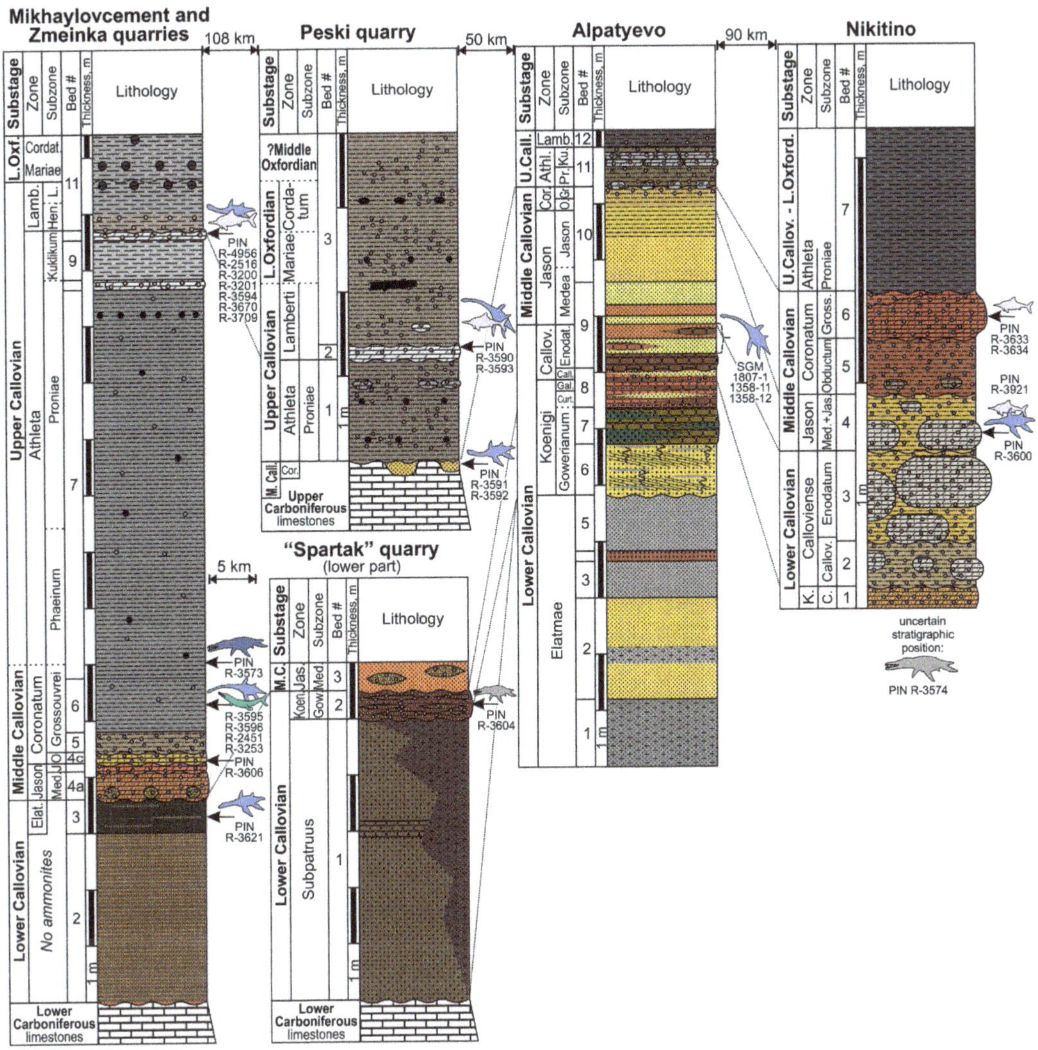

Figure 3. Stratigraphic sections of the Callovian marine reptile localities in Moscow and Ryazan regions and distribution of marine reptile remains. For lithology and other symbols of the legend see Figure 2. Abbreviations of ammonite zones and subzones: Athl, Athleta; C/Curt, Curtilobum; Call/Callov, Calloviense; Cor, Coronatum; Cordat, Cordatum; Enod/Enodat, Enodatum; Gal, Galilaeii; Gow, Gowerianum; Gr/Gross, Grossouvrei; Hen, Henrici; J/Jas, Jason; K/Koen, Koenigi; Ku, Kuklikum; L/Lamb, Lamberti; M/Med, Medea; O/Obd, Obductum; Ph, Phaeinum; Pr, Proniae. Mikhaylovcement section modified after [79]; Peski after [80]; Alpatyevo after [81].

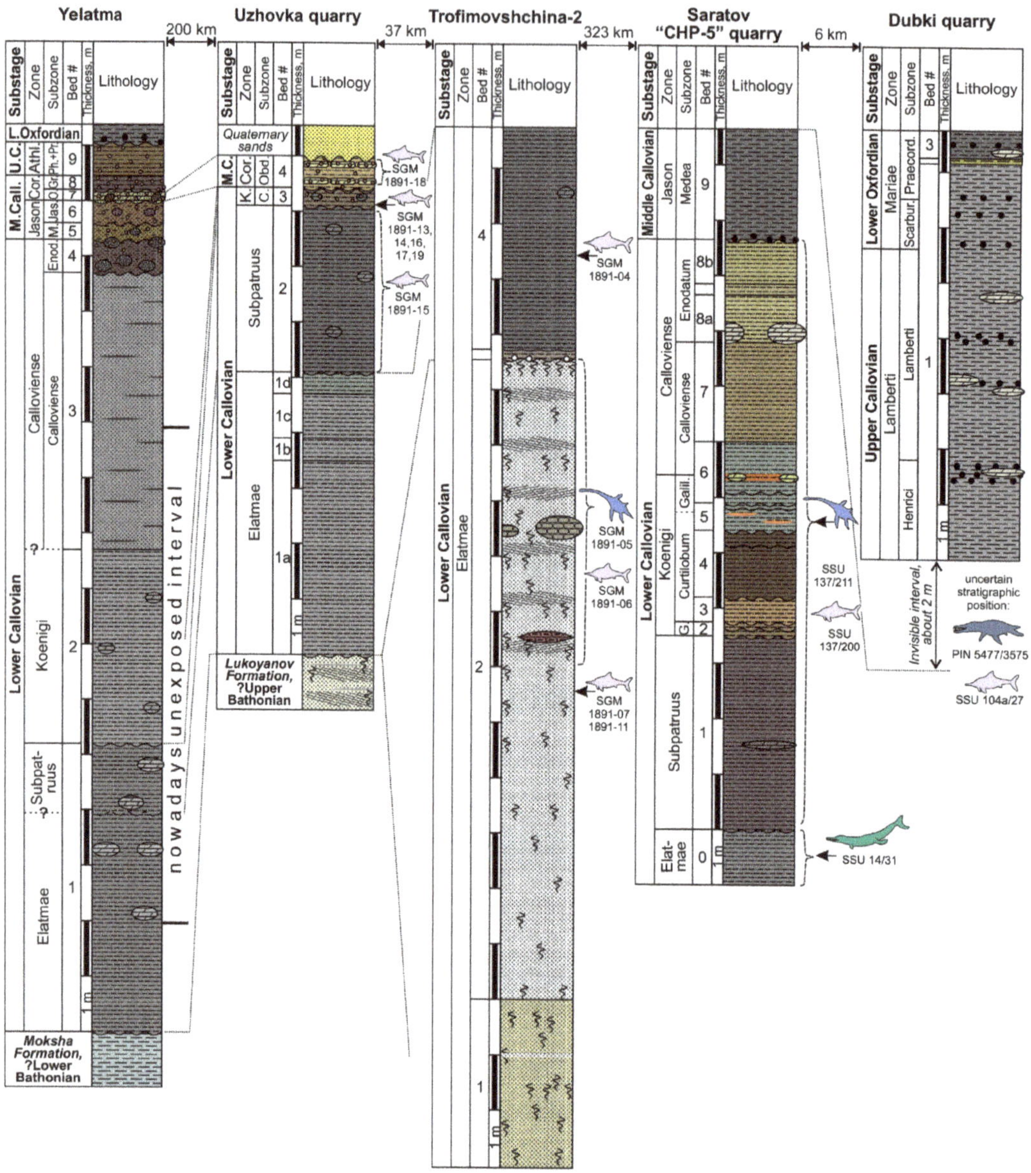

Figure 4. Stratigraphic sections of the Callovian marine reptile localities in Ryazan and Nizhny Novgorod regions, Republic of Mordovia, and Saratov Region. For lithology and other symbols of the legend, see Figure 2. Abbreviations of ammonite zones and subzones: Athl, Athleta; C, Curtilobum; Call/Callov, Calloviense; Cor, Coronatum; Enod, Enodatum; G, Gowerianum; Galil, Galilaeii; Gr–Grossouvrei; Jas, Jason; K, Koenigi; M, Medea; O/Obd, Obductum; Ph, Phaeinum; Pr, Proniae; Praecord, Praecordatum; Scarbur, Scarburgense. Yelatma locality section modified after [82–84]; Uzhovka and Trofimovshchina-2 originally drawn herein based on A.I.'s field descriptions; CHP-5 after [85]; Dubki after [86,87].

3.3. Geological Features of the Callovian of European Russia and Patterns of Stratigraphical Distributions of Marine Reptile Remains

The deposits of the Callovian age crop out in riverbanks, ravines, and quarries across the European part of Russia, commonly transgressively overlying the Late Paleozoic–Early Triassic and/or coastal–continental pre-Callovian Jurassic deposits in the western and eastern peripheries, and coastal-marine and marine pre-Callovian Jurassic in the central parts of European Russia [88,89]. Typically, bones and teeth of marine reptiles occur in condensed beds or above erosional surfaces, therefore, it might be difficult to date them to ammonite zone or even to the substage/stage level if they are found loose. The important feature derived from such a confinement is that, in most cases, vertebrate fossils are discovered only as disarticulated elements. Below, we list the most remarkable condensed intervals, which yielded the findings described in the present paper and have the highest potential for further discoveries of the Callovian marine reptiles.

At the north of the territory under consideration, in the Sysola River basin, there is a slight condensation at the boundary of the sandy Sysola Formation (Bathonian) and the clayey "Churkino" Formation (Callovian) [90–93], but the most prospective interval is the uppermost lower–middle Callovian, which is strongly condensed and packed with plenty of fossil invertebrates and vertebrates [90,91,93]. It is likely that most Callovian vertebrate fossils ever found in the region originate from this interval. Another example from the northern part of the region is shallow-water, strongly condensed sandy sections of the Callovian, located along the eastern shore of the paleobasin, adhering to the Ural Mountains (Adzvavom).

To the south, in the lower-middle reaches of the Unzha River Basin [84,94–96] (and D.G.'s unpublished data; see Figure 2), reptile remains can be associated with condensed sandy and sandy-clayey deposits of the lower Callovian (upper part of *Paracadoceras elatmae*, *Proplanulites koenigi,* and *Sigaloceras calloviense* zones), as well as to the disconformity between the middle Callovian and middle Oxfordian [41], and condensation sometimes covering the stratigraphic interval from the top of the *Kosmoceras jason* Zone to the bottom of the *Cardioceras densiplicatum* Zone.

Within the Callovian of the central part of European Russia (Figure 3), the condensed interval encompasses most of the lower Callovian and the middle Callovian in full. A very productive condensed interval is present at the transition from the *Cadochamoussetia subpatruus* Zone to the base of the *Pr. koenigi* Zone. There is a well-defined erosional surface at the base of the middle Callovian (represented by the *K. jason* or *Erymnoceras coronatum* Zone), overlying different horizons of the lower Callovian. Finally, the remarkable erosional event occurred around the Callovian/Oxfordian boundary. In many places, middle and late Callovian strata are eroded to a significant depth, whereas the fauna of this age can be redeposited at the base of the lower/middle Oxfordian.

In the Middle Volga region [61,84,89,94,97–100] (Figure 4), reptile remains are often confined to the usually eroded and chronostratigraphically sliding boundary (upper Bathonian–lower Callovian) of the sandy Lukoyanov and clayey Yelatma formations. Thus, some of the finds originate from the upper Bathonian *Paracadoceras infimum* Zone, while some are from the lower part of the lower Callovian *P. elatmae* Zone. Furthermore, in the same region, well-defined condensed horizons are confined to the base of the lower Callovian *Pr. koenigi* Zone and to the base of the middle Callovian *E. coronatum* Zone. In the first case, the condensation interval may cover the top of the *P. elatmae* Zone, the entire *C. subpatruus* Zone, and the basal part of the *Pr. koenigi* Zone, while in the second, it expands from the top of the *P. elatmae* Zone to the base of the *E. coronatum* Zone. At the same time, the preserved deposits of the *Pr. koenigi*–*E. coronatum* zones are condensed and represented by oolitic facies [69]. In well-studied sections near Saratov [65,94], the main Callovian condensed levels fall on the boundary interval of the *C. subpatruus* and *Pr. koenigi* zones, and *S. calloviense*/*K. jason* zones, and the largest one, between the middle and upper Callovian, includes a biostratigraphic interval from the upper half of the *K. jason* Zone to the base of the *Lamberticeras lamberti* Zone.

In the western part of the former Middle Russian Sea, only the lower and middle Callovian are known [65,101–103]. In some sections (Mikhaylovsky Quarry, Kursk Region), there is a remarkable hiatus between the coarse sands of the subcontinental Arkino Formation (Bathonian?) and marine clayey deposits in the middle part of the *P. elatmae* Zone. Here, a lensing bed of rolled belemnite rostra is observed, among which fragments of hexacoral colonies occur (observed in 1997 by A.B. Guzhov, PIN RAS). In the same section, significant erosion occurs at the boundary of the *P. elatmae* and *C. subpatruus* zones. The middle Callovian is also present but not well-studied there [94]. The uppermost lower to middle Callovian interval contains multiple minor disconformities, each containing reworked fauna, and thus has a potential for marine reptile remains search.

4. Materials and Methods

More than a hundred of ichthyosaur, plesiosaur, and thalattosuchian specimens stored in 13 institutions are known from the Callovian of Russia. These are summarized in Table 1. The "Regional collection" of the Paleontological Institute RAS has a common catalogue number, 5477, which is abbreviated as "R" herein.

Prior to photography, teeth were coated with ammonium chloride to highlight the patterns of their enamel ornamentation.

The taxonomic frameworks used herein for reptilian taxa follow Benson and Druckenmiller [104] for plesiosaurians, Zverkov [105] for ichthyosaurians, and Johnson et al. [106] and Young et al. [107,108] for thalattosuchians. Ammonoid taxonomy and biostratigraphy follow Gulyaev and Ippolitov [65], and Kiselev [91].

Institutional abbreviations: **IG**, A.A. Chernov Geological Museum, Institute of Geology of the Komi Scientific Center of Ural Branch of RAS, Syktyvkar, Russia; **MM**, Mining Museum, St Petersburg, Russia; **MRUM**, Mordovian Republican United Museum of Local Lore, named after I.D. Voronin, Saransk, Russia; **NHMUK**, Natural History Museum, London, UK; **NNGASU**, Museum of Nizhny Novgorod State University of Architecture, Building and Civil Engineering, Nizhny Novgorod, Russia; **PIN**, Borissiak Paleontological Institute, Russian Academy of Sciences, Moscow, Russia; **PSM**, Penza State Museum of Local Lore, Penza, Russia; **SGM**, V.I. Vernadsky State Geological Museum of the Russian Academy of Sciences, Moscow, Russia; **SSU**, Saratov State University, Regional Museum of Earth Sciences, Saratov, Russia; **SSTU MEZ**, Museum of Natural History of Saratov State Technical University named after Y.A. Gagarin, Saratov, Russia; **TsNIGR**, F.N. Chernyshev Central Research Geological Survey Museum, St Petersburg, Russia; **UMLH**, Ukhta Museum of Local History, Ukhta, Russia; **YSPU**, Museum at K.D. Ushinsky Yaroslavl State Pedagogical University, Yaroslavl, Russia; **ZIN PH**, Paleoherpetological Collection, Zoological Institute of the Russian Academy of Sciences, St. Petersburg, Russia.

Table 1. Catalogue of marine reptile remains from the Callovian of European Russia.

Repository and Catalogue No, Status	Identification (Historical in Parentheses)	Material	Locality	Stratigraphic Position	Reference
ZIN PH 1/215	Ichthyosauria indet.	posterior dorsal centrum	Zamezhnaya, "Outcrop 14" (label info; but "outcrop 16" in [109]), Komi Republic	middle/upper Callovian boundary strata, beds with *Longaeviceras*	collected by V.S. Kravets
Specimen location unknown	Ichthyosauria indet.	centrum	Adzvavom, Inta District, Komi Republic, "outcrop 14"in [110]	Upper Callovian, *Longaeviceras nikitini* Zone (Boreal standard)	[110]
UMLH MPZ KP 1982/1	Ichthyosauria indet.	anterior presacral centrum	Aiyuva-6, Sosnogorsk District, Komi Republic	middle Callovian?	[111]
IG 722/6	Ichthyosauria indet.	posterior caudal centrum	Aiyuva-6	middle Callovian, *Rondiceras milaschevici* Zone (Boreal standard)	[111]
IG 93/13	Ichthyosauria indet.	anterior presacral centrum	Votcha, Sysola District, Komi Republic	ex situ, likely from the middle Callovian, *K. jason–E. coronatum* zones	[112]
SGM 1358-53	Plesiosauria indet. ("*Cryptoclidus* (?) sp." in [33])	caudal vertebra	Kargort (Yb) at Sysola River, Komi Republic	middle Callovian	[33] (p. 230)
B.A. Mal'kov's private collection	Plesiosauria indet., Pliosauridae indet., Ichthyosauria indet., *Muraenosaurus* sp.	cervical and dorsal vertebrae, plesiosaurian distal propodial	Kargort (Yb)	middle Callovian?	[112,113]
SGM uncatalogued	Ichthyosauria indet.	cast of posterior dorsal centrum	Mytets River, Sychugovy Village (abandoned), Nagorsk District, Kirov Region	Callovian?	[114] collected by Khabakov, 1924
NNGASU 147/2080	Pliosauridae indet.	proximal portion of a large propodial bone	Vyatka–Kama phosphorite field, Verkhnekamsky District, Kirov Region	middle Jurassic, Callovian?	collected by Yu. S. Tamoykin
MM 219	Pliosauridae indet.	partial postcranial skeleton	Gradulevo, Manturovo District, Kostroma Region	middle Callovian or middle Oxfordian	[41]
SGM 1891-02, 03 and 20	cf. *Ophthalmosaurus*	anterior dorsal vertebra, indet. bone fragments, and partial mandible with teeth.	Znamenka, Manturovo District, Kostroma Region	lower Callovian, *Pr. koenigi* Zone, *G. galilaeii* Subzone	collected by V.V. Mitta, 2015

Table 1. *Cont.*

Repository and Catalogue No, Status	Identification (Historical in Parentheses)	Material	Locality	Stratigraphic Position	Reference
SGM 1807	*Liopleurodon ferox*	isolated posterior tooth and partial postcranial skeleton; small plesiosaurian vertebra found in association	Makariev North locality, Makariev District, Kostroma Region	lower Callovian, upper part of *Pr. koenigi Zone*, *G. galilaeii* Subzone	collected by A.V. Stupachenko
SGM 2007-6,7	*Liopleurodon* sp.	three associated teeth	Makariev, Makariev District, Kostroma Region	lower Callovian, *S. calloviense* Zone	collected by A.V. Stupachenko, 2016
SGM 569-1	*Simolestes* sp.	propodial, likely femur	Makariev	lower Callovian, *S. calloviense* Zone, *C. enodatum* Subzone	collected by A.V. Stupachenko and V.V. Mitta
YSPU M/F-45	Pliosauridae indet.	articular surface of a large vertebral centrum	Makariev North	lower Callovian, *S. calloviense* Zone	collected by D.N. Kiselev
YSPU M/F-44	cf. *Cryptoclidus*	nearly complete dorsal vertebra	Makariev North	same as above	collected by D.N. Kiselev
SGM 1574-1	*Simolestes* sp.	left maxilla	Mikhalenino, Makariev District, Kostroma Region	lower Callovian, horizon of reworked fauna of *P. elatmae* and *C. subpatruus* zones at the base of *G. gowerianum* Subzone	[37] collected by A.V. Stupachenko
SGM 1891-10	cf. *Muraenosaurus*	posterior cervical vertebra	Mikhalenino	lower Callovian, *S. calloviense* Zone	collected by A.V. Stupachenko, 2001
SGM 1891-09	cf. *Muraenosaurus*	posterior cervical vertebra	Mikhalenino	same as above	collected by A.V. Stupachenko
SGM 1891-08	*Muraenosaurus* sp.	anterior cervical vertebra	Mikhalenino	same as above	collected by A.V. Stupachenko,
SGM 2007-3(1–5)	*Muraenosaurus* sp.	five isolated teeth	Mikhalenino	lower Callovian, horizon of reworked fauna of *P. elatmae* and *C. subpatruus* zones at the base of *G. gowerianum* Subzone	collected by A.V. Stupachenko, 1999
PIN 5819/3	*Muraenosaurus sp.*	partial tooth	Mikhalenino	same as above	collected by A.V. Stupachenko, 1999
SGM 2007-4	cf. *Tricleidus*	tooth crown	Mikhalenino	same as above	collected by A.V. Stupachenko, 1999
SGM 2007-5	Pliosauridae indet.	fragmant of a large tooth	Mikhalenino	same as above	collected by A.V. Stupachenko, 2016
SGM 2007-8	cf. *Ophthalmosaurus*	tooth	Mikhalenino	same as above	collected by A.V. Stupachenko,
PIN 5819/4	cf. *Ophthalmosaurus*	tooth crown	Mikhalenino	same as above	collected by A.V. Stupachenko, 1999

Table 1. *Cont.*

Repository and Catalogue No, Status	Identification (Historical in Parentheses)	Material	Locality	Stratigraphic Position	Reference
SGM 1891-22	cf. *Ophthalmosaurus*	left jugal	Mikhalenino	same as above	collected by A.V. Stupachenko, 1999
SSU uncatalogued	cf. *Ophthalmosaurus*	left quadrate	Mikhalenino	same as above	collected by A.V. Stupachenko, 2001
SSU uncatalogued	Ophthalmosauria indet.	right nasal, anterodorsal centrum, rib	Mikhalenino	same as above	collected by A.V. Stupachenko, 1998
SGM 1891-21	Ophthalmosauria indet.	a series of nine tail fluke centra with neural arches	Mikhalenino	same as above	collected by A.V. Stupachenko, 1999
SGM 1891-23	Ichthyosauria indet.	long dorsal rib (50 cm proximoditally)	Mikhalenino	same as above	collected by A.V. Stupachenko, 1999
SGM 1960-01	Geosaurini indet.	tooth crown	Mikhalenino	same as above	[40] (as "SGM BX-12") collected by A.V. Stupachenko
SGM 1891-01	Thalattosuchia indet.	autopodial element	Mikhalenino	same as above	Collected by A.V. Stupachenko
PIN 5819/1, 2, 6	Metriorhynchidae indet.	three miscellaneous teeth	Mikhalenino	same as above	[40]
SGM 1891-25	Ichthyosauria indet.	posterior presacral centrum	Mikhalenino	lower Callovian, *S. calloviense* Zone	collected by A.V. Stupachenko,
SGM 1891-24	Ichthyosauria indet.	caudal centrum	Mikhalenino	same as above	collected by A.V. Stupachenko,
SGM 2000-1	*Ophthalmosaurus icenicus*	left premaxilla	Mikhalenino	clays at the boundary of the lower and middle Callovian	collected by A.V. Stupachenko, 1999
SGM 1961	*Ophthalmosaurus icenicus*	skull fragments including quadrate, angular and tooth; atlas-axis and vertebrae from all regions of the column; fragments of pectoral girdle and partial forelimb	Perebory, Rybinsk District, Yaroslavl Region	middle Callovian; *K. jason* Zone, *K. medea* Subzone	collected by K.K. Kotov and N.Z., 2021.
specimen is lost, holotype	*Simolestes* sp. (*Thaumatosaurus calloviensis*)	isolated tooth crown	Rechitsy near Gzhel village, Ramenskoe District, Moscow Region	middle Callovian	[33] (p. 200, pl. II, figs. 1 and 6)
PIN R-3589	cf. *Simolestes*	partial crown	Gzhel, Ramenskoe District, Moscow Region	middle Callovian	collected by M.S. Boiko, 1990
PIN 5818/8	Pliosauridae indet.	partial crown	Rechitsy	middle Callovian	collected by P.A. Gerasimov, 1928

Table 1. *Cont.*

Repository and Catalogue No, Status	Identification (Historical in Parentheses)	Material	Locality	Stratigraphic Position	Reference
PIN 5818/9	*Tyrannoneustes* sp.	tooth crown	Rechitsy	middle Callovian, likely *E. coronatum* Zone	[40] collected by P.A. Gerasimov, 1928
PIN 5819/7	Metriorhynchidae indet.	tooth crown	Rechitsy	middle to lower upper Callovian	[40]
PIN R-3590	cf. *Muraenosaurus*	posterior cervical centrum	Peski Quarry, Kolomensk District, Moscow Region	upper Callovian, *L. lamberti* Zone	[39] (fig. 5b) collected by A. Kuzmenko, 2015
PIN R-3591, PIN R-3592	Cryptoclididae indet.	cervical of juvenile	Peski Quarry	middle Callovian, *E. coronatum* Zone	[39] (fig. 5d); collected by N. Denisova and M. Sushko, 2015
PIN R-3593	Ichthyosauria indet.	four posterior caudal centra	Peski Quarry	upper Callovian, *L. lamberti* Zone	collected by I.V. Ilyasov, 1990s
SGM 1358-11	Plesiosauria indet.	two articulated pectoral or sacral centra	Gorky (nowadays Fruktovaya Station), Lukhovitsy District, Moscow Region	lower Callovian, likely *S. calloviense* Zone, *C. enodatum* Subzone	[33] (p. 149, pl. I)
SGM 1358-12	Plesiosauria indet.	proximal fragment of propodial	Alpatyevo, Lukhovitsy District, Moscow Region	lower Callovian?	[33] (p. 151)
SGM 1358-37	*Muraenosaurus* cf. *leedsi*	cervical vertebra	Alpatyevo	lower Callovian, likely *S. calloviense* Zone, *C. enodatum* Subzone	[33] (p. 235, pl. III, figs. 1 and 2)
PIN R-3600	*Cryptoclidus eurymerus*	partial skeleton, including several cervical, pectoral, dorsal and caudal vertebrae, dorsal ribs and gastralia, partial pubis, humerus, radii, femora	Nikitino, Spasskiy District, Ryazan Region	middle Callovian, *K. jason* Zone	collected by the Club of Junior Paleontologists of the Paleontological Museum, lead by A.S., in 2014–2015
PIN R-3574	Pliosauridae indet.	partial tooth crown	Nikitino	middle Callovian	collected by D.N. Kasantsev, 2011
PIN R-3633, 3634	Ichthyosauria indet.	caudal centrum and jawbone fragments	Nikitino	middle Callovian, *E. coronatum* Zone, *K. grossouvrei* Subzone	collected by the Club of Junior Paleontologists of the Paleontological Museum, lead by A.S., in 2014-2015
PIN R-3573	*Liopleurodon ferox*	large caniniform tooth crown	Mikhaylovcement Quarry, Mikhaylov District, Ryazan Region	same as above	[39] collected by A. Kuraev and A.S., 2012
V. Bakhtin private collection	*Simolestes* sp.	partial tooth crown	Zmeinka Quarry, Mikhaylov District, Ryazan Region	upper Callovian	[39] (fig. 1C)

Table 1. *Cont.*

Repository and Catalogue No, Status	Identification (Historical in Parentheses)	Material	Locality	Stratigraphic Position	Reference
PIN R-3604	Pliosauridae indet.	tooth crown fragment	Spartak Quarry, Mikhaylov District, Ryazan Region	lower Callovian, *Pr. koenigi* Zone	collected by A.S.
PIN R-3595	cf. *Muraenosaurus*	posterior cervical vertebra	Mikhaylovcement Quarry	middle Callovian, *E. coronatum* Zone, *K. grossouvrei* Subzone	[39] (fig. 5c) collected by A. Churkin
PIN R-3621	*Cryptoclidus* sp.	tooth crown	Mikhaylovcement Quarry	lower Callovian, upper part of *P. elatmae* Zone	collected by D.G., 2017
PIN R-3670	Cryptoclididae indet.	cervical vertebra	Mikhaylovcement Quarry	upper Callovian, *P. athleta* Zone, *K. kuklikum* Subzone	collected by S.V. Grishin, 2017
PIN R-3709	Cryptoclididae indet.	pectoral vertebra	Mikhaylovcement Quarry	same as above	collected by Ershova O.G.
PIN R-3606	Cryptoclididae indet.	vertebra	Mikhaylovcement Quarry	middle Callovian, *E. coronatum* Zone, *K. obductum* Subzone	collected by Ershova O.G.
PIN R-3201, PIN R-3202	Plesiosauria indet.	caudal centrum and dorsal neural arch fragment	Mikhaylovcement Quarry	upper Callovian, *P. athleta* Zone, *K. kuklikum* Subzone	collected by K. Nazarov, 2011
PIN R-3596	Plesiosauia indet.	posterior caudal centrum	Mikhaylovcement Quarry	middle Callovian, *E. coronatum* Zone, *K. grossouvrei* Subzone	collected by S. Rossiysky, 2016
PIN R-4956	*Ophthalmosaurus icenicus*	humerus, radius, intermedium, radiale	Zmeinka Quarry	upper Callovian, *P. athleta* Zone *K. kuklikum* Subzone	collected by L.V. Kulagina, 2023
PIN R-2516	cf. *Ophthalmosaurus*	anterior accessory epipodial element	Mikhaylovcement Quarry	same as above	[39] (fig. 12D) collected by A.S., 2010
PIN R-3200	Ichthyosauria indet.	posterodorsal centrum	Mikhaylovcement Quarry	same as above	collected by N. Ushakov, 2011
PIN R-3594	Ichthyosauria indet.	dorsal centrum	Zmeinka Quarry	upper Callovian, *L. lamberti* Zone	collected by I.A. Dadykin, 2016
PIN R-2451	*Tyrannoneustes* sp.	tooth crown	Mikhaylovcement Quarry	middle Callovian, *E. coronatum* Zone, *K. grossouvrei* Subzone	[39,40] collected by A.S., 2010
PIN R-3253	cf. *Thalattosuchus*	tooth crown	Mikhaylovcement Quarry	same as above	[39,40] collected by A.S. Shmakov, 2012
SGM w/o number	Geosaurinae indet.	cervical vertebra	Mikhaylovcement Quarry	middle to upper Callovian; collected ex situ	collected by K. Volkov
TsNIGR 157a/649	Pliosauridae indet.	sacral vertebra	Yelatma, Kasimov District, Ryazan Region	indet. Callovian	collected by N.A Bogoslovsky

Table 1. *Cont.*

Repository and Catalogue No, Status	Identification (Historical in Parentheses)	Material	Locality	Stratigraphic Position	Reference
TsNIGR 144/1712	cf. *Muraenosaurus*	dorsal vertebra	Yelatma	indet. Callovian	collected by E.M. Lutkevich, 1925
SGM 1891-13, 14, 16, 17, 19	Ichthyosauria indet.	weathered centrum (17), anterior to middle caudal centrum (19); posterior caudal centrum (14); coracoid lateral fragment (13); small rib fragment (16)	Uzhovka Quarry, Pochinki District, Nizhny Novgorod Region	lower Callovian, lower part of *Pr. koenigi* Zone	collected by D.G., 1995
SGM 1891-18	Ichthyosauria indet.	small caudal vertebra	Uzhovka Quarry	middle Callovian, *K. jason* Zone, *K. obductum* Subzone	collected by D.G., 1995
SGM 1891-15	Ichthyosauria indet.	posterior dorsal centrum	Uzhovka Quarry	lower Callovian, *P. elatmae or C. subpatruus* zones	collected by D.G., 1995
SGM 1891-05	Plesiosauria indet.	dorsal centrum of juvenile	Trofimovshchina-2, Romodanovo District, Republic of Mordovia	lower Callovian, *P. elatmae* Zone	collected by N.Z., 2018
SGM 1891-06	*Ophthalmosaurus* cf. *calloviensis*	radius	Trofimovshchina-2	same as above	collected by N.Z., 2018
SGM 1891-04	Ichthyosauria indet.	middle dorsal centrum	Trofimovshchina-2	lower Callovian, *P. elatmae* Zone, upper clayey part	collected by A.P. Ippolitov, 2016
SGM 1891-07	Ichthyosauria indet.	rib	Trofimovshchina-2	same as above	collected by A. I., 2016
SGM 1891-11	Ichthyosauria indet.	weathered vertebral centrum and proximal rib fragments	Trofimovshchina-2	same as above	collected by A. I. and N.Z., 2018
PSM 3999-4004	*Ophthalmosaurus* cf. *calloviensis*	jaw fragments with teeth; sclerotic ring; dorsal and caudal centra; neural arches; proximal fragment of humerus; radius, ulna and phalanges	Moksha River near Rybkino Village, Kovylkino District, Republic of Mordovia	lower Callovian, *P. elatmae* to *Pr. koenigi* zones	collected by A.A. Stuckenberg, 1925
Collection of V.M. Efimov	*Ophthalmosaurus* sp.	anterior part of the skeleton including skull, ribs and limb elements	Sinyakovo Village, Krasnoslobodsk District, Republic of Mordovia	lower Callovian, *P. elatmae* Zone	[38]
MRUM 1315/1	cf. *Thalattosuchus*	tooth crown	Gumny, Krasnoslobodsk District, Republic of Mordovia	lower Callovian, *P. elatmae* to *Pr. koenigi* Zones	[40]
SGM 1891-27	Plesiosauria indet.	scapular fragment	Lebedinsky Quarry; Belgorod Region	lower Callovian, *P. elatmae* Zone	collected by N.Y. Bragin, 2000

Table 1. *Cont.*

Repository and Catalogue No, Status	Identification (Historical in Parentheses)	Material	Locality	Stratigraphic Position	Reference
SSU 137/211	cf. *Muraenosaurus*	cervical vertebra	CHP-5 (TETs-5), Saratov, Saratov Region	lower Callovian, *C. subpatruus–S. calloviense* zones	collected 1985
SSU 137/200	Ichthyosauria indet.	caudal centrum	sovkhoz Leninskiy Put′ near CHP 5 power station (TETs-5), Saratov, Saratov Region	lower Callovian, *P. elatmae* Zone	collected by M.A., 1984
SSU 14/31 (historical No 104a/29)	Geosaurini indet.	tooth crown	sovkhoz Leninskiy Put′ near CHP 5	lower Callovian, *P. elatmae* Zone	[35,40] collected by M.A., 1984
SSU 104a/27, holotype	*Ophthalmosaurus calloviensis* (*Khudiakovia calloviensis*)	partial forelimb	construction pit near Dubki Village, Saratov District	upper Callovian? (or lower Callovian)	[35] collected 1977 by the Children's Club of Local History of Saratov
PIN R-3575	Pliosauridae indet.	Crown fragment	Dubki, Saratov District, Saratov Region	upper Callovian, *L. lamberti* Zone	collected by R. Yu. Stredinin, 2011
SGM 1445-97−120	Rhomaleosauridae indet.	Ilium, partial ischium and pubis; dorsal and caudal centra; rib, metatarsals and phalanges	Verkhnaya Dobrinka, Zhirnovsk district, Volgograd Region	lower Callovian, *P. elatmae* Zone	[34,36] collected by E.V. Milanovsky, 1920
SSTU MEZ 3/4	cf. *Ophthalmosaurus*	partial skeleton	Gvardeyskoe Village, Krasnoarmeysk District, Saratov Region	lower Callovian, *P. elatmae* Zone	donated to SSTU by S.A. Bratashova in early 2000s
Whereabouts of specimen unknown	Ichthyosauria indet.	Partial skeleton	0.5 km south of Bartolomeevka, Saratov District, Saratov Region	lower Callovian, *P. elatmae* Zone	[115] collected by V. Grizbovskiy, 1998

5. Results of the Study of the Marine Reptile Material

5.1. Pliosaurids

The majority of marine reptile remains collected from the Callovian deposits of European Russia are plesiosaurian teeth and vertebrae. Among these, robust teeth of pliosaurids are quite common.

5.1.1. *Liopleurodon ferox*

The specimen, PIN R-3573, is a very large caniniform tooth crown with a partial root. The apicobasal length of the crown is 123 mm, and its basal diameter is 51 mm. This tooth crown is the largest ever reported for Callovian pliosaurids, among which no specimens with crowns exceeding 110 mm in apicobasal height were hitherto reported [85,116]. PIN R-3573 can be referred to the genus *Liopleurodon*, as it is consistent with the holotypic crown in its large size, circular cross-section, coarse enamel ridges triangular in cross-section, rare on the labial surface, numerous on the lingual surface, and all reaching the base of the crown [12,116]. The crown of PIN R-3573 is slightly curved and terminates in a pointed, sharp apex. The labial (outer convex) side bears only two small ridges in its basal part (Figure 5A); the rest of its surface is subtly rugose, with numerous small ridglets. The lingual, mesial, and distal surfaces bear numerous sharp and nearly straight ridges, which emerge at the base of the crown and extend towards the apex to a varying extent. Only five of these reach the tip (Figure 5D). The ridges enlarge towards the apex, being approximately 0.5 mm wide at the base of the crown and more than doubling in width towards the apex.

In addition to PIN R-3573, the crown morphology described above is present in three associated teeth, SGM 2007-6,7 (Figure 5J–M), from the lower Callovian of the Unzha River. Furthermore, one tooth with such morphology (Figure 5E–I) was found in association with the postcranial skeleton SGM 1807, which is described below. The largest of these teeth (SGM 2007-7) is similar to the above-described PIN R-3573, whereas smaller teeth are posterior "ratchet"-type teeth with stout and strongly curved crowns bearing frequent ridges around their entire circumference (Figure 5E–L).

A partial skeleton of a large pliosaurid, SGM 1807, unearthed by Andrey V. Stupachenko in the Makariev North locality at the Unzha River, Kostroma Region, is among the most significant marine reptile discoveries in the Callovian of European Russia to date. This specimen can be referred to the genus *Liopleurodon* based on the following features: a tooth crown that is conical in shape, bearing well-spaced and coarse ridges, all reaching the crown base, and the femur with elongated diaphysis giving the bone "*almost parallel sides for the essential part of its length*", as diagnosed by Tarlo [6] (p. 167). No cranial remains were found for SGM 1807, except a single tooth (Figure 5E–I). The total length of the tooth, including the crown and root, is 56 mm. The crown apicobasal height is 18 mm, and the basal diameter is 13 mm.

The vertebrae are not preserved in SGM 1807; however, the cervical, dorsal, and sacral ribs are present. Anterior cervical ribs are robust, double-headed, with widely spaced capitulum and tuberculum (Figure 6P–V). Their anterolateral edge is tapered, their posterior surface bears a pronounced trough (Figure 6T,U), and their distal end is expanded, forming anterior and posterior processes, with the posterior process being more prominent than the anterior (Figure 6P–S). Five complete dorsal ribs and multiple rib fragments are preserved. The dorsal ribs are large and massive (the longest preserved rib is 550 mm long and 45 mm in cross-section diameter). The dorsal ribs are curved and circular in cross-section (Figure 6W–Y). One sacral rib is preserved; it is short and robust, possessing anteroposteriorly expanded proximal and distal ends, which are twisted relative to one another (Figure 6Z).

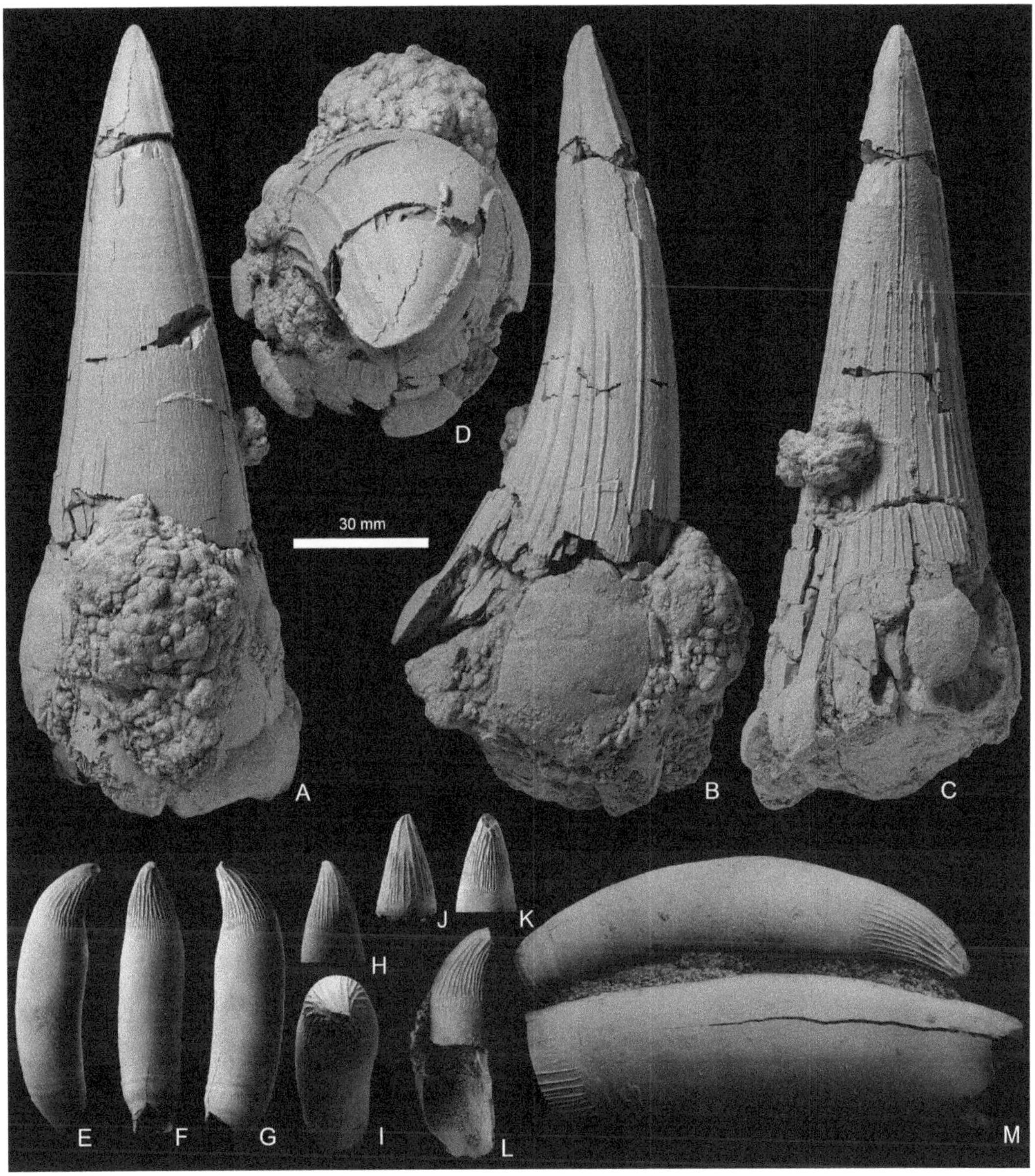

Figure 5. Teeth of *Liopleurodon ferox* from the Callovian of European Russia. Mesial ("caniniform" type) teeth PIN R-3573 (**A–D**) from the middle Callovian of Ryazan Region, and SGM 2007-7 (**M**) from the lower Callovian of Makariev, Kostroma Region. Distal ("ratchet" type) teeth SGM 1807 (**E–I**) and SGM 2007-6 (**J–L**) from the lower Callovian of Makariev, Kostroma Region. Views: anterior or posterior (**D**,**E**,**G**,**L**), lingual (**C**,**F**,**K**), labial (**B**,**H**,**J**), and apical (**D**,**I**).

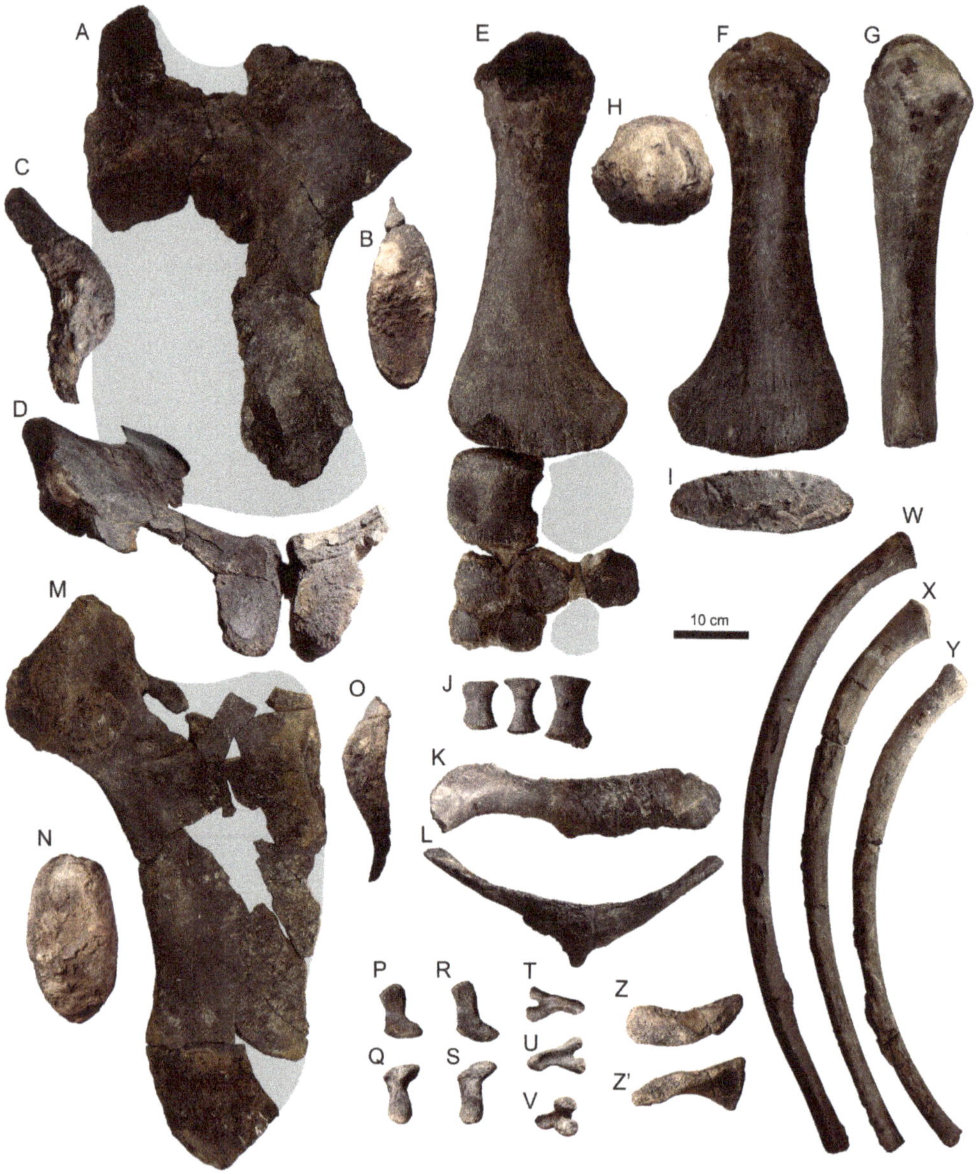

Figure 6. Postcranial skeleton of *Liopleurodon ferox*, SGM 1807. Right coracoid (**A–D**) in dorsal (**A**), glenoidal (**B**), mesial symphysial (**C**), and anterior (in articulation with the medial portion of left coracoid) (**D**) views. Partial right hindlimb in ventral view (**E**); right femur in dorsal (**F**), posterior (**G**), proximal (**H**), and distal (**I**) views. Phalanges (**J**). Interclavicle (?) in dorsal (**K**) and anterior (**L**) views. Left ischium in dorsal (**M**), acetabular (**N**), and mesial, symphysial (**O**) views. Cervical ribs (**P–V**) in dorsal (**P**,**R**), ventral (**Q**,**S**), anterior (**T**), posterior (**U**), and proximal (**V**) views. Dorsal ribs (**W–Y**) and sacral rib (**Z**,**Z′**). Reconstructed outlines of broken parts are shown in gray.

Both coracoids are partially preserved, lacking posteromedial portions of the plates. They are large and thin elements with a mediolateral width of 450 mm. The thickest part of the bone is the glenoid region. The angle between the glenoid and scapular surfaces is obtuse and measures about 150°. The glenoid contribution is strongly inclined anteriorly, with its posterior portion markedly protruding laterally, unlike in any other Callovian pliosaurid, except for *Liopleurodon* (N.Z.'s pers. obs. on NHMUK PV R2738; see text-fig. 5 in [5]). This is an additional feature supporting our generic referral of SGM 1807. The scapular facet is triangular in outline and gradually transitions medially into the anterior coracoid shelf (Figure 6B,D). The medial articular facet of the coracoid is S-curved with convex dorsal and concave ventral surfaces in the thickest part (Figure 6C). The anteromedial process is thus directed anteroventrally. A bilateral element of problematic identity may represent an interclavicle. It consists of two slender and compressed rami and a short posterior process (Figure 6K,L). It is 40 cm wide. If such identification is correct, SGM 1807 is among the very few post-Liassic pliosaurids known to have a clavicle and/or interclavicle (clavicles and/or interclavicles are reported for *Peloneustes* [5,10], *Marmornectes* [9], and *Eardasaurus* [11]).

The pelvic girdle is preserved as two fragmental pubes and an almost complete left ischium. The ischium measures 750 mm anteroposteriorly, from its posterior end to the anterior surface of the acetabular facet. It is 445 mm wide mediolaterally, from the acetabulum to the medial symphysis, and therefore has a length-to-width ratio of 1.69. The medial and lateral surfaces of the ischiadic blade are approximately parallel posteriorly, converging only weakly. The ischial neck, connecting the acetabular process to the ischial blade, is relatively wide due to shallow excavations of anterior and posterior edges, and its anterior border is sharply edged. The acetabular process of the ischium is 230 mm long anteroposteriorly and 110 mm high dorsoventrally. It is divided into facets for the ilium (facing posterolaterally), acetabulum (facing laterally), and pubis (facing anteriorly). Overall, this element is in agreement with the ischia of *Liopleurodon* from the Oxford Clay [5].

A partial right hindlimb is preserved, including the femur in articulation with the tibia and mesopodial elements, as well as several isolated phalanges (Figure 6E,J). The femur is 535 mm long proximodistally and 238 mm wide anteroposteriorly at its distal end. The femur has a long shaft with almost parallel anterior and posterior sides for the essential part of its length (Figure 6E,F). The proximal portion is expanded to form a capitulum and dorsal trochanter (Figure 6H). The distal end bears two subequal facets for the tibia and fibula, making an obtuse angle to one another. The tibia is isometric, as long as it is wide (120 mm). It has a convex and tapered anterior edge and concave posterior edge that contribute to the extensive spatium interosseum between the tibia and fibula.

The distal expansion of the femur in SGM 1807 is less pronounced than that in any known Oxford Clay *Liopleurodon* specimen [5,117], which may suggest that SGM 1807 represents a distinct species of this genus.

5.1.2. *Simolestes* sp.

The left maxilla SGM 1574-1 is anteroposteriorly short and massive; it bears a small rounded external naris and its posterior edge is irregularly digitated for contacts with other cranial elements (i.e., frontal, prefrontal and lacrimal), which is typical for pliosaurids [118]. It lacks the orbital contribution known for rhomaleosaurids and plesiosauroids [119]. The posterior edge of the bone is incomplete ventrally. The preserved length of SGM 1574-1 is 260 mm, maximum dorsoventral height is 90 mm. There are nine alveoli on the preserved part, and several smaller alveoli could have been present on the broken posteroventral portion. The first two alveoli do not exceed 20 mm in diameter (Figure 7B), and the diameter of the subsequent alveoli increases abruptly, reaching 33 mm at the third alveolus. Starting from the sixth, the diameter of the alveoli decreases (Figure 7B). The body of the bone expands laterally in the region of the third to fifth alveoli, which had the largest caniniform teeth. The significant anteroposterior shortening, robustness, and circular outline of the external naris and alveolar pattern allow the referral of SGM 1574-1 to *Simolestes*, the only Callovian pliosaurid genus with such a short and robust snout [5,6,116].

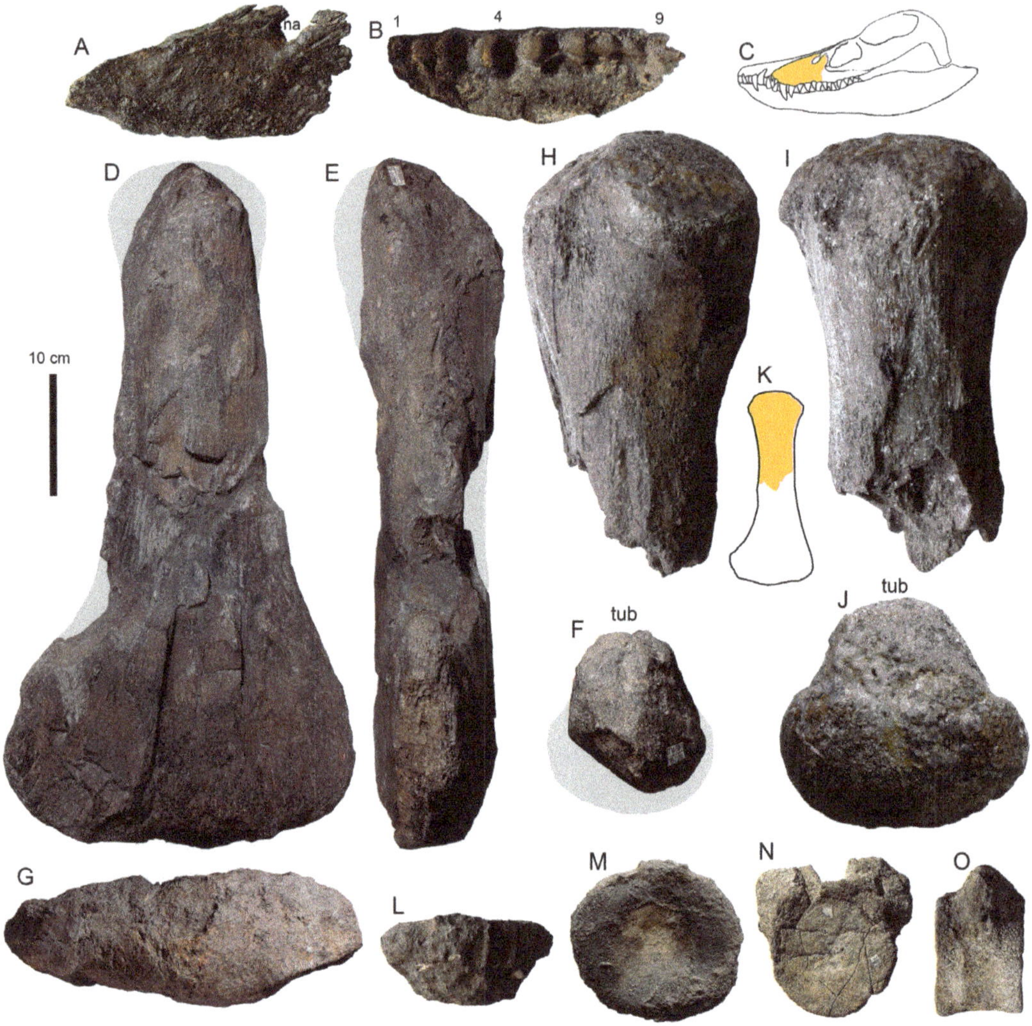

Figure 7. Pliosaurid remains. Left maxilla of *Simolestes* sp. SGM 574-01 (**A–C**), in lateral (**A**) and ventral (**B**) views. (**C**) Position of the element in the skull (in orange). Femur of *Simolestes* sp. SGM 569-1 (**D–G**), in dorsal (**D**), posterior (**E**), proximal (**F**), and distal (**G**) views. Proximal portion of a large propodial (likely femur), NNGASU 147/2080 (**H–K**), in anterior or posterior (**H**), dorsal (**I**), and proximal (**J**) views. (**K**) Position of the fragment (in orange). Articular portion of the vertebral centrum, YSPU M/F-45, in (?) dorsal (**L**) and articular (**M**) views. Pliosaurid sacral vertebra, TsNIGR 157a/649, in articular (**N**) and right? lateral (**O**) views. Abbreviations: 1–9, positions of maxillary alveoli; na, external nares; tub, dorsal trochanter/tuberosity.

The teeth of *Simolestes* are distinctive in having fine apicobasal enamel ridges, more numerous on the lingual surface of the crown [116]. The ridges on the labial surface always begin some distance from the base, which is a constant feature of the teeth of this taxon used in its recognition [5,6,116]. The largest teeth of *Simolestes* have no ornamentation on the labial surface of the crown; the individual enamel ridges are narrower, less raised from the surface of the enamel, and apparently do not coarsen upwards in comparison to *Liopleurodon*, the individual ridges can be irregularly roughened and wavy [116]. A tooth crown described by Bogolubov [33] from the middle Callovian of Gzhel greatly matches these characteristics, and we concur with Tarlo [6] in its referral to *Simolestes*. Two other teeth from the middle and upper Callovian of Moscow and Ryazan regions (PIN R-3589 from Gzhel and specimen in the private collection of V. Bakhtin from the Zmeinka Quarry, see [39]) match some, or all of, the above characteristics (Figure 8A–C); therefore, they are here referred to *Simolestes* sp.

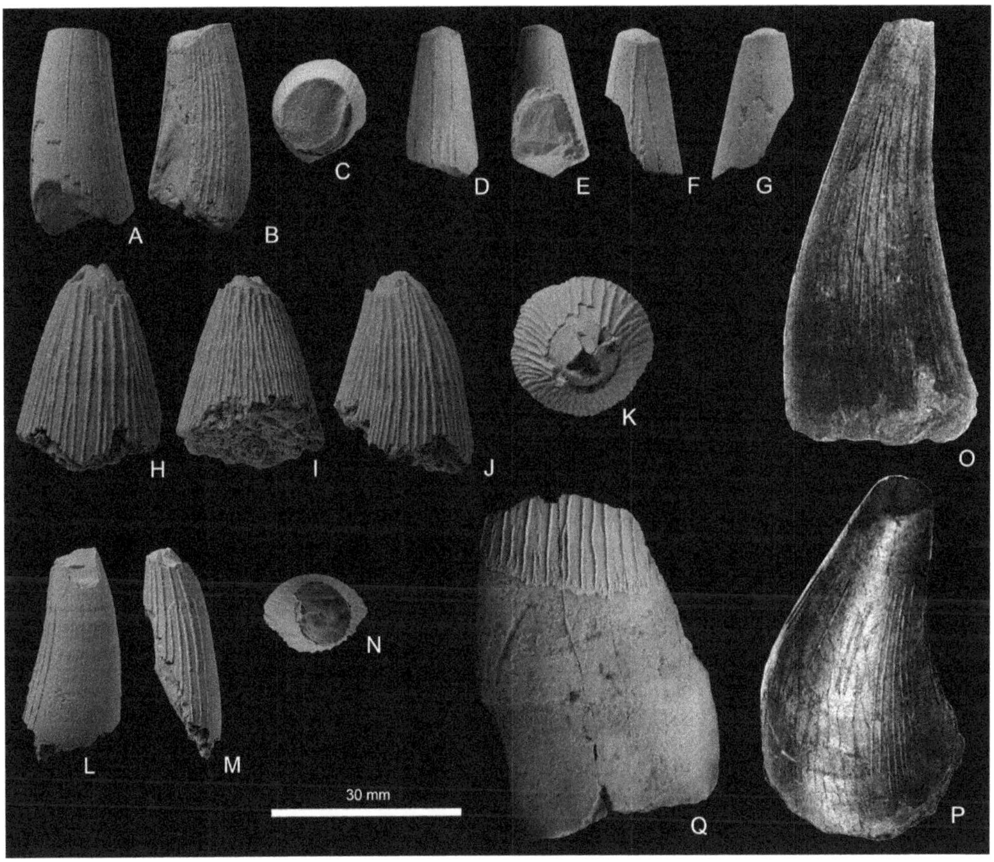

Figure 8. Pliosaurid teeth from the Callovian of European Russia. Pliosauridae cf. *Simolestes* PIN R-3589 (**A–C**) from the middle Callovian of Gzhel; Pliosauridae indet., PIN 5818/8 (**D–G**) from the middle Callovian of Rechitsy; Pliosauridae indet. PIN R-3574 (**H–K**) from the middle Callovian of Nikitino; Pliosauridae indet. PIN R-3575 (**L–N**) from the upper Callovian of Dubki; lost holotype of "*Thaumatosaurus calloviensis*" (**O,P**), referred to *Simolestes* sp. herein, from the middle Callovian of Rechitsy, reproduced from Bogolubov [33] (pl. II, figs. 1 and 6); Pliosauridae indet. SGM 2007-5 (Q) from the lower Callovian of Mikhalenino, Kostroma Region. Views: labial (**A,E,H,L**), lingual (**D,I**), mesial or distal (**B,F,G,J,M,P,Q**), apical (**C,K,N**), and oblique labial from the apex (**O**).

Of interest is a large propodial, SGM 569-1 (Figure 7D–G), from the lower Callovian of the Unzha River. It is 58 cm long proximodistally, 31 cm wide anteroposteriorly at the distal end, and 11 cm thick distally. Its capitulum is largely broken off, so the complete length of the bone could have reached 60 cm. The great distal expansion of the propodial clearly marked off from the shaft allows the referral of SGM 569-1 to *Simolestes*, which shares such propodial configuration [5], differing it from other Middle and Late Jurassic pliosaurids [5,6]. If this referral is correct, SGM 569-1 is among the largest representatives of the genus, as it is 16% longer than the femur of the holotype (NHMUK PV R 3319).

5.1.3. Pliosauridae gen. et sp. indet.

There are several specimens, represented by fragmentary tooth crowns and some postcranial remains attributable to pliosaurids, but not diagnostic at the genus level.

Tooth crown fragments from Dubki (PIN R-3575), Nikitino (PIN R-3574), Gzhel (PIN 5818/8), and Mikhalenino (SGM 2007-5) localities (Figure 8) do not show distinct features for referral to a particular pliosaurid genus, although the relatively large size of PIN R-3575 and PIN R-3574 and the semicircular cross-section of enamel ridges suggest the referral to *Simolestes*, rather than *Liopleurodon*.

NNGASU 147/2080 is the proximal portion of a very large propodial bone from the Kama River basin (Figure 7H–J). Its preserved length is 37 cm, and the diameter of the capitulum is 23 cm. The precise locality and age of this specimen is not recorded. Based on the museum label, it is Middle Jurassic in age, which implies the Callovian age, although the Bathonian age cannot be ruled out, considering the stratigraphy of the region, e.g., [71,114]. This specimen is equally large to the "Stewartby Pliosaur" NHMUK PV R 8322 from the *E. coronatum* Zone of the UK [120] and indicates the presence of large pliosaurids, around 8–9 m long, in the Middle Jurassic of European Russia.

A sacral vertebra TsNIGR 157a/649 (Figure 7N,O), from Yelatma, Ryazan Region, has moderate size with a 115 mm maximum diameter. A slightly larger (143 mm in maximum diameter) is the articular surface of a centrum, YSPU M/F-45, from the lower Callovian of the Makariev locality at the Unzha River (Figure 7L,M).

A partial postcranial skeleton MM 219 was described and figured by Riabinin [41], who somewhat conventionally referred it to *Peloneustes philarchus*. However, apart from the relatively small size for a pliosaurid, no distinct character supports this referral. Therefore, we follow the opinion of other researchers [43,44] and consider this specimen as Pliosauridae indet.

5.2. *Plesiosauroids*

5.2.1. *Cryptoclidus eurymerus*

The specimen PIN R-3600 is a partial postcranial skeleton comprising several cervical, pectoral, dorsal, and caudal vertebrae, rib fragments, some gastralia, fragments of girdle, and limb elements (Figures 9 and 10). This specimen originates from the middle Callovian of Nikitino, Ryazan Region. It was excavated by the Club of Junior Paleontologists of the Paleontological Museum, Moscow, led by one of us (A.S.) in the summer of 2014 and 2015. This specimen can be positively referred to *Cryptoclidus eurymerus* based on anterior expansion of the humeral distal portion bearing the radial facet and the respectively enlarged anterior expansion of the radius portion bearing the humeral facet, which is more than twice as long as the facet for the radiale (autapomorphies of *C. eurymerus* [7]). Neural arches fused with the centra and radii angular and tightly articulated with humeri indicate that the specimen is an osteologically mature individual [7]. This is also supported by the size of preserved elements; the humerus is 34.5 cm long and the femur 33.5 cm long, which agrees with the dimensions of large *C. eurymerus* specimens reported by Andrews [4], and indicates the total length of PIN R-3600 being around 4.2 m.

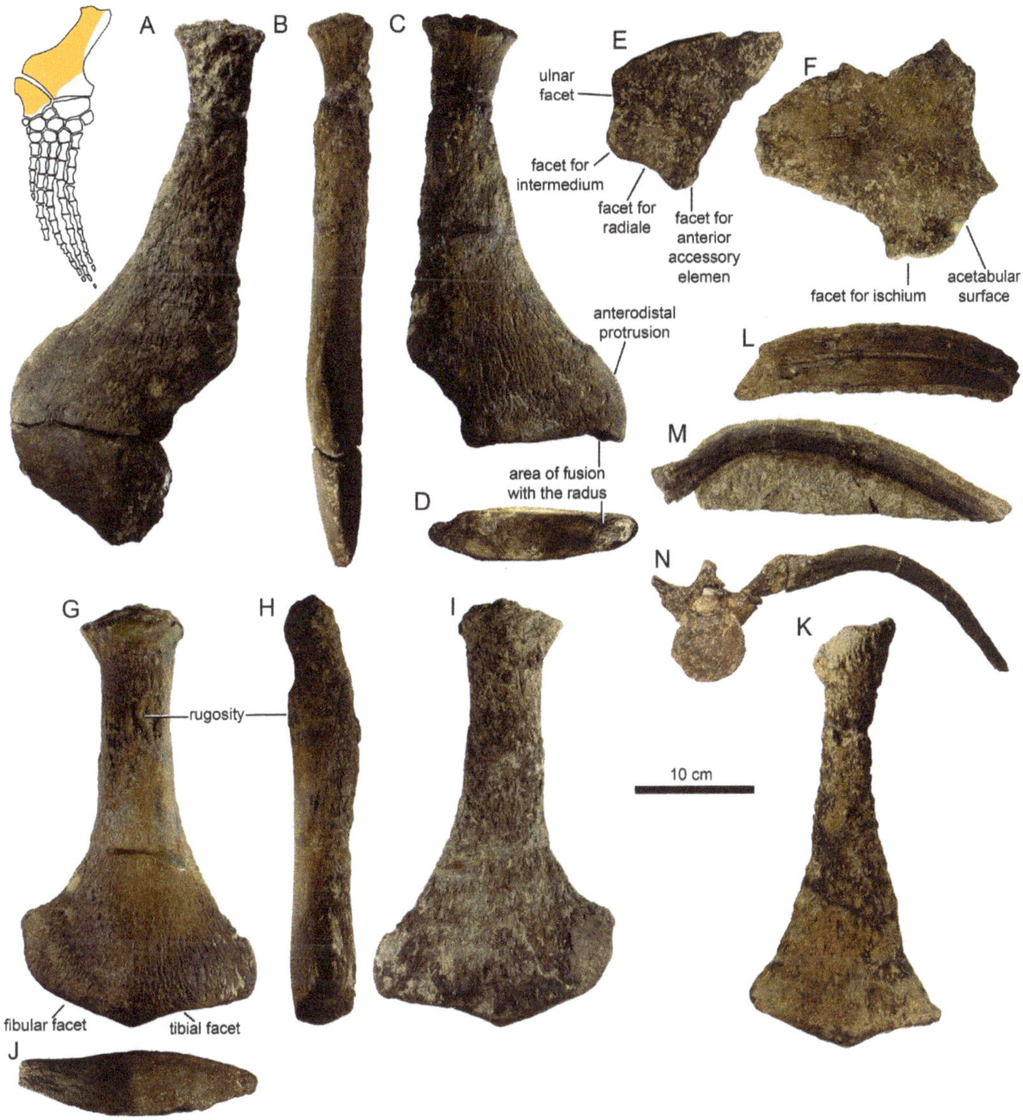

Figure 9. Postcranial elements of *Cryptoclidus eurymerus*, PIN R-3600, from the middle Callovian of Nikitino, Ryazan Region. Left humerus in articulation with the radius in dorsal (**A**) and anterior (**B**) views; reconstruction of their position in the limb (in orange), left humerus in ventral (**C**) and distal (**D**) views. Right radius in dorsal view (**E**). Partial pubis in dorsal view (**F**). Left femur in ventral (**G**), anterior (**H**), dorsal (**I**), and distal (**J**) views. Partial right femur in dorsal view (**K**). Two gastralia (**L**), dorsal rib in anterior view (**M**), and in articulation with dorsal vertebra (**N**).

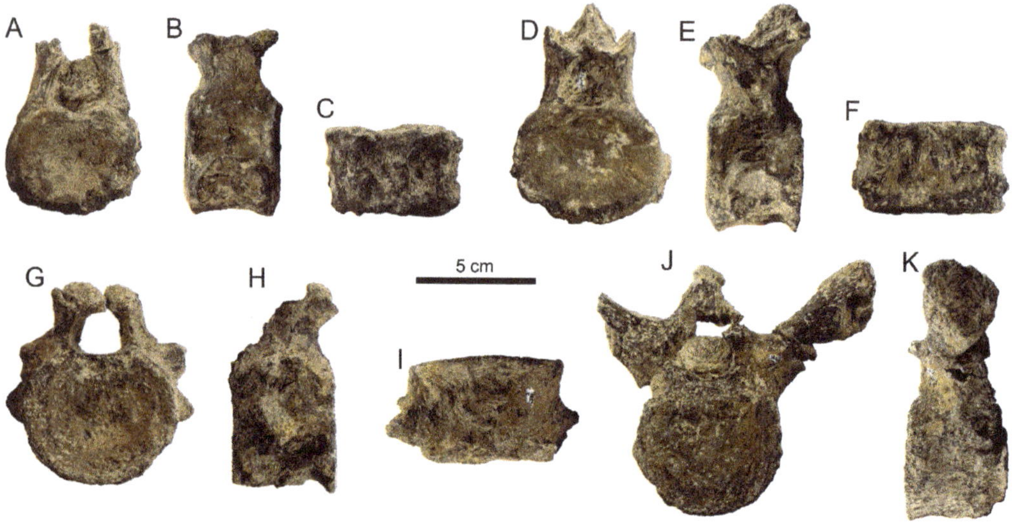

Figure 10. Vertebrae of *Cryptoclidus eurymerus* PIN R-3600, from the middle Callovian of Nikitino, Ryazan Region. Cervical vertebrae (**A–F**) in anterior (**A**,**D**), left lateral (**B**,**E**), and ventral (**C**,**F**) views. Pectoral vertebra (**G–I**) in anterior (**G**), right lateral (**H**), and ventral (**I**) views. Dorsal vertebra in anterior (**J**) and left lateral (**K**) views.

The preserved cervical and pectoral vertebrae have concave articular surfaces (commonly referred to as amphicoelous type), oval in outline (Figure 10A,D,G). The cervical and pectoral centra are mediolaterally wider than they are long anteroposteriorly and high dorsoventrally (Figure 10A–I). There are no ventral and lateral keels. The prezygapophyses are trough-like, with articular surfaces faced anteromedially, they are not fused medially and their combined width is narrower than the centrum width. The preserved middle dorsal vertebra has circular in outline articular facets (Figure 10J), its transverse processes are raised high dorsally, and the long axis of the rib facet is nearly vertical (Figure 10K). When articulated with the rib, it results in a marked posterior inclination of the dorsal rib and a more dorsoventrally compressed body cross-section (Figure 9N) than that reconstructed by O'Keefe et al. [121] (fig. 7B). This indicates that *Cryptoclidus* was more similar to the Oxfordian *Tatenectes* in its low body shape than was previously thought [121].

From the forelimbs, the left humerus and both radii are preserved, although heavily weathered (Figure 9A–E). As noted above, the humeri and radii are diagnostic in *Cryptoclidus eurymerus* in forming a strong anterior protrusion of the leading edge of the forelimb, which is well pronounced in PIN R-3600. In the left forelimb, the radius and humerus are so tightly articulated that they become partially fused anteriorly.

The left femur is nearly completely preserved, with its proximal portion only slightly weathered (Figure 9G,H), whereas the right femur is heavily weathered (Figure 9K). The femora are well consistent with the femora described and figured for *Cryptoclidus eurymerus* [4,7]. The femur has a slender and long shaft and is not as greatly expanded distally as the humerus (Figure 9F–H). The distal end of the femur is dorsoventrally thicker than that of the humerus (45 vs. 37 mm).

Other plesiosaurian specimens known from the Callovian of European Russia are represented by isolated bones (mostly vertebrae) and teeth. The teeth of the Oxford Clay cryptoclidids have been demonstrated to be diagnostic [7,122]. In the available plesiosaurian teeth from the Callovian of European Russia, three morphotypes occur, which are in agreement with the teeth of *Muraenosaurus*, *Cryptoclidus*, and *Tricleidus* ([7] and N.Z.'s pers. obs. on NHMUK specimens).

5.2.2. Isolated Plesiosauroid Teeth

Several teeth (SGM 2007-3/1–3/5; PIN 5819/3) are referable to *Muraenosaurus*. The largest of these (SGM 2007-3/1) has a tooth crown 38 mm high. The tooth crown is slender and gently curved, subcircular in cross-section, becoming more oval to the apex (Figure 11E,F,R). Its labial side is largely smooth, with a band of short and discontinuous ridges present at the base (Figure 11D,J,U). The lingual side is ornamented by numerous ridges, some of which extend almost to the apex, but none reach it (Figure 11B,H,O). The ridges are straight and sparse in the upper half of the crown; to the base, they became numerous and form a vermiculation pattern (Figure 11A,B,G,H,S,T). This morphology perfectly matches that described for *Muraenosaurus leedsii* by Brown [7] and among Callovian plesiosaurians found only in this species (*Cryptoclidus* has small teeth with rare ridges, *Tricleidus* has ridges on the labial surface [7]).

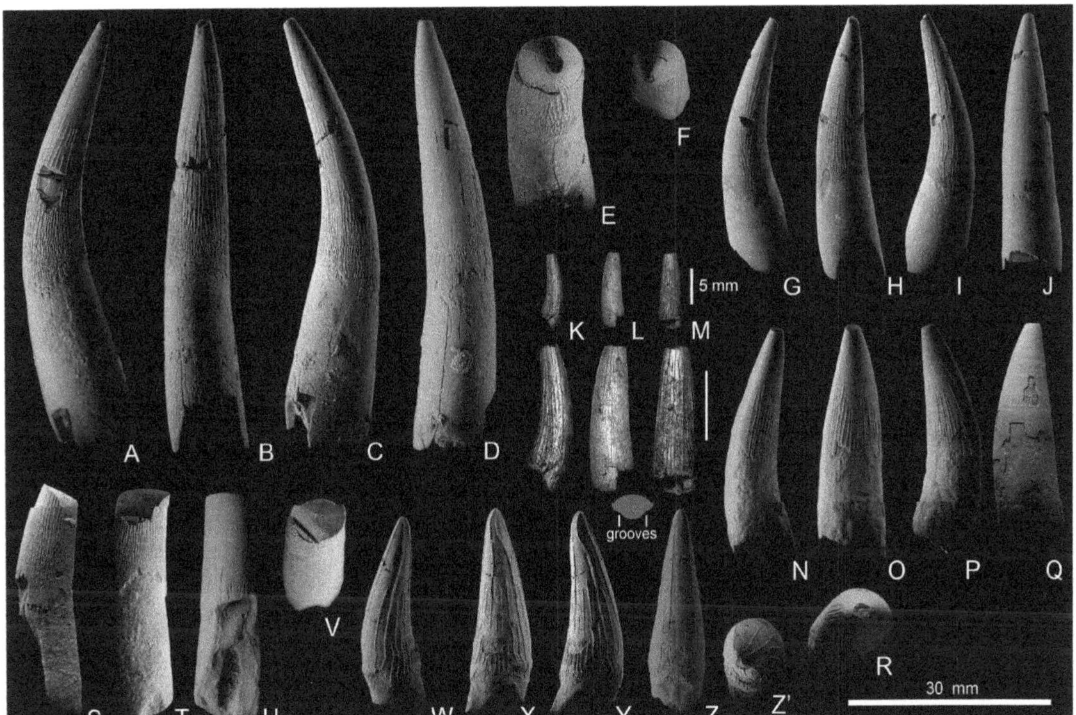

Figure 11. Isolated cryptoclidid teeth from the lower Callovian of Kostroma (**A–J**, **N–Z′**) and Ryazan (**K–M**) regions. *Muraenosaurus* sp. SGM 2007-3/1 (**A–E**), SGM 2007-3/2 (**F–J**), SGM 2007-3/3 (**N–Q**), PIN 5819/3 (**S–V**). *Cryptoclidus* sp. PIN R-3621 (**K–M**). cf. *Tricleidus* SGM 2007-4 (**W–Z′**). Views: mesial or distal (**A,C,G,I,K,N,P,S,W,Y**), labial (**D,J,L,Q,U,Z**), lingual (**B,H,M,O,X**), apical (**E,F,R,V,Z′**). Occlusal wear facets are visible at the tooth root on (**C**) and (**G,H**). K–M are SEM photographs with the bottom row magnified x2 relative to the top row and schematic cross-section of the crown with apicobasal lingual grooves shown below.

One small tooth crown, PIN R-3621, from the lower Callovian of Mikhaylovcement Quarry has a peculiar cross-section, slightly flattened on the labial side and undulating on the lingual side, due to the two apicobasal grooves separating the lingual surface onto three lobes. The ridges are rare; the two longest mesial and distal ridges separate the smooth labial surface from the ridged lingual side, and between them, there are ten fine ridges on the lingual side, none of which reach the apex (Figure 11M). This morphology perfectly matches that of *Cryptoclidus eurymerus*, described by Brown [7], and although Brown had not described the trilobate cross-section of the teeth in *C. eurymerus*, this is confirmed by our personal observations (N.Z.'s pers. obs, on NHMUK PV R3730, April 2019). Therefore, we refer this tooth crown to *Cryptoclidus* sp.

A tooth crown, SGM 2007-4, from the lower Callovian of the Unzha River differs from other teeth in its more robust ridges, not that sparse, as in *Cryptoclidus*, and not that frequent, as in *Muraenosaurus*. Six of the ridges reach the apex. The ridges on the labial surface are rarer and finer than those on the lingual side. Brown [7] provided restricted information on the teeth of *Tricleidus seeleyi*, and in his fig. 24 [7], the tooth of *T. seeleyi* appears superficially similar to that of *M. leedsii*. However, personal examination of the type specimen (N.Z.'s pers. obs, on NHMUK PV R3539, April 2019) confirms that the teeth of *T. seeleyi* differ from those of other Oxford Clay cryptoclidids by having more robust ridges, lack of dense vermiculation pattern at the base (characteristic of *M. leedsii*), the presence of sparse but prominent ridges on the labial side, and several (more than two) ridges reaching the apex (no ridges reach the apex in *M. leedsii* and only two reach the apex in *C. eurymerus*). All these features are present in SGM 2007-4, therefore, we refer it to as cf. *Tricleidus*. An interesting feature of SGM 2007-4 is that the ridges, reaching the apex, curve and form a spiral pattern (Figure 11X,Y,Z'). To our knowledge, such a condition has not been reported for plesiosauroids and may represent some abnormality of tooth development.

5.2.3. Isolated Plesiosauroid Vertebrae

We identify two morphotypes of plesiosauroid cervical vertebrae in the available material.

To Morphotype 1 (Figure 12) we refer vertebrae with flat or only slightly convex articular surfaces (platycoelous to acoelous type). Among Oxford Clay cryptoclidids, this condition is present in *Muraenosaurus* and *Picrocleidus* ([7] and N.Z.'s pers. obs.).

In *Muraenosaurus*, the articular surfaces are only slightly wider than they are high, and are semicircular in cross-section, whereas in *Picrocleidus*, the articular surfaces are markedly wider than they are high, oval in outline, and bear distinct depression under the neural canal floor [4]. Furthermore, *Picrocleidus* is a small taxon, with the longest cervical vertebral centra of the largest known specimen being 39 mm long [4], whereas most of the vertebral centra of Morphotype 1 are larger (44 mm and longer). Therefore, these vertebrae are referred to cf. *Muraenosaurus*. In the smaller anterior to middle cervical vertebrae, the anteroposterior length equals or slightly exceeds the mediolateral width (Figure 12B,E). In larger posterior cervical vertebrae, mediolateral width (W) exceeds the dorsoventral height (H) and anteroposterior length (L), so that W > H > L (Figure 12G–T). There are no lateral and ventral keels. A slight depression on the dorsal side of the articular surface under the neural canal is present. The cervical ribs are dorsoventrally flattened at the base and fused with the centrum without visible sutures in all vertebrae of Morphotype 1 (Figure 12B,E,J,N,Q,T), indicating the maturity of the specimens (sensu Brown [7]). The neural arches are also fused to the centrum, however sutures are noticeable in some specimens (Figure 12B). At present, six vertebrae of Morphotype 1 are identified from the Callovian of European Russia, they all are depicted in Figure 12.

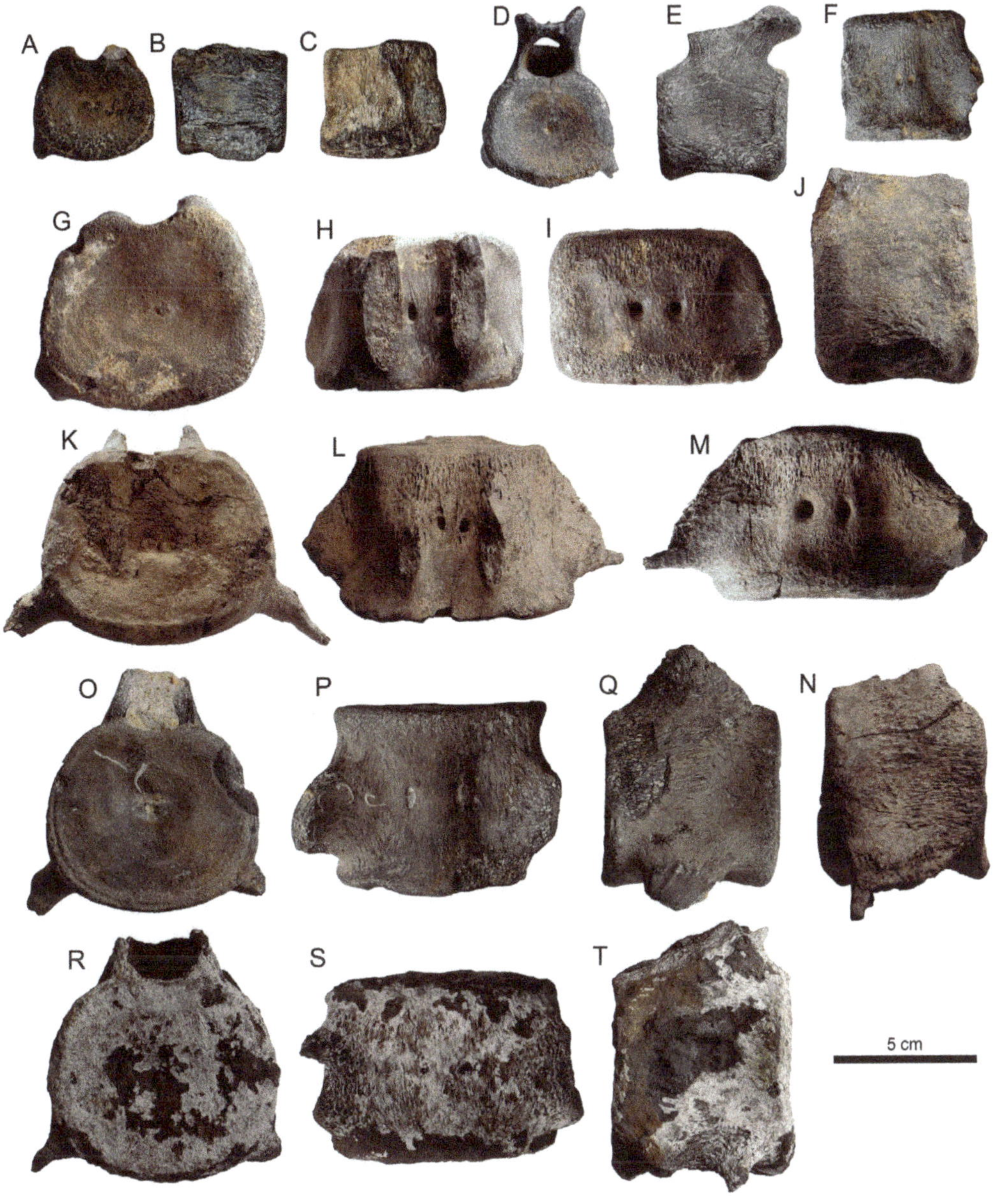

Figure 12. Cervical vertebrae of cf. *Muraenosaurus* (Morphotype 1). Anterior cervical vertebra SGM 1358-37 from the (?)lower Callovian of Alpatyevo (**A–C**), originally described by Bogolubov [33]. Anterior cervical vertebra SGM 1891-08 (**D–F**) from the lower Callovian of Mikhalenino. Posterior cervical vertebrae SGM 1891-10 (**G–J**) and SGM 1891-9 (**K–N**) from the lower Callovian of Mikhalenino. Middle to posterior cervical vertebra PIN R-3590 (**O–Q**) from the upper Callovian of Peski. Posterior cervical vertebra PIN R-3595 (**R–T**) from the middle Callovian of Mihailovcement. Views: anterior articular (**A**,**D**,**G**,**K**,**R**), lateral (**B**,**E**,**J**,**N**,**Q**,**T**), ventral (**C**,**F**,**I**,**M**,**P**,**S**), and dorsal (**H**,**L**).

To Morphotype 2 (Figure 13A–H) we refer proportionally shortened (W > H > L) cervical vertebrae with markedly concave articular surfaces (amphicoelous type), which are known for Callovian cryptoclidid genera *Cryptoclidus* and *Tricleidus* [7], and for the Oxfordian cryptoclidid *Tatenectes* [123]; therefore, these vertebrae are referred to Cryptoclididae indet. There are only three isolated vertebrae of this morphotype. Two associated centra from the middle Callovian of Peski Quarry, PIN R-3591-3592, belong to a small, osteologically immature individual (Figure 13A–D), with their neural arches and ribs not fused to the centrum (criterion of immaturity, sensu Brown [7]). Specimen PIN R-3670, from Mikhaylovcement Quarry, belongs to a larger, osteologically mature individual with the neural arch and ribs fused to the centrum without any visible suture (Figure 13E–H). Furthermore, there are proportionally short pectoral and sacral vertebrae with concave articular facets from the Callovian of Mikhaylovcement Quarry (Figure 13U–B′).

A small anterior dorsal vertebra (Figure 13I–K) was collected in association with the skeleton of *Liopleurodon* SGM 1807 and possibly represents its gastric contents. This vertebra has neural ach fused to the centrum, thus it belongs to an osteologically mature individual, although its small size (length 33 mm) indicates that it belonged to a small-bodied taxon, such as *Picrocleidus* or *Tricleidus* [4].

Another dorsal vertebra TsNIGR 144/1712 belongs to a large and osteologically mature plesiosaurian. Its articular surfaces are nearly flat, transverse processes are horizontal, and the long axis of the fib facet is nearly vertical (Figure 13M), all of which is characteristic of *Muraenosaurus* [4,121]. Therefore, we identify this specimen as cf. *Muraenosaurus*.

The dorsal vertebra YSPU M/F-44, in contrast, has concave articular surfaces and dorsally deflected transverse processes. These characters, together with the relatively large size allow the referral of YSPU M/F-44 to cf. *Cryptoclidus*.

Several isolated dorsal and caudal vertebral centra cannot be identified further than Plesiosauria indet. These are shown in Figure 13Q–T and C′–I′. The dorsal centrum, SGM 1891-05, belonging to a juvenile individual, is interesting in terms of its great anteroposterior shortness. A poorly preserved anterior caudal centrum, PIN R-3201, lacks chevron facets and has a rib facet confluent with the neural arch facet. A small posterior caudal centrum, PIN R-3596, has large chevron facets, and its rib and neural arch facets are also confluent.

Figure 13. Plesiosauroid vertebrae from the Callovian of European Russia. Cervical vertebrae of short-necked cryptoclidids (Morphotype 2) (**A–H**); juvenile PIN R-3591 (**A–D**) and osteologically mature PIN R-3670 (**E–H**). Anterior dorsal vertebra of osteologically mature small cryptoclidid collected in association with *Liopleurodon* skeleton SGM 1807. Dorsal vertebra cf. *Muraenosaurus* TsNIGR 144/1712 (**L–N**). Dorsal vertebra cf. *Cryptoclidus* YSPU M/F-44 (**O,P**). Dorsal centrum of osteologically immature plesiosaurian SGM 1891-05 (**Q–T**). Pectoral PIN R-3709 (**U–X**), sacral PIN R-3606 (**Y–B′**), and caudal PIN R-3201(**C′–E′**) and PIN R-3596(**F′–I′**) vertebrae. Views: articular (**A,E,I,L,O,Q,U,Y,C′,F′**), dorsal (**B,G,J,S,W,E′,H′**), ventral (**C,H,K,T,X,B′,I**), lateral (**D,F,M,P,V,Z,D′,G′**), cross-section (**A′**).

5.3. Ichthyosaurians

5.3.1. Ophthalmosaurian Ichthyosaurs

The first documented finding of an ichthyosaur in the Callovian of Russia was made in 1925, near Rybkino Village (at present, the Republic of Mordovia), by a prominent Russian geologist and paleontologist, Alexander A. Stuckenberg (according to PSM archive records). The excavated specimen is represented by a nearly complete sclerotic ring, small jaw fragments with associated teeth, several vertebral centra, fragmental neural arches, proximal portion of the humerus, and complete radius and ulna (Figure 14). However, this specimen, PSM 3999-4004, has not been described and figured until now. It is worth mentioning that in 2014, another skeleton referable to *Ophthalmosaurus* was excavated not far from this locality [38], in the lowermost Callovian strata, near Sinyakovo Village, Krasnoslobodsk District (Figure 14N). The preserved portions of the jaws of PSM 3999-4004 bear small teeth with roots circular in cross-section and slender crowns ornamented by rather sparse striations (Figure 14B–E). The crowns are 10–12 mm high and 5–7 mm at the basal diameter; they are very similar to those of *Ophthalmosaurus icenicus*, both in size and morphology [52]. The ulna and radius of PSM 3999-4004 are quite large (6 and 6.5 cm anteroposteriorly long, respectively), indicating a moderately large individual of approximately 3 m in length (estimated compared to known *O. icenicus* specimens; e.g., radius and ulna are each ~7–8 cm long in the holotype NHMUK PV R2133; N.Z.'s pers. obs.). The ulna of PSM 3999-4004 is proximally wider than the radius (Figure 14L). Its posterior edge is convex and demonstrates unfinished ossification, compared to the straight or concave and proximodistally long posterior edge of the ulna in *O. icenicus* from the Oxford Clay [52]. In this aspect, the ulna of PSM 3999-4004 is more similar to those of Late Jurassic *Arthropterygius* spp. [124], *Nannopterygius borealis* [125], and to the yet undescribed Bajocian ophthalmosaurian from Luxembourg [126]. The radius of PSM 3999-4004 has an oval outline with poorly demarcated facets, unlike a pentagonal outline in many ophthalmosaurids, including all the known specimens of *O. icenicus*, regardless of their ontogenetic state (N.Z.'s pers. obs. on numerous specimens in NHMUK and CAMSM collections, 2018–2019). This outline of the radius suggests the presence of two anterior surfaces, one for articulation with the preaxial accessory element and another free of contact with other elements. Among ophthalmosaurids, such a condition is present in *Ophthalmosaurus natans*, e.g., [127], some specimens of *Nannopterygius enthekiodon* [125], and *Ophthalmosaurus calloviensis* [35] (redescribed below). The sclerotic ring of PSM 3999-4004 (Figure 14A) is similar to that of many other ophthalmosaurids, with peripheral portions of the individual plates rather thin (Figure 14A). The preserved posterior caudal vertebrae are similar to those of *O. icenicus* [52], and the apical and fluke centra lack chevron facets (Figure 14H–K). In summary, despite the similarity of teeth to *O. icenicus*, the epipodial elements of PSM 3999-4004 are dissimilar to this species and resemble those of some other ophthalmosaurids, thus allowing its identification as Ophthalmosauridae indet. However, given its provenance and similarity to *O. calloviensis* from the Callovian of Saratov Region, PSM 3999-4004 may belong to this species, which is characterized below.

Figure 14. *Ophthalmosaurus* specimens from the lower Callovian of the Republic of Mordovia. Partial skeleton of *Ophthalmosaurus* cf. *calloviensis* PSM 3999-4004 (**A–M**) from Rybkino locality. Sclerotic ring (**A**). Dentigerous bone fragment (likely maxilla) in lateral (**B**) and ventral (**C**) views. Magnified teeth (**D**,**E**). Caudal preflexural (**F**,**G**), apical (**H**,**I**) and postflexiral, "fluke" (**J**,**K**), centra in articular (**F**,**I**,**J**) and lateral (**G**,**H**,**K**) views. Radius and ulna in proximal (**L**) and dorsal/ventral (**M**) views. Skull of a partial skeleton of *Ophthalmosaurus* sp. (**N**) from the lower Callovian of Sinyakovo locality reported by [38], in oblique anterolateral view. Isolated radius of *Ophthalmosaurus* cf. *calloviensis* (SGM 1891-06) from Trofimovshchina-2 locality in dorsal/ventral (**O**) and proximal (**P**) views. Abbreviations: faae, facet for anterior accessory element; fe, free surface; fim, facet for intermedium; fre, facet for radiale.

The first Callovian ichthyosaur that was formally described from Russia [35] is an incomplete forelimb, SSU104a/27, found in a trench near Dubki Village, Saratov Region. It was initially referred to a new genus and species, *Khudiakovia calloviensis*, by Arkhangelsky, 1999, and consequently synonymized with *Ophthalmosaurus icenicus* [50–52]. Arkhangelsky [53], however, retains this taxon as a valid species of *Ophthalmosaurus*. Our re-examination of this holotype specimen allows the definition of some difference to *O. icenicus* specimens from the Oxford Clay. With the humerus being 176 mm long, SSU104a/27 is a moderately large ophthalmosaurid within the size range of *Ophthalmosaurus* [4,52]. The humerus of SSU104a/27 is overall similar to that of *O. icenicus* [52] and lacks the constriction between the radial and ulnar facets known for *Arthropterygius* [124]. The presence of an accessory epipodial element, posterior to the ulna, suggested by Arkhangelsky [35], cannot be confirmed. However, the posterior edge of the ulna in SSU104a/27 is proximo-distally short, convex, and has unfinished ossification; the radius of SSU104a/27 has an oval outline with poorly demarcated facets (Figure 15A,B), implying the presence of two anterior surfaces rather than one convex facet, as interpreted by Arkhangelsky [35]. Both the radius and ulna of SSU104a/27 are similar to those of the above-described PSM 3999-4004. The intermedium of SSU 104a/27 has two distal facets of subequal size for articulation with the fourth and third distal carpals, as in many ophthalmosaurids with "latipinnate" condition [52,124].

The morphology of the humerus in SSU104a/27 supports its referral to *Ophthalmosaurus*. However, the morphology of its epipodial elements questions the assignment to *O. icenicus* and demonstrates affinities to a wider range of ophthalmosaurids. The similarity of epipodial elements in SSU104a/27 and PSM 3999-4004 suggests that PSM 3999-4004 can represent the same taxon as SSU104a/27, for which the species name *O. calloviensis*, Arkhangelsky, 1999, is available. However, due to the fragmentary nature of both specimens, we consider *Ophthalmosaurus calloviensis* as species inquirenda, pending more complete materials.

An isolated radius, SGM 1891-06 (Figure 14O,P), from the lower Callovian (*P. elatmae* Zone) of the Trofimovshchina-2 locality, Republic of Mordovia, is also similar to the above-described specimens in being oval in dorsal view and having two anterior surfaces, further supporting this feature a as characteristic of early Callovian ophthalmosaurians of the Middle Russian Sea.

Another specimen of interest is an incomplete and poorly preserved skeleton, SSTU MEZ 3/4, found from the lower Callovian near Gvardeyskoe Village, Saratov Region (Figure 15F–W). The specimen has "lateral wings" over the external nares (Figure 15F), tapered anterior extremity of the dentary (Figure 15H), a scapula with a well-developed acromial process and short, mediolaterally compressed shaft (Figure 15I,J), intermedium rhomboid in dorsal view bearing two distal facets subequal in size (Figure 15R), and dorsoventrally thickened and rounded phalanges (Figure 15S–W). All this makes it possible to attribute the specimen to cf. *Ophthalmosaurus*.

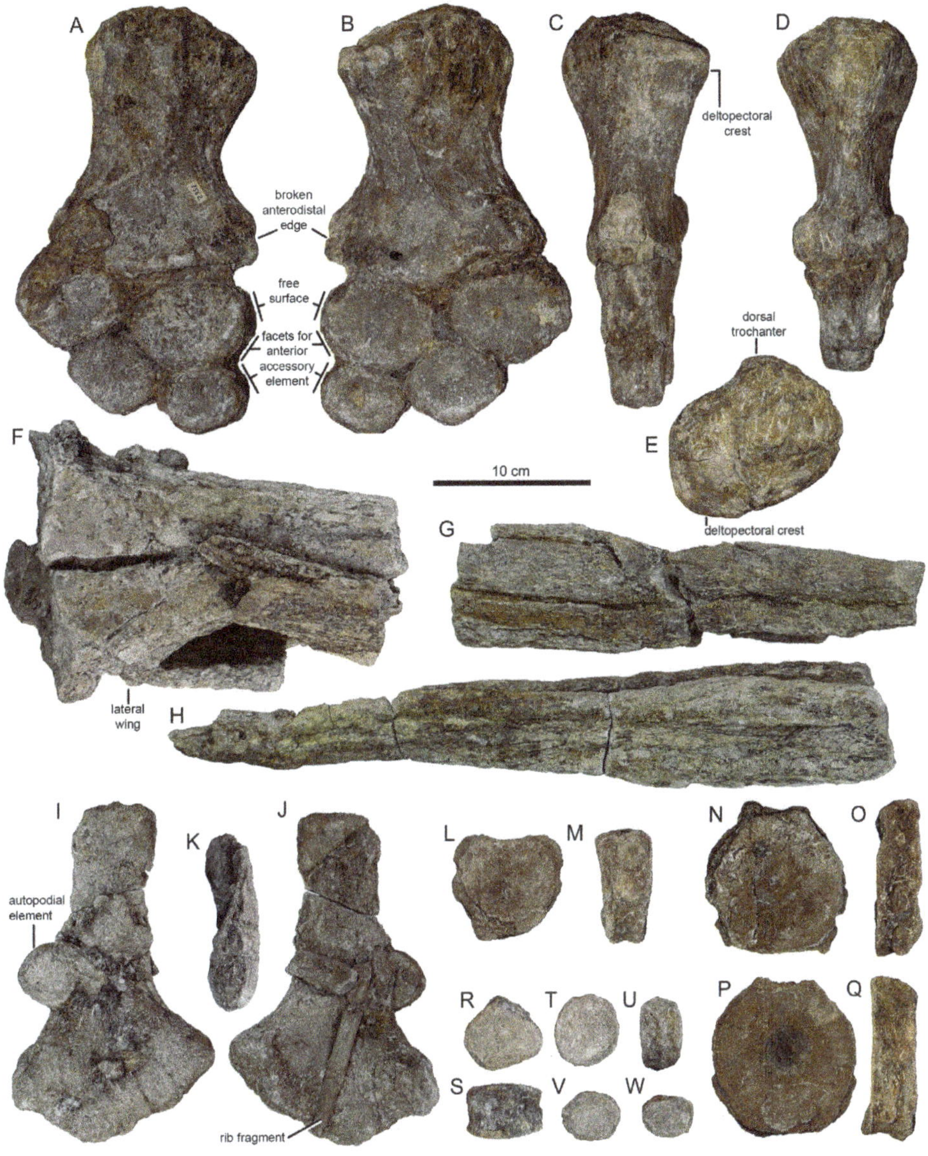

Figure 15. Ophthalmosaurids from the Callovian of Saratov Region. Holotype forelimb of *Ophthalmosaurus calloviensis* SSU 104a/27 (**A–E**) in dorsal (**A**), ventral (**B**), anterior (**C**), posterior (**D**), and proximal (**E**) views. Fragments of a partial skeleton SSTU MEZ 3/4 (**F–W**). Articulated nasals in dorsal view (**F**), partial premaxilla (**G**) and dentary (**H**) in lateral views. Right scapula (with associated rib fragments and distal limb elements) in dorsal (**I**), ventral (**J**), and proximal (**K**) views. Anterior dorsal (**L,M**), and posterior dorsal to anterior caudal (**N–Q**) centra in articular (**L,N,P**) and lateral (**M,O,Q**) views. Intermedium in dorsal/ventral (**R**) and proximal (**S**) views. Anterior accessory element in dorsal/ventral (**T**) and posterior (**U**) views. Phalanges in dorsal/ventral view (**V,W**).

In May 2022, a partial skeleton, SGM 1961, referable to *Ophthalmosaurus icenicus*, was discovered from the middle Callovian of the Perebory locality, Yaroslavl Region. The skull is poorly preserved and fragmented. Among the available parts, the most informative is the condylar portion of the left quadrate (Figure 16A–C), which has a shape and size typical for the Oxford Clay *O. icenicus* specimens. Vertebral centra from all regions of the spinal column are preserved, including the atlas–axis complex (Figure 17A–E). They also agree with the vertebral morphologies described for *O. icenicus* [52]. Among the interesting traits is that the third centrum has diapophysis and parapophysis fused into the dorsoventrally elongated synapophysis (Figure 17C,F), which is probably a malformation. Of interest is that the surface of the floor of the neural canal is roughened, bearing undulating longitudinal striations (Figure 17I,M,P,T). This condition is present in all Oxford Clay *O. icenicus* specimens, but is not seen in any other ophthalmosaurian (N.Z.'s pers. obs.), thus it likely represents an autapomorphy of *O. icenicus*. In every aspect, the humerus is similar to that of the *O. icenicus* paratype, NHMUK PV R2134, although the anterodistal edge in SGM 1961 is broken. The proximodistal length of the humerus is 14 cm, which corresponds to the size of young adult *O. icenicus* specimens [4]. Epipodial and autopodial elements are all typical for *O. icenicus*; the ulna is proximodistally elongated with a straight and dorsoventrally compressed posterior edge, the radius has a pentagonal outline dorsally, the intermedium bears two distal facets of subequal size, and the distal autopodial elements are rounded and dorsoventrally thick.

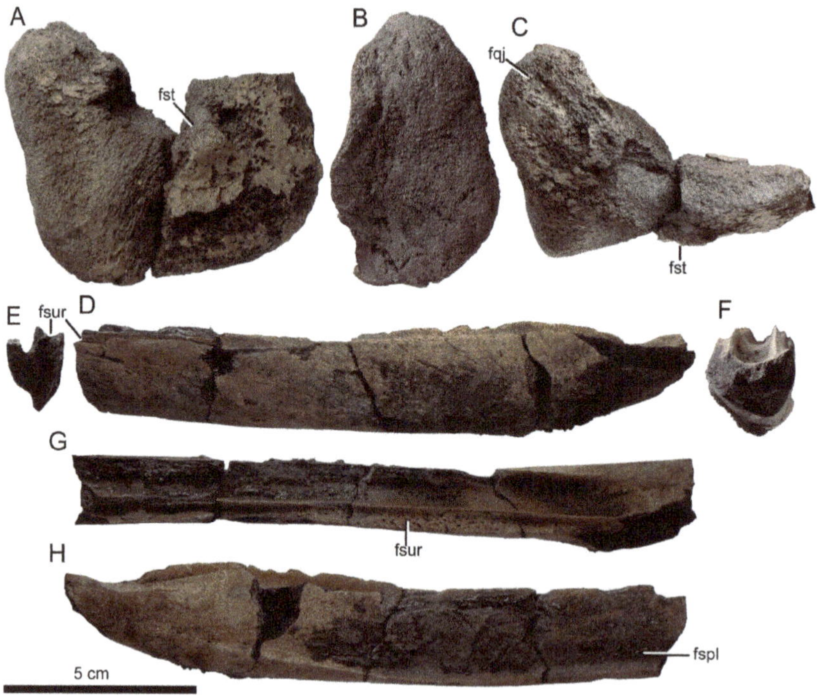

Figure 16. Cranial elements of *Ophthalmosaurus icenicus*, SGM 1961, from the middle Callovian of Perebory, Yaroslavl Region. Ventral portion of the left quadrate in posteromedial (**A**), condylar (**B**) and dorsal (**C**) views. Partial left angular in anterior cross-sectional (**E**), lateral (**D**), posterior (**F**), dorsal (**G**), and medial (**H**) views. Abbreviations: fqj, facet for quadratojugal; fspl, facet for splenial; fst, facet for stapes; fsur, facet for surangular.

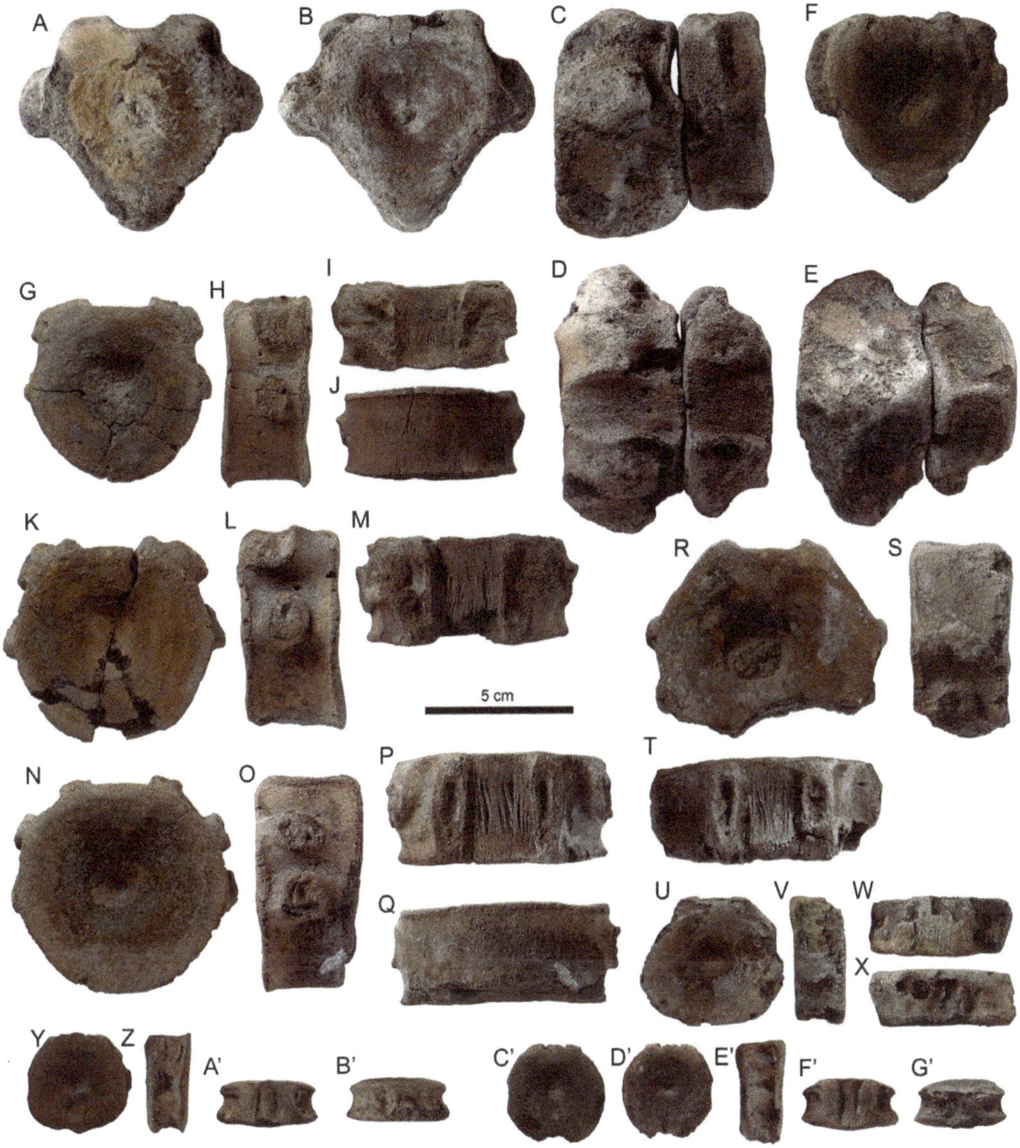

Figure 17. Vertebral centra of *Ophthalmosaurus icenicus*, SGM 1961, from the middle Callovian of Perebory, Yaroslavl Region. Atlas–axis complex in anterior (**A**) and posterior (**B**) views; articulated with the third centrum in left lateral (**C**), dorsal (**D**), and ventral (**E**) views. Third centrum in anterior articular view (**F**). Anterior dorsal (**G–M**), posterior dorsal (**N–T**), and caudal (**U–G′**) centra, in articular (**G,K,N,R,U,Y,C′,D′**), lateral (**H,L,O,S,V,Z,E′**), dorsal (**I,M,P,T,W,A′,F′**), and ventral (**J,Q,X,B′,G′**) views.

Another partial forelimb, PIN R-4956 (Figure 18I–N), from the upper Callovian of Mikhaylovcement Quarry, with the humerus 16 cm in proximodistal length, belongs to a larger individual. The humerus is nearly completely preserved and is indistinguishable in its shape from the *O. icenicus* NHMUK PV R2134 paratype humerus. Most noticeable is the anteriorly acute, triangular in outline facet for the preaxial accessory epipodial element (Figure 18N). This condition occurs only in *O. icenicus* and allows for the robust referral of the specimen to this species [128]. Associated with the humerus, the radius, intermedium, and a carpal element are also typical of *O. icenicus* in their shape and proportions.

The isolated premaxilla, SGM 2000-1 (Figure 19P,Q,R), is 50 cm long and 6 cm in maximum height. It has a characteristically anteriorly tapered and acute tip, which diverges from the sagittal plane anteriorly (Figure 19P). This anterior divergence of the premaxillae was reported to be unique to *O. icenicus* among Jurassic ichthyosaurians [52].

Some other less complete specimens are also attributable to ophthalmosaurids and likely belonged to *Ophthalmosaurus*. Among these are several specimens from the lower Callovian of the Unzha River, including a partial mandible with teeth (SGM 1891-20; Figure 19A–D), and collected in association with it, an anterior dorsal centrum (SGM 1891-02; Figure 20A–C). The teeth are small (crowns 14 mm in maximum height and 7 mm at basal diameter) with slender curved crowns (Figure 19B–D), and weak roots of semicircular cross-section (Figure 19A–D), similar to *Ophthalmosaurus* [52] and *Arthropterygius* [124]. The anterior dorsal centrum has undulating longitudinal striations on the neural canal floor, which implies its affinity to *O. icenicus* (see above). The jugal SGM 1891-22 (Figure 19N,O) is a slender J-shaped element with anteroposteriorly narrow dorsal portion, very similar to that of *O. icenicus* [52]. The isolated left quadrate (SSU uncatalogued; Figure 19K–M) is also most similar in its morphology to *O. icenicus*, see [4,52]. An isolated tooth (SGM 2007-8; Figure 19E–H) and tooth crown (PIN 5819/4; Figure 19I,J) are small and slender, similar to the above-described teeth of SGM 1891-20 and PSM 3999-4004.

An isolated limb element, PIN R-2516 (Figure 18O,P), from the upper Callovian of Mikhaylovcement Quarry is rounded in outline and compressed at one of the edges. It likely represents an anterior accessory element, or one of the elements from the anterior or posterior digits of the limb. In its rounded outline and large size (6 cm in diameter), it is similar to the respective elements of *Ophthalmosaurus* [52] and *Arthropterygus* [124].

5.3.2. Indeterminate Ichthyosaurians

Numerous isolated ichthyosaurian vertebrae are known from the Callovian strata of European Russia. Some of these are the only finds from certain localities, including the northernmost occurrence in European Russia, Adzvavom [110] (see Table 1 for details). Some of these specimens are depicted in Figure 20. We omit the description of these specimens here, as in the present state of knowledge, it is not possible to identify them more precisely than Ichthyosauria indet., although they most likely belong to ophthalmosaurians. Therefore, these specimens only demonstrate the ubiquitous presence of ichthyosaurians in the Middle Russian Sea during the Callovian.

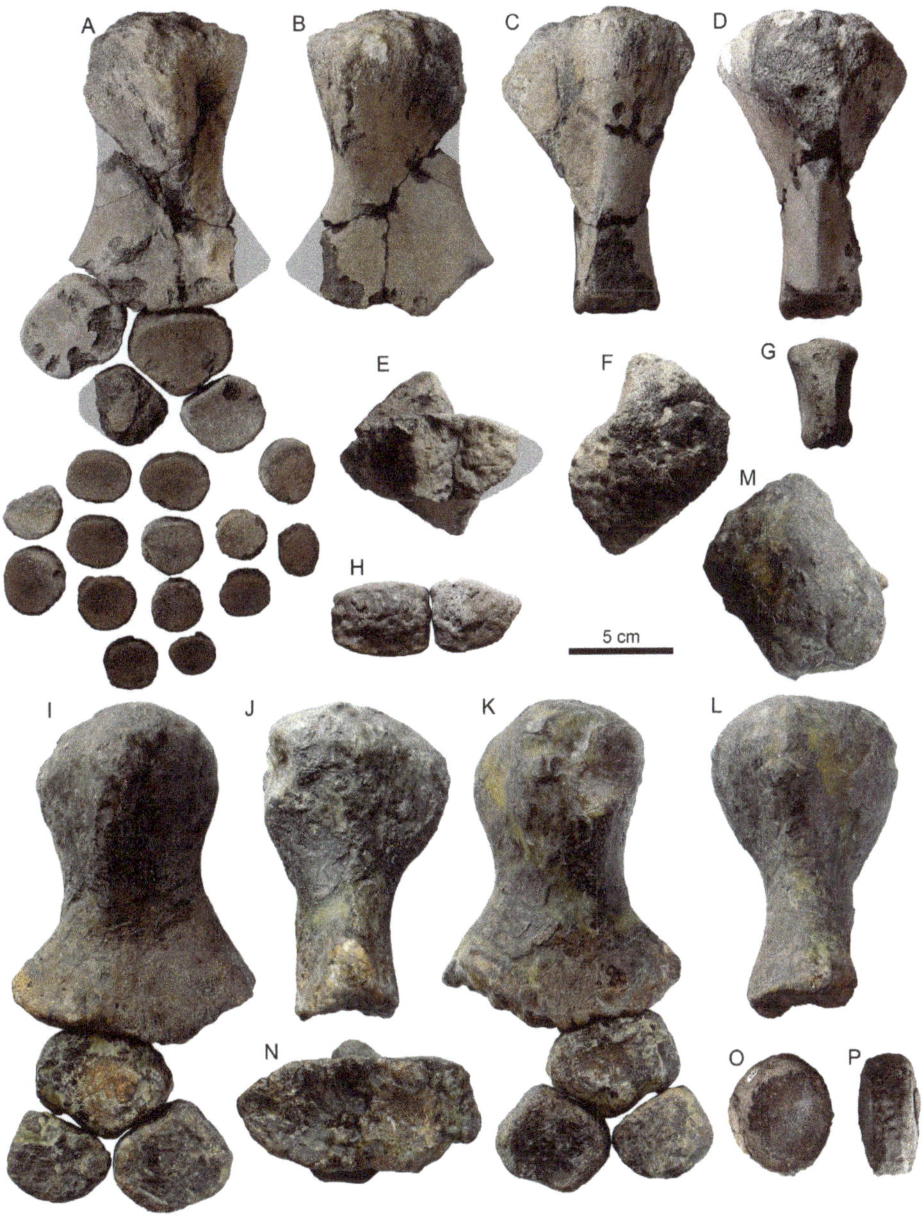

Figure 18. Forelimb remains of *Ophthalmosaurus*. Partial right forelimb of *Ophthalmosaurus icenicus*, SGM 1961, in dorsal view (**A**). Humerus of SGM 1961 in ventral (**B**), anterior (**C**), posterior (**D**), distal (**E**), and proximal (**F**) views. Ulna of SGM 1961 in posterior view (**G**), and proximal surfaces of the ulna and radius (**H**). Partial left forelimb of *Ophthalmosaurus icenicus* PIN R-4956 (**I–N**) from the upper Callovian of Zmeinka Quarry, Ryazan Region, in dorsal (**I**) and ventral (**K**) views. Humerus of PIN R-4956 in anterior (**J**), posterior (**L**), proximal (**M**), and distal (**N**) views. Anterior accessory epipodial element (?) of cf. *Ophthalmosaurus* PIN R-2516 (**O,P**), from the upper Callovian of Mikhaylovcement Quarry, Ryazan Region in dorsal/ventral (**O**) and anterior (**P**) views.

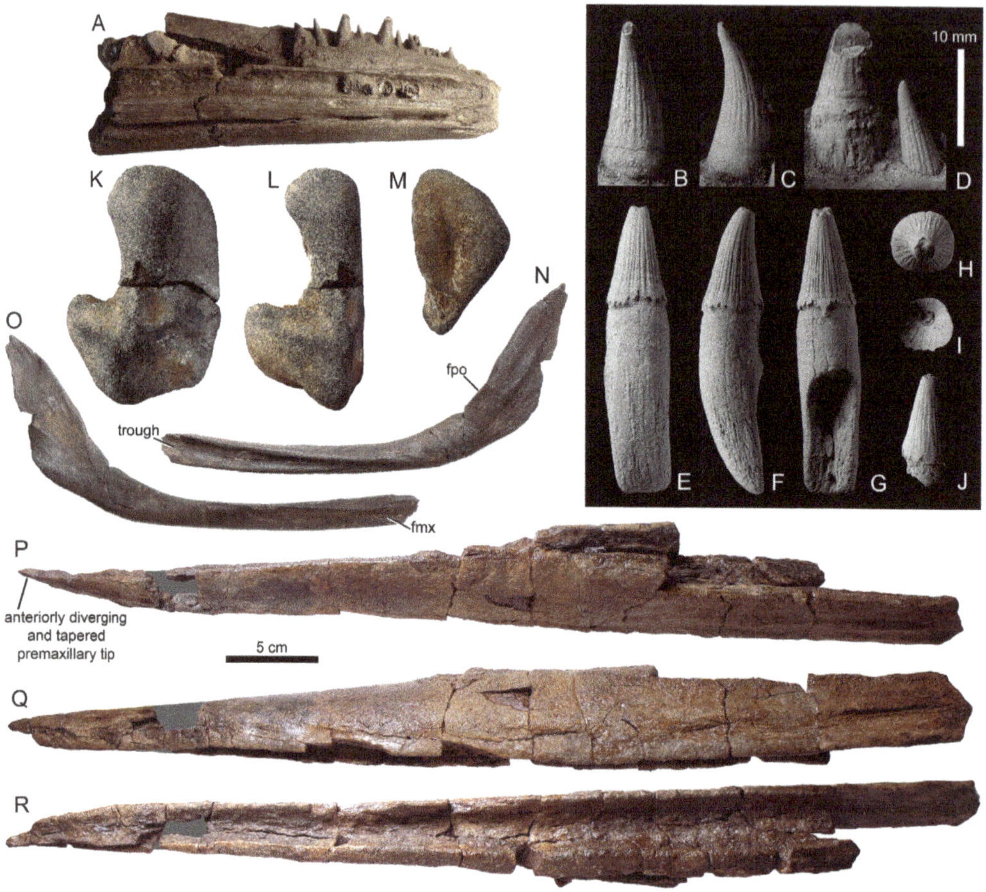

Figure 19. Ophthalmosaurid cranial remains from the Callovian of the Unzha River basin, Kostroma Region. Partial mandible with teeth SGM 1891-20 (**A–D**), on (**B–D**) magnified teeth. Isolated tooth SGM 2007-8 (**E–H**) and tooth crown PIN 5819/4 (**I,J**) in labial (**E,J**), anterior or posterior (**F,C**), lingual (**B,G,D**), and apical (**H,I**) views. Left quadrate, SSU uncatalogued, (**K–M**) in posteromedial (K), posterior (**L**), and condylar (**M**) views. Left jugal SGM 1891-22 in lateral (**N**) and medial (**O**) views. Left premaxilla SGM 2000-1 (**P–R**), in dorsal (**P**), lateral (**Q**), and ventral (**R**) views.

5.4. Thalattosuchian Crocodylomorphs

Recently Young et al. [40] described isolated thalattosuchian teeth from the Callovian of European Russia, therefore, we refer the reader to that contribution for details. Young et al. [40] identified three teeth from the lower Callovian of Unzha Village as Metriorhynchidae indet., and one more crown, from nearby Mikhalenino Village, as Geosaurini indet. (NB localities "Unzha" and "Mikhalenino" may represent different records of one natural outcrop near the villages Unzha, Mikalenino, and Popovo). A tooth crown from the lower Callovian of CHP-5 locality in Saratov, Saratov Region, originally described and figured by Arkhangelsky [35], was referred to Geosaurini indet., although it represents a taxon different to the Unzha geosaurin. One tooth crown from the lower Callovian of Gumny village, Republic of Mordovia, and another from the middle Callovian of the Mikhaylovcement Quarry, Ryazan Region, were referred to cf. *Thalattosuchus*. Two crowns from the middle Callovian of Gzhel and Mikhaylovcement were referred to emphTyrannoneustes sp. [40].

Figure 20. Ichthyosaurian veretebral centra from the Callovian of European Russia. Anterior dorsal centra SGM 1891-02 (**A–C**) and IG 93/13 (**D**). Middle dorsal centra ZIN PH 1/215 (**E–G**) and SGM 1891-04 (**I–K**). Posterior dorsal centra PIN R-3200 (**H**), SGM 1891-25 (**N–P**), SGM 1891-15 (**Q–S**). Anterior caudal centrum SGM 1891-19 (**L,M**). Posterior preflexural caudal centra SGM 1891-14 (**T–V**), PIN R-3593 (**W–Z**) and SGM 1891-24 (**A′–D′**). Series of postflexural (fluke) vertebrae SGM 1891-21 (**E′,F′**). Views: articular (**A,D,G,H,I,L,N,Q,T,W,A′,F′**), lateral (**B,F,J,O,R,U,X,B′,E′**), dorsal (**C,E,K,V,Y,C′**), ventral (**M,P,S,Z,D′**).

In autumn 2023, a cervical vertebra (SGM 2007-02) from the Callovian of Mikhaylovcement Quarry was donated to SGM by Konstantin Volkov. The vertebral centrum (Figure 21A–F) has approximately equal height and width (50 and 49 mm, respectively), and its length (53 mm) only slightly exceeds the width. The anterior articular surface is flat to slightly convex, and the posterior articular surface is concave. The ventral hypapophyseal keel is well-developed and somewhat convex ventrally (Figure 21D). It is narrowest in the central part and expands into triangular surfaces anteriorly and posteriorly. These surfaces have a conspicuously rugose, mammilated surface (Figure 21E). The parapophyses are protruding and oval in outline, located closer to the anterior surface. The anterior surfaces of parapophyses are rugose (Figure 21D). The subequal length and width of the centrum suggest its referral to a geosaurine metriorhynchid, as metriorhynchine metriorhynchids have posterior cervicals that are slightly shorter than wide ([78]; M.T. Young pers. comm. 2023).

Furthermore, in SGM collection, there is a long bone from the lower Callovian of the Unzha River basin (Figure 21G–K), which likely represents a thalattosuchian metacarpal or metatarsal. It is a slender, 65 mm long element with a triangular to teardrop cross-section.

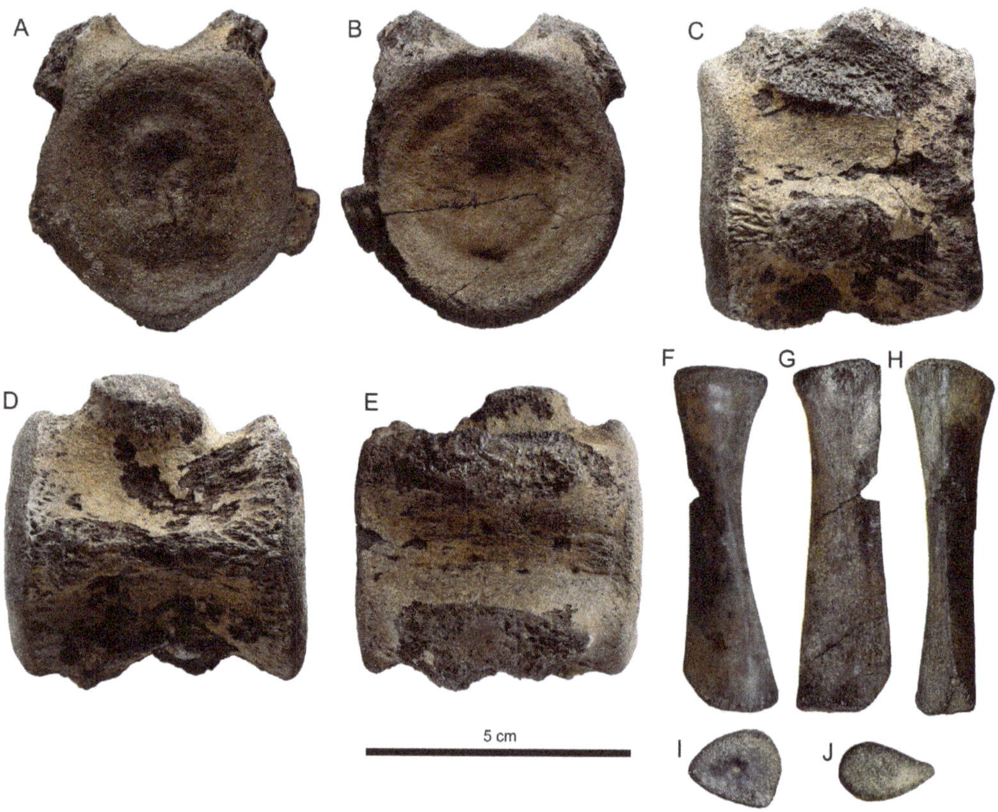

Figure 21. Metriorhynchid cervical vertebra SGM 2007-02 from the Callovian of Mikhaylovcement Quarry in anterior (**A**), posterior (**B**), left lateral (**C**), ventral (**D**), and dorsal (**E**) views. Thalattosuchian metacarpal or metatarsal SGM 1891-01 (**F–J**) from the lower Callovian of Mikhalenino.

6. Discussion

The fossil record of marine reptiles since the middle Toarcian and to the middle Callovian is extremely poor globally [1,2,126,129–136], which makes it hardly possible to assess any turnovers during this time interval, or trace patterns of marine reptile distribution [126]. Furthermore, our knowledge of marine reptile fauna of the only well-characterized middle Jurassic stage, the Callovian, is largely based on the Oxford Clay Formation of England, with less sound records from coeval strata of other countries of Western Europe [12–26], and extremely limited data on Callovian marine reptiles outside of Europe [27–32]. Moreover, for many Callovian marine reptile specimens from the Oxford Clay, precise information on their stratigraphic position was not recorded [4,5,137–139]. According to the existing records and later collections, the majority of marine reptile remains from the Oxford Clay Formation, particularly articulated skeletons and bone associations, appear to come from the basal beds of the Peterborough Member (Middle Callovian, lower part of the *Kosmoceras jason* Subzone of *K. jason* Zone; Figure 22) [8,139–141]. This is partially due to the peculiarities of clay extraction, both manual and mechanical, such that *Gryphaea* beds and concretionary nodules, especially abundant in Bed 10 [141], as well as the poorer quality of these basal beds due to frequent shells, resulted them becoming the predominant area for fossil collecting [139], even though a few finds from the below (uppermost lower Callovian *Catasigaloceras enodatum* Subzone of *S. calloviense* Zone [9]) and above (middle to upper Callovian *Erymnoceras coronatum*, *Peltoceras athleta* and *Lamberticeras lamberti* zones [8,83,139]) intervals of the Oxford Clay Formation are also known. Therefore, the taxonomically rich Callovian herpetofauna of England largely characterizes a short episode in the Middle Callovian. Brown and Keen [122] also described the marine reptile fauna from the uppermost lower Callovian (*C. enodatum* Subzone of *S. calloviense* Zone) sands of the Kellaways Formation of the UK. They identified *Liopleurodon*, *Cryptoclidus*, *Muraenosaurus*, as well as teleosaurid and metriorhynchid teeth. Thus, they demonstrated that at least some reptile genera of the Oxford Clay were present already in the lower Callovian Kellaways Sands. Early Callovian marine reptile fauna of European Russia provides additional data for this poorly characterized time interval and for older intervals within the Callovian. Similarly to Brown and Keen [122], we identified *Liopleurodon*, *Muraenosaurus*, and *Cryptoclidus* in the lower Callovian of Russia, but also documented *Simolestes* and cf. *Tricleidus* herein. Thus, the lower Callovian plesiosaurians of European Russia are similar to those from the Kellaways and Oxford Clay formations of the UK, at least at the genus level. The fragmentary nature of the available lower Callovian specimens in both the UK and European Russia complicates further comparisons of these faunas with each other and with younger Oxford Clay fauna. However, it is possible that new findings will reveal some anatomic differences to characterize the temporal and/or geographic separation of the early Callovian marine reptile fauna of the Middle Russian Sea.

Of interest is the presence of a relict rhomaleosaurid in the lower Callovian of Russia, which is in line with the finds from the lower Callovian of Arctic Canada, as well as Callovian records from Argentina and the UK [36]. This probably shows that early Callovian marine reptile faunas particularly retained stratigraphically older taxa from poorly characterized faunas of the Aalenian–Bathonian [126].

Among metriorhynchids, Young et al. [40] reported several indeterminate metryorhynchid teeth, cf. *Thalattosuchus*, and two morphotypes of Geosaurini indet. from the lowermost Callovian of Russia. The tooth crowns referred to Geosaurini represent two distinct morphotypes, different from known European taxa, and imply that thalattosuchian fauna in the earliest Callovian of European Russia was partially different from that of the middle Callovian of Western Europe.

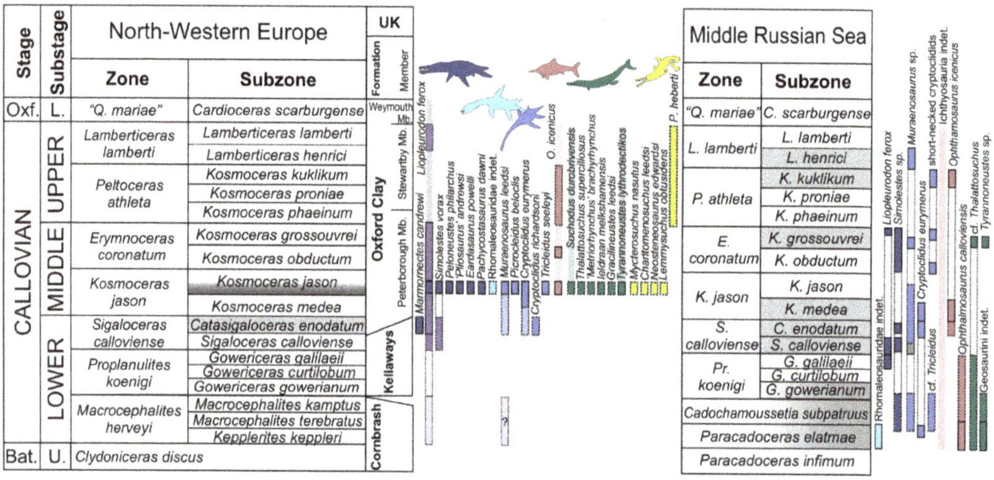

Figure 22. Stratigraphic distribution of marine reptiles in Western Europe (**left**) and European Russia (**right**). Ammonite zones/subzones highlighted in gray are the main levels with marine reptile fossils. For Western Europe, data on stratigraphic distribution of reptiles follow [8,9,73,122,139,142], and partially follow [78,143] for thalattosuchians. Occurrences with some uncertainties are shown in boxes with dashed margins; light shades indicate potential ranges based on ambiguous evidence or uncertainties in the age of specimens. Some ranges for early and late Callovian plesiosaurians in Western Europe are expanded according to Bardet [19,20] and Sachs and Nyhuis [25].

Similarly, early Callovian ophthalmosaurians of European Russia demonstrate some differences from the only known Callovian ophthalmosaurian taxon, *Ophthalmosaurus icenicus*, from the middle Callovian of Western Europe; whereas diagnostic specimens of middle to late Callovian ichthyosaurians in European Russia are referable to *O. icenicus*, supporting the connection of herpetofaunas of the Middle Russian Sea and seas of Western Europe. For this time interval, we also observe other typical representatives of the Oxford Clay herpetofauna in the Middle Russian Sea; plesiosaurians *Cryptoclidus eurymerus*, *Liopleurodon ferox*, *Simolestes* sp., and metriorhynchids cf. *Thalattosuchus* and *Tyrannoneustes* sp. [40], all of which have no marked differences from their contemporary congeners of Western Europe. This is not surprising, as the invertebrate faunas of the middle and late Callovian are also highly similar in these basins [91,144]. However, the studied area is still quite close to Western Europe on a global scale, and further data on Callovian marine reptiles from other regions of the world, especially those showing significant differences in invertebrate fauna, are required.

Some differences between the early Callovian ichthyosaurian and thalattosuchian taxa from European Russia and the known Callovian taxa of Western Europe could be explained by the older age of the Russian fauna, rather than its geographic position. However, the presence of several common genera implies that younger middle–late Callovian herpetofaunas inherited most of it. Thus, the earliest Callovian marine reptiles of European Russia expand our knowledge of the global diversity of marine reptiles of this age and are likely to represent the fauna ancestral to that of the middle Callovian age. Post-Callovian marine reptile faunas of Europe are less well known, as reptile remains are very rare and fragmentary in the Oxfordian of European Russia and Western Europe [145]. In this respect, it is still difficult to trace the patterns of marine reptile fauna evolution and geographic distribution during the Middle and early-Late Jurassic epochs in Europe and globally, and the Callovian remains a narrow, 4-million-year-long "window" into the Middle Jurassic marine reptile world, surrounded by the Aalenian–Bathonian and Oxfordian intervals of blurred marine reptile fossil records.

Author Contributions: Conceptualization, N.Z., M.A. and A.S.; Methodology, investigation, resources, data curation, all authors; writing—original draft preparation, N.Z. (most text), D.G. and A.I. (sections on paleobiogeography and stratigraphy); writing—review and editing, all authors; visualization, N.Z. and D.G. (stratigraphic columns). All authors have read and agreed to the published version of the manuscript.

Funding: The work of N.Z. was funded by the Geological Institute of RAS (Program FMMG-2021-0003), and by a grant for young researchers of the Geological Institute (in 2021), while for A.I., it was supported by the Kazan Federal University Strategic Academic Leadership Program (PRIORITY-2030).

Institutional Review Board Statement: Not applicable.

Data Availability Statement: All data used in the present paper are published in the paper.

Acknowledgments: This work is a tribute to Andrey Vadimovich Stupachenko, who assembled an important collection of marine reptiles from the lower Callovian of Kostroma Region and generously donated it to SGM, and earlier partially to PIN and SSU. We ask for his pardon, that this project, started as far as in 2015, took such a long time to finish. Other important specimens described in this contribution were provided by Kirill K. Kotov (Moscow) and Vasily V. Mitta (PIN). Mikhail Shekhanov (Yaroslavl) helped with fieldworks of N.Z. in Perebory, Yaroslavl Region. We thank Dmitry N. Kiselev (YSPU) for consultation on the stratigraphy of Pereory locality and ammonite identifications, as well as for the photographs and measurements of YSPU specimens. Dmitry V. Varenov provided the photograph of *Ophthalmosaurus* specimen reported by [38]. Participants of the Club of Junior Paleontologists of the Paleontological Museum, I.A. Dadykin, M.A. Nikiforov, E.D. Orlova, and A.D. Voronkina took part in the preparation of specimens collected by the Club, PIN R collection (see Table 1). Arina Voronkina further took part in the study of PIN R-3600 as her school project. We thank S.V. Bagirov for the high-resolution photographs of teeth from PIN R collection. We thank Mark Young for his comments on the metriorhynchid vertebra. We thank Sandra Chapman (NHMUK) and Matt Riley (CAMSM) for their help during N.Z. work with the collections of Oxford Clay marine reptiles under their care. Two anonymous reviewers provided valuable comments and corrections. Finally, we thank the editor, Nathalie Bardet, for her kind invitation to contribute to this Special Issue.

Conflicts of Interest: The authors declare no conflicts of interest.

References

1. Benson, R.B.J.; Butler, R.J.; Lindgren, J.; Smith, A.S. Mesozoic marine tetrapod diversity: Mass extinctions and temporal heterogeneity in geological megabiases affecting the vertebrates. *Proc. R. Soc. Lond. B, Biol. Sci.* **2010**, *277*, 829–834. [CrossRef] [PubMed]
2. Benson, R.B.J.; Butler, R.J. Uncovering the diversification history of marine tetrapods: Ecology influences the effect of geological sampling biases. *Geol. Soc. Spec. Publ.* **2011**, *358*, 191–208. [CrossRef]
3. Raja, N.B.; Dunne, E.M.; Matiwane, A.; Khan, T.M.; Nätscher, P.S.; Ghilardi, A.M.; Chattopadhyay, D. Colonial history and global economics distort our understanding of deep-time biodiversity. *Nat. Ecol. Evol.* **2022**, *6*, 145–154. [CrossRef] [PubMed]
4. Andrews, C.W. *A Descriptive Catalogue of the Marine Reptiles of the Oxford Clay. Part 1*; British Museum (Natural History): London, UK, 1910.
5. Andrews, C.W. *A Descriptive Catalogue of the Marine Reptiles of the Oxford Clay. Part 2*; British Museum (Natural History): London, UK, 1913.
6. Tarlo, L.B. A review of the Upper Jurassic pliosaurs. *Bull. Br. Mus. Geol.* **1960**, *4*, 147–189.
7. Brown, D.S. The English Upper Jurassic Plesiosauroidea (Reptilia) and a review of the phylogeny and classification of the Plesiosauria. *Bull. of the Brit. Mus. Geol.* **1981**, *35*, 253–347.
8. Martill, D.M. The preservation of marine vertebrates in the Lower Oxford Clay (Jurassic) of central England. *Philos. Trans. R. Soc. Lond. B* **1985**, *311*, 155–165.
9. Ketchum, H.F.; Benson, R.B.J. A new pliosaurid (Sauropterygia, Plesiosauria) from the Oxford Clay Formation (Middle Jurassic, Callovian) of England: Evidence for a gracile, longirostrine grade of Early–Middle Jurassic pliosaurids. *Spec. Pap. Palaeontol.* **2011**, *86*, 109–129.
10. Ketchum, H.F.; Benson, R.B.J. The cranial anatomy and taxonomy of *Peloneustes philarchus* (Sauropterygia, Pliosauridae) from the Peterborough Member (Callovian, Middle Jurassic) of the United Kingdom. *Palaeontology* **2011**, *54*, 639–665. [CrossRef]
11. Ketchum, H.F.; Benson, R.B.J. A new pliosaurid from the Oxford Clay Formation of Oxfordshire, UK. *Acta Palaeontol. Pol.* **2022**, *67*, 297–315. [CrossRef]
12. Sauvage, H.-E. Notes sur les reptiles fossiles. *Bull. Soc. Géol. Fr.* **1873**, *3*, 365–386.

13. Sauvage, H.-E. Synopsis des poisons et des reptiles des terrains jurassiques de Boulogne-sur-Mer. *Bull. Soc. Géol. Fr.* **1880**, *8*, 524–547.
14. Sauvage, H.-E. Catalogue des Reptiles Jurassiques du Boulonnais. *Bull Sc Acad de l'Arrondissement de Boulogne-sur-Mer* **1914**, *10*, 253–264.
15. Bigot, A. Sauroptérygiens du Jurassique du Calvados. *Bull. Soc. Géol. Fr.* **1938**, *5*, 631–636.
16. Brunet, M. Note préliminaire sur une faune de vertébrés du Callovien des environs de Poitiers. *C. R. Hebd. Séanc. Acad. Sci. Série D Sci. Nat.* **1969**, *268*, 2667–2670.
17. Bardet, N.; Pennetier, G.; Pennetier, E. *Muraenosaurus leedsii* Seeley (Plesiosauria, Elasmosauridae) dans le Callovien du Calvados (France). *Neues Jahrb. Geol. Paläontol. Monatshefte* **1991**, *7*, 402–408. [CrossRef]
18. Bardet, N. Pliosaurs and plesiosaurs from the Middle Jurassic (Callovian) of Normandy. *Rev. Paléobiol.* **1993**, *7*, 1–7.
19. Bardet, N.; Pennetier, G.; Pennetier, E.; Queromain, J. Présence du pliosaure *Liopleurodon ferox* Sauvage dans le Callovien de Villers-sur-Mer (Normandie). *Bull. Trim. Soc. Géol. Normandie Amis Mus. Havre* **1993**, *80*, 11–14.
20. Bardet, N. Les plésiosaures du Callovien de Normandie. *L'Écho Des Falaises* **1996**, *2*, 11–20.
21. Bardet, N. Les ichthyosaures et les plésiosaures du Jurassique et du Crétacé des falaises des Vaches-Noires (Normandie, France). *Fossiles* **2013**, *Hors-Série 4*, 98–104.
22. Hermann, R. *Pliosaurus* sp. aus dem Ornatenton des fränkischen Jura. *Cbl Mineral Geol. Palaontol.* **1907**, *1907*, 667–669.
23. Von Huene, F. Ein großer *Pliosaurus* aus dem schwäbischen Ornatenton. *Jahresh. Ges. Naturkunde Wurtt.* **1934**, *90*, 31–46.
24. Michelis, I.; Sander, P.M.; Metzdorf, R.; Breitkreutz, H. Die Vertebratenfauna des Calloviums (Mittlerer Jura) aus dem Steinbruch Störmer (Wallücke, Wiehengebirge). *Geol. Paläontol. Westfal.* **1996**, *44*, 1–66.
25. Sachs, S.; Nyhuis, C.J. Plesiosaurier-Funde aus dem Mittleren Jura von Hildesheim. *Naturhistorica* **2018**, *160*, 115–128.
26. Waskow, K.; Grzegorczyk, D.; Sander, P.M. The first record of *Tyrannoneustes* (Thalattosuchia: Metriorhynchidae): A complete skull from the Callovian (late Middle Jurassic) of Germany. *PalZ* **2018**, *92*, 457–480. [CrossRef]
27. Gasparini, Z.; Chong-Díaz, G. *Metriorhynchus casamiquelai* n. sp. (Crocodilia, Thalattosuchia) a marine crocodile from the Jurassic (Callovian) of Chile, South America. *Neues Jahrb. Geol. Paläontol. Abh.* **1977**, *153*, 341–360.
28. Gasparini, Z.B. Un nuevo cocodrilo marino (Crocodylia, Metriorhynchidae) del Caloviano del norte de Chile. *Ameghiniana* **1980**, *17*, 97–103.
29. Gasparini, Z. Marine reptiles from the circum-Pacific region. In *The Jurassic of the Circum Pacific*; Westermann, G.E.G., Ed.; Cambridge University Press: Cambridge, UK, 1992; pp. 361–364.
30. Gasparini, Z.; Spaletti, L. First Callovian plesiosaurs from the Neuquén Basin, Argentina. *Ameghiniana* **1993**, *30*, 245–254.
31. Russell, D.A. Jurassic marine reptiles from Cape Grassy, Melville Island, Arctic Canada. *Bull. Geol. Surv. Can.* **1994**, *450*, 195–201.
32. Sato, T.; Wu, X.-C. A new Jurassic pliosaur from Melville Island, Canadian Arctic Archipelago. *Can. J. Earth Sci.* **2008**, *45*, 303–320. [CrossRef]
33. Bogolubov, N.N. *On the History of Plesiosaurs in Russia*; Tipografia Imperatorskogo Moskovskogo Universiteta: Moscow, Russia, 1911. (In Russian)
34. Milanovsky, E.V. A find of plesiosaur in the Jurassic deposits of Saratov province. *Geol. Vestn.* **1921**, *4*, 118–120. (In Russian)
35. Arkhangelsky, M.S. On an ichthyosaur from the Callovian stage of the Volga Region near Saratov. *Paleontol. J.* **1999**, *33*, 88–91.
36. Benson, R.B.J.; Zverkov, N.G.; Arkhangelsky, M.S. Youngest occurrences of rhomaleosaurid plesiosaurs indicate survival of an archaic marine reptile clade at high palaeolatitudes. *Acta Palaeontol. Pol.* **2015**, *60*, 769–780. [CrossRef]
37. Zverkov, N.G.; Arkhangelsky, M.S. On a find of a Callovian short-snouted pliosaur in Kostroma Region (Russia). In *Jurassic System of Russia: Problems of Stratigraphy and Paleogeography, Proceedings of the Sixth All-Russian Meeting*; Zakharov, V.A., Rogov, M.A., Ippolitov, A.P., Eds.; ALEF: Makhachkala, Russia, 2015; pp. 135–138. (In Russian)
38. Efimov, V.M.; Meleshin, I.A. The first find of an ichthyosaur in the Jurassic deposits of Mordovia. In Proceedings of the all-Russian Scientific Conference "Treshnikovskie Chteniya-2017", Ulyanovsk, Russia, 30–31 March 2017; pp. 22–23. (In Russian).
39. Zverkov, N.G.; Shmakov, A.S.; Arkhangelsky, M.S. Jurassic Marine Reptiles of Moscow and Surroundings. In *Jurassic Deposits of the Southern Part of the Moscow Syneclyse and Their Fauna*; Rogov, M.A., Zakharov, V.A., Eds.; GEOS: Moscow, Russia, 2017; pp. 230–263. (In Russian)
40. Young, M.T.; Zverkov, N.G.; Arkhangelsky, M.S.; Ippolitov, A.P.; Meleshin, I.A.; Mirantsev, G.V.; Shmakov, A.S.; Stenshin, I.M. Thalattosuchian crocodylomorphs from European Russia, and new insights into metriorhynchid tooth serration evolution and their palaeolatitudinal distribution. *PeerJ* **2023**, *11*, e15781. [CrossRef]
41. Riabinin, A.N. Zwei Plesiosaurier aus den Jura- und Kreideablagerungen Russlands. *Mém. Com. Géol. St. Petersburg* **1909**, *43*, 1–49. (In Russian)
42. Bogolubov, N.N. Sur quelques restes de deux reptiles (*Cryptodidus simbirskensis* n. sp. et *Ichthyosaurus steleodon* n. sp.), trouvés par Mr. le Profes. A.P. Pavlow sur les bords de la Volga dans les couches mésozoiques de Simbirsk. *Ann. Géol. Minéral. Russ.* **1909**, *11*, 42–64. (In Russian)
43. Arkhangelsky, M.S.; Sennikov, A.G. Subclass Synaptosauria. In *Fossil Vertebrates of Russia and Neighboring Countries. Fossil Reptiles and Birds. Part 1*; Ivakhnenko, M.F., Kurochkin, E.N., Eds.; GEOS: Moscow, Russia, 2008; pp. 224–243. (In Russian)
44. Storrs, G.W.; Arkhangel'sky, M.S.; Efimov, V.M. Mesozoic marine reptiles of Russia and other former Soviet republics. In *The Age of Dinosaurs in Russia and Mongolia*; Benton, M.J., Shishkin, M.A., Unwin, D.M., Kurochkin, E.N., Eds.; Cambridge University Press: Cambridge, UK, 2000; pp. 187–210.

45. Knutsen, E.M.; Druckenmiller, P.S.; Hurum, J.H. A new plesiosauroid (Reptilia: Sauropterygia from the Agardhfjellet Formation (middle Volgian) of central Spitsbergen, Norway. *Nor. J. Geol.* **2012**, *92*, 213–234.
46. Smith, A.S. The back-to-front plesiosaur *Cryptoclidus* (*Apractocleidus*) *aldingeri* from the Kimmeridgian of Milne Land, Greenland. *Bull. Geol. Soc. Den.* **2007**, *55*, 1–7. [CrossRef]
47. Riabinin, A.N. Class Reptilia. In *Atlas of the Guide Forms of the Fossil Faunas of USSR. Vol. IX. Upper Series of the Jurassic System*; Krymgoltz, G.Y., Ed.; Gosgeolizdat: Moscow, Russia, 1949; pp. 290–298. (In Russian)
48. Zverkov, N.G.; Fischer, V.; Madzia, D.; Benson, R.B.J. Increased pliosaurid dental disparity across the Jurassic-Cretaceous transition. *Palaeontology* **2018**, *61*, 825–846. [CrossRef]
49. Nesov, L.A.; Ivanov, A.O.; Khozatskii, L.I. On the finds of the remains of ichthyosaurs in the USSR and the problem of faunal change in the middle Cretaceous. *Vestn. Leningr. Universiteta. Seriya 7 Geol. Geogr.* **1988**, *1*, 15–25. (In Russian)
50. Maisch, M.W.; Matzke, A.T. The Ichthyosauria. *Stutt Beitr. Naturkd B* **2000**, *298*, 1–159.
51. McGowan, C.; Motani, R. *Handbook of Paleoherpetology. Part 8. Ichthyopterygia*; Verlag Dr. Friedrich Pfeil: München, Germany, 2003.
52. Moon, B.C.; Kirton, A.M. Ichthyosaurs of the British Middle and Upper Jurassic Part 1. *Palaeontogr. Soc. Monogr.* **2016**, *170*, 1–84. [CrossRef]
53. Arkhangelsky, M.S. Subclass Ichthyopterygia. In *Fossil Vertebrates of Russia and Neighboring Countries. Fossil Reptiles and birds. Part 1*; Ivakhnenko, M.F., Kurochkin, E.N., Eds.; GEOS: Moscow, Russia, 2008; pp. 244–262. (In Russian)
54. Mitta, V.V.; Seltzer, V.B. First finds of Arctocephalitinae (Ammonoidea) in Jurassic of the southeastern Russian Platform and correlation of the boreal Bathonian with the standard scale. *Trans. Sci. Res. Geol. Inst. Saratov State Univ.* **2002**, *10*, 12–39. (In Russian)
55. Seltzer, V.B. Boreal event in the Early Bathonian basin of the southeastern part of the Russian Plate. In Proceedings of the LV session of the Paleontological Society of RAS "Paleontology and Development of the Stratigraphic Basis for Geologic Mapping", St. Petersburg, Russia, 6–10 April 2009. (In Russian).
56. Mitta, V.V.; Zakharov, V.A.; Barskov, I.S.; Sel'tser, V.B.; Ivanov, A.V. The Upper Bajocian–Lower Bathonian boundary section in the outskirts of Saratov: Molluscan characteristics and biostratigraphy. *Stratigr. Geol. Correl.* **2011**, *19*, 502–514. [CrossRef]
57. Mitta, V.; Kostyleva, V.; Dzyuba, O.; Glinskikh, L.; Shurygin, B.; Seltzer, V.; Ivanov, A.; Urman, O. Biostratigraphy and sedimentary settings of the Upper Bajocian–Lower Bathonian in the vicinity of Saratov (Central Russia). *Neues Jahrb. Geol. Paläontol. Abh.* **2014**, *271*, 95–121. [CrossRef]
58. Ippolitov, A.P. Belemnites and biostratigraphy of the Lower Bathonian of central and southern parts of East European platform. Part 2. Cylindroteuthididae and Belemnotheutididae. *Stratigr. Geol. Correl.* **2018**, *26*, 433–458. [CrossRef]
59. Gulyaev, D.B. Ammonites and Infrazonal Stratigraphy of the Lower Bathonian Besnosovi Zone of the Russian Plate. *Stratigr. Geol. Correl.* **2019**, *27*, 95–117. [CrossRef]
60. Ippolitov, A.P.; Gulyaev, D.B. Palaeogeography and marine connections of the Middle Russian Sea across the Middle Jurassic: Cephalopod-based model. In Proceedings of the 10th International Symposium "Cephalopods—Present and Past", Fes, Morocco, 26 March–3 April 2018; pp. 53–54.
61. Gulyaev, D.B. Infrazonal Ammonite Scale for the Upper Bathonian–Lower Callovian of Central Russia. *Stratigr. Geol. Correl.* **2001**, *9*, 65–92.
62. Gulyaev, D.B. New data on ammonite biostratigraphy of the Upper Batonian and Lower Callovian deposits of Churkinskaya Stchelya reference stratigraphic section (Pizhma River, river basin of Pechora). In *Jurassic System of Russia: Problems of Stratigraphy and Paleogeography. Second All-Russian Meeting: Scientific Materials*; Zakharov, V.A., Dzyuba, O.S., Kiselev, D.N., Rogov, M.A., Eds.; Yaroslavl State Pedagogical University: Yaroslavl, Russia, 2007; pp. 49–58. (In Russian)
63. Kiselev, D.N.; Rogov, M.A. Stratigraphy of the Bathonian–Callovian boundary deposits in the Prosek secion (Middle Volga Region), Article 1. Ammonites and Infrazonal biostratigraphy. *Stratigr. Geol. Correl.* **2007**, *15*, 485–515. [CrossRef]
64. Gulyaev, D.B.; Rogov, M.A. Macrocephalitids (Sphaeroceratidae, Ammonoidea) in the Early Callovian of the East-European subboreal sea. In *Contributions to current cephalopod research: Morphology, Systematics, Evolution, Ecology and Biostratigraphy, Issue 2*; Leonova, T.B., Barskov, I.S., Mitta, V.V., Eds.; PIN RAS: Moscow, Russia, 2009; pp. 72–74. (In Russian)
65. Gulyaev, D.B.; Ippolitov, A.P. Lower Callovian of Kanev dislocations, Cherkasy oblast, Ukraine: Ammonites and stratigraphy. *Stratigr. Geol. Correl.* **2021**, *29*, 767–847. [CrossRef]
66. Mönnig, E. *Der Macrocephalen Oolith von Hildesheim*; Georg Olms Verlag: Hildesheim, Germany, 1995; pp. 1–77.
67. Mönnig, E.; Dietl, G. The systematics of the ammonite genus *Kepplerites* (upper Bathonian and basal Callovian, Middle Jurassic) and the proposed basal boundary stratotype (GSSP) of the Callovian Stage. *Neues Jahrb. Geol. Paläontol. Abh.* **2017**, *286*, 235–287. [CrossRef]
68. Mitta, V.V.; Dietl, G.; Callomon, J.H.; Schweigert, G.; Dietze, V. The ammonite genus *Cadoceras* (Cardioceratidae) in the Lower Callovian (Middle Jurassic) of the Swabian Alb and the Wutach area (Southern Germany), *Neues Jahrb. Geol. Paläontol. Abh.* **2015**, *278*, 303–321. [CrossRef]
69. Sasonova, I.G.; Sasonov, N.T. *Paleogeography of the Russian Platform during Jurassic and Early Cretaceous Time*; Nedra: Leningrad, Russia, 1967. (In Russian)
70. Ippolitov, A.P. The history of Middle Jurassic belemnites in the Middle Russian Sea, in the light of recent discoveries. In *Contributions to Current Cephalopod Research: Morphology, Systematics, Evolution, Ecology and Biostratigraphy. Vol. 5, Proceeding of*

Conference, Moscow, 29–31 October 2018; Leonova, T.B., Barskov, I.S., Mitta, V.V., Eds.; PIN RAS: Moscow, Russia, 2018; pp. 49–52. (In Russian)

71. Vinogradov, A.P. (Ed.) *Atlas of the Lithological-Paleogeographic Maps of the Russian Platform and its Geosynclinal Margins. Volume II. Mesozoic and Cenozoic*; Gosudarstvennoe Nauchno-Tehnicheskoe Izdatelstvo Literatury po Geologii i Ohrane nedr: Moscow–Leningrad, Russia, 1961. (In Russian)
72. Wierzbowski, H.; Rogov, M. Reconstructing the palaeoenvironment of the Middle Russian Sea during the Middle–Late Jurassic transition using stable isotope ratios of cephalopod shells and variations in faunal assemblages. *Palaeogeogr. Palaeoclimatol. Palaeoecol.* **2011**, *299*, 250–264. [CrossRef]
73. Benton, M.J.; Spencer, P.S. British Mid Jurassic fossil reptile sites. In *Fossil Reptiles of Great Britain*; Benton, M.J., Spencer, P.S., Eds.; Chapman and Hall: New York, NY, USA, 1995; pp. 124–164.
74. Riou, B.; Bennourine, M.; Klee, N.; Lutz, M. Jurassic marine crocodiles in the Monts d'Ardèche UNESCO Global Geopark. *Geoconservation Res.* **2021**, *4*, 368–377.
75. Parrilla-Bel, J.; Young, M.T.; Moreno-Azanza, M.; Canudo, J.I. The first metriorhynchid crocodylomorph from the Middle Jurassic of Spain, with implications for evolution of the subclade Rhacheosaurini. *PLoS ONE* **2013**, *8*, e54275. [CrossRef] [PubMed]
76. Madzia, D.; Sachs, S.; Klug, C. Historical significance and taxonomic status of *Ischyrodon meriani* (Pliosauridae) from the Middle Jurassic of Switzerland. *PeerJ* **2022**, *10*, e13244. [CrossRef] [PubMed]
77. Madzia, D.; Szczygielski, T.; Wolniewicz, A.S. The giant pliosaurid that wasn't-revising the marine reptiles from the Kimmeridgian, Upper Jurassic, of Krzyżanowice, Poland. *Acta Palaeontol. Pol.* **2021**, *66*, 99–129. [CrossRef]
78. Young, M.T.; Andrade, M.B.; Brusatte, S.L.; Sakamoto, M.; Liston, J. The oldest known metriorhynchid super-predator: A new genus and species from the Middle Jurassic of England, with implications for serration and mandibular evolution in predacious clades. *J. Syst. Palaeontol.* **2013**, *4*, 475–513. [CrossRef]
79. Kiselev, D.N.; Rogov, M.A. Detailed biostratigraphy of the Middle Callovian–lowest Oxfordian in the Mikhaylov reference section (Ryazan region, European part of Russia) by ammonites. *Vol. Jurass.* **2018**, *16*, 73–186. [CrossRef]
80. Tesakova, E.M. Callovian and Oxfordian Ostracodes from the Central Region of the Russian Plate. *Paleont J.* **2003**, *37*, S107–S227.
81. Gulyaev, D.B. New data on stratigraphy of the classical Callovian section near the village Alpat'evo (Oka river, Moscow Region). In *Jurassic System of Russia: Problems of Stratigraphy and Paleogeography, Proceedings of the Fifth all-Russian Meeting*; Zakharov, V.A., Rogov, M.A., Schurygin, B.N., Eds.; IzdatNaukaServis LLC: Yekaterinburg, Russia, 2013; pp. 53–57. (In Russian)
82. Sasonov, N.T. *Jurassic Deposits of the Central Areas of the Russian Platform*; Gosgeoltehizdat: Leningrad, Russia, 1957. (In Russian)
83. Meledina, S.V. Ammonites and zonal stratigraphy of the Callovian of Subboreal regions of USSR. *Trans. Inst. Geol. Geophys. Sib. Branch. USSR Acad. Sci.* **1987**, *691*, 1–182. (In Russian)
84. Kiselev, D.N. *Zones, Subzones, and Biohorizons of the Middle Callovian of Central Russia*; Pedagogical University of Yaroslavl: Yaroslavl, Russia, 2001; pp. 1–38. (In Russian)
85. Gulyaev, D.B.; Ippolitov, A.P. Detailed biostratigraphy of the Lower Callovian of the Kanev dislocations area (Cherkassy region, Ukraine). In *Jurassic System of Russia: Problems of Stratigraphy and Paleogeography, Proceedings of the Fifth all-Russian meeting*; Zakharov, V.A., Rogov, S.B.N., Eds.; IzdatNaukaServis LLC: Yekaterinburg, Russia, 2013; pp. 65–72. (In Russian)
86. Tesakova, E. Late Callovian and Early Oxfordian ostracods from the Dubki section (Saratov area, Russia): Implications for stratigraphy, paleoecology, eustatic cycles and palaeobiogeography. *Neues Jahrb. Geol. Paläontol. Abh.* **2008**, *249*, 25–45. [CrossRef]
87. Kiselev, D.; Rogov, M.; Glinskikh, L.; Guzhikov, A.; Pimenov, M.; Mikhailov, A.; Dzyuba, O.; Matveev, A.; Tesakova, E. Integrated stratigraphy of the reference sections for the Callovian/Oxfordian boundary in European Russia. *Vol. Jurass.* **2013**, *11*, 59–96.
88. Mitta, V.V.; Alekseev, A.S.; Shik, S.M. (Eds.) *Unified Regional Stratigraphic Scheme of the Jurassic of East European Platform*; PIN RAS—VNIGNI: Moscow, Russia, 2012. (In Russian)
89. Gulyaev, D.B. Stratigraphy of the Bathonian-Callovian nearboundary deposits of European Russia. In *Jurassic System of Russia: Problems of Stratigraphy and Paleogeography, Proceedings of the Sixth All-Russian Meeting*; Zakharov, V.A., Rogov, M.A., Ippolitov, A.P., Eds.; ALEF: Makhachkala, Russia, 2015; pp. 94–101. (In Russian)
90. Kiselev, D.N. Ammonites and biostratigraphy of the Callovian beds in the Votcha section on the Sysola river. *News Paleontol. Strat.* **2006**, *9*, 47–69.
91. Kiselev, D.N. *Ammonites and Infrazonal Stratigraphy of the Boreal and Subboreal Bathonian and Callovian*; GEOS: Moscow, Russia, 2022. (In Russian)
92. Gulyaev, D.B.; Ippolitov, A.P. On the Bathonian-Callovian boundary deposits in the reference sections of the north of European Russia (Komi Republic). In *Golden Age of Russian Malacology. Collective Volume of the All-Russia Research Conference Dedicated to the 100th Anniversary of the Birth of Professor Viktor Nikolaevich Shimansky*; Barskov, I.S., Ivanov, A.V., Leonova, T.B., Nikolayeva, S.V., Yashkov, I.A., Eds.; Borissiak Paleontological Institute of RAS–Yuri Gagarin Saratov State Technical University: Moscow–Saratov, Russia, 2016; pp. 235–248. (In Russian)
93. Kiselev, D.N.; Ippolitov, A.P.; Rogov, M.A. Jurassic section near the Votcha village. In *Jurassic Deposits of the Sysola River Basin in the Vicinities of Syktyvkar (Komi Republic) Guidebook of the Field Geological Excursion of the IX All-Russian Meeting "Jurassic System of Russia: Problems of Stratigraphy and Paleogeography"*; Kiselev, D.N., Beznosov, P.A., Rogov, M.A., Ippolitov, A.P., Zverkov, N.G., Lyyurov, S.V., Eds.; Institute of Geology, Komi SC UB RAS: Syktyvkar, Russia, 2023; pp. 5–20. (In Russian)
94. Mitta, V.V. Ammonites and stratigraphy of the Lower Callovian of the Russian Platform. *Bull. VNIGRI* **2000**, *3*, 1–144. (In Russian)

95. Gründel, J.; Mitta, V. Gastropoden aus dem Untercallovium des Unzha-Beckens (Zentralrussland). *Freib. Forschungshefte* **2013**, *C545*, 107–139.
96. Keupp, H.; Mitta, V. Cephalopod jaws from the Middle Jurassic of Central Russia. *Neues Jahrb. Geol. Paläontol. Abh.* **2013**, *270*, 23–54. [CrossRef]
97. Mitta, V.V. The genus *Cadochamoussetia* in the phylogeny of the Jurassic Cardioceratidae (Ammonoidea). In *Advancing Research on Living and Fossil Cephalopods. Development and Evolution Form, Construction, and Function Taphonomy, Palaeoecology, Palaeobiogeography, Biostratigraphy, and Basin Analysis, Proceedings of the IV International Symposium on Cephalopods: Present and Past, Granada, Spain, 14–18 July 1996*; Olóriz, F., Rodríguez-Tovar, F.J., Eds.; Kluwer Acad./Plenum Publ.: New York, NY, USA, 1999; pp. 125–136.
98. Mitta, V.V. Late Bathonian Cardioceratidae (Ammonoidea) from the middle reaches of the Volga River. *Paleontol. J.* **2005**, *39*, 629–644.
99. Gulyaev, D.B. Macrocephalitinae and Gowericeratinae (Ammonoidea) from the Elatmae Zone and the stratigraphy of the Lower Callovian of central areas of Russian Platform. In *Problems of the Mesozoic Stratigraphy and Palaeontology. Lecturing in Memory of M.S. Mesezhnikov*; Kozlova, G.E., Prozorovskyi, V.A., Eds.; VNIGRI: St.-Petersburg, Russia, 1999; pp. 63–85. (In Russian)
100. Kiselev, D.N. The zone and subzone ammonite assemblages from the Middle Callovian in Central Russia. In *Problems of the Mesozoic Stratigraphy and Palaeontology. Lecturing in memory of M.S. Mesezhnikov*; Kozlova, G.E., Prozorovskyi, V.A., Eds.; VNIGRI: St.-Petersburg, Russia, 1999; pp. 87–115. (In Russian)
101. Gerasimov, P.A.; Mitta, V.V.; Kochanova, M.D.; Tesakova, E.M. *Fossils of the Callovian Stage of Central Russia*; VNIGNI: Moscow, Russia, 1996; pp. 1–127. (In Russian)
102. Tesakova, E.M.; Strezh, A.S.; Gulyaev, D.B. New lower Callovian ostracodes from the Kursk Region. *Paleontol. J.* **2009**, *43*, 258–271. [CrossRef]
103. Ippolitov, A.P.; Gulyaev, D.B. Belemnite stratigraphy of the Lower Callovian of south-west of the Western Europe platform: Preliminary results. In *Jurassic System of Russia: Problems of Stratigraphy and Paleogeography, Proceedings of the Fifth All-Russian Meeting*; Zakharov, V.A., Rogov, S.B.N., Eds.; IzdatNaukaServis LLC: Yekaterinburg, Russia, 2013; pp. 85–89. (In Russian)
104. Benson, R.B.J.; Druckenmiller, P.S. Faunal turnover of marine tetrapods during the Jurassic–Cretaceous transition. *Biol. Rev.* **2014**, *89*, 1–23. [CrossRef]
105. Zverkov, N.G. A Problem of naming of the families of Late Jurassic and Cretaceous ichthyosaurs. *Paleontol. J.* **2022**, *56*, 463–470. [CrossRef]
106. Johnson, M.M.; Young, M.T.; Brusatte, S.L. The phylogenetics of Teleosauroidea (Crocodylomorpha, Thalattosuchia) and implications for their ecology and evolution. *PeerJ* **2020**, *8*, e9808. [CrossRef]
107. Young, M.T.; Brignon, A.; Sachs, S.; Hornung, J.; Foffa, D.; Kitson, J.J.N.; Johnson, M.M.; Steel, L. Cutting the Gordian knot: A historical and taxonomic revision of the Jurassic crocodylomorph *Metriorhynchus*. *Zool. J. Linn. Soc.* **2021**, *192*, 510–553. [CrossRef]
108. Young, M.T.; Wilberg, E.W.; Johnson, M.M.; Herrera, Y.; de Andrade, M.B.; Brignon, A.; Sachs, S.; Abel, P.; Foffa, D.; Fernández, M.S.; et al. The history, systematics, and nomenclature of Thalattosuchia (Archosauria: Crocodylomorpha). *Zool. J. Linn. Soc.* **2024**, *200*, 547–617. [CrossRef]
109. Meledina, S.V. Boreal Middle Jurassic of Russia. *Trans. Inst. Geol. Geophys. Sib. Branch. Russ. Acad. Sci.* **1994**, *819*, 1–184. (In Russian)
110. Repin, Y.S.; Zakharov, V.A.; Meledina, S.V.; Nalnyaeva, T.I. Atlas of the mollusks of the Pechora Jurassic. *Bull. VNIGRI* **2006**, *3*, 1–262. (In Russian)
111. Beznosov, P.A.; Zverkov, N.G. Fossil vertebrates in the Jurassic of the Ukhta area. In *Jurassic deposits of the Ukhta area (Komi Republic)*; Rogov, M.A., Ed.; Institute of Geology Komi SC UB RAS: Syktyvkar, Russia, 2023; pp. 88–93.
112. Beznosov, P.A.; Zverkov, N.G. Vertebrate remains from the Jurassic deposits of the Sysola River basin. In *Jurassic Deposits of the Sysola River Basin in the Vicinities of Syktyvkar (Komi Republic). Guidebook of the Field Geological Excursion of the IX All-Russian Meeting "Jurassic System of Russia: Problems of Stratigraphy and Paleogeography"*; Institute of Geology Komi SC UB RAS: Syktyvkar, Russia, 2023; p. 29. (In Russian)
113. Malkov, B.A.; Lysuk, A.Y.; Zhornyak, A.V. New finds of fossil bones of Jurassic plesiosaurs and ichthyosaurs in the Callovian clays of Sysola depression. *Vestn. Inst. Geol. Komi SC UB RAS.* **2004**, *9*, 28–30. (In Russian)
114. Kassin, N. *Carte Géologique Générale de la Partie Européenne De l'U.R.S.S. Feuille 107. Région Viatka—Slobodskoi—Omoutninsk—Kai*; Comité Géologique: Leningrad, Russia, 1928. (In Russian)
115. Arkhangelsky, M.S. On a find of incomplete ichthyosaur skeleton in the lower Callovian deposits of Saratov Region. In *Materials of the First All-Russian Meeting "Jurassic System of Russia: Problems of Stratigraphy and Paleogeography"*; Zakharov, V.A., Rogov, M.A., Dzyuba, O.S., Eds.; GIN RAS: Moscow, Russia, 2005; p. 4. (In Russian)
116. Noè, L.F. A Taxonomic and Functional Study of the Callovian (Middle Jurassic) Pliosauroidea (Reptilia; Sauropterygia). Ph.D. Thesis, School of Environmental and Applied Sciences, University of Derby, Derby, UK, 2001.
117. Linder, H. Beitrage zur Kenntnis der Plesiosaurier-Gattungen *Peloneustes* und *Pliosaurus*. *Geol. Palæontol. Abh.* **1913**, *11*, 339–401.
118. Benson, R.; Evans, M.; Smith, A.S.; Sassoon, J.; Moore-Faye, S.; Ketchum, H.F.; Forrest, R. A giant pliosaurid skull from the Late Jurassic of England. *PLoS ONE* **2013**, *8*, e65989. [CrossRef] [PubMed]
119. Smith, A.S.; Benson, R.B.J. Osteology of *Rhomaleosaurus thorntoni* (Sauropterygia: Rhomaleosauridae) from the Lower Jurassic (Toarcian) of Northamptonshire, England. *Monogr. Palaeontogr. Soc.* **2014**, *168*, 1–40. [CrossRef]
120. Newman, B.; Tarlo, L.B. A giant marine reptile from Bedfordshire. *Animals* **1967**, *10*, 61–63.

121. O'Keefe, F.R.; Street, H.P.; Wilhelm, B.C.; Richards, C.D.; Zhu, H. A new skeleton of the cryptoclidid plesiosaur *Tatenectes laramiensis* reveals a novel body shape among plesiosaurs. *J. Vertebr. Paleontol.* **2011**, *31*, 330–339. [CrossRef]
122. Brown, D.S.; Keen, J.A. An extensive marine vertebrate fauna from the Kellaways Sand (Callovian, Middle Jurassic) of Lincolnshire. *Mercian Geol.* **1991**, *12*, 87–96.
123. O'Keefe, F.R.; Street, H.P. Osteology of the cryptocleidoid plesiosaur *Tatenectes laramiensis*, with comments on the taxonomic status of the Cimoliasauridae. *J. Vertebr. Paleontol.* **2009**, *29*, 48–57. [CrossRef]
124. Zverkov, N.G.; Prilepskaya, N.E. A prevalence of *Arthropterygius* (Ichthyosauria: Ophthalmosauridae) in the Late Jurassic—Earliest Cretaceous of the Boreal Realm. *PeerJ* **2019**, *7*, e6799. [CrossRef] [PubMed]
125. Zverkov, N.G.; Jacobs, M.L. Revision of *Nannopterygius* (Ichthyosauria: Ophthalmosauridae): Reappraisal of the "inaccessible" holotype resolves a taxonomic tangle and reveals an obscure ophthalmosaurid lineage with a wide distribution. *Zool. J. Linn. Soc.* **2021**, *191*, 228–275. [CrossRef]
126. Fischer, V.; Weis, R.; Thuy, B. Refining the marine reptile turnover at the Early–Middle Jurassic transition. *PeerJ* **2021**, *9*, e10647. [CrossRef] [PubMed]
127. Gilmore, C.W. Osteology of *Baptanodon*. *Mem. Carnegie Mus.* **1905**, *2*, 77–129. [CrossRef]
128. Arkhangelsky, M.S.; Zverkov, N.G.; Spasskaya, O.S.; Evgrafov, A.V. On the first reliable record of the ichthyosaur *Ophthalmosaurus icenicus* Seeley in the Oxfordian–Kimmeridgian beds of European Russia. *Paleontol. J.* **2018**, *52*, 49–57. [CrossRef]
129. Bardet, N. Extinction events among Mesozoic marine reptiles. *Hist. Biol.* **1994**, *7*, 313–324. [CrossRef]
130. Fernández, M. A new long-snouted ichthyosaur from the early Bajocian of Neuquén basin (Argentina). *Ameghiniana* **1994**, *31*, 291–297.
131. Fernández, M. A new ichthyosaur from the Los Molles Formation (Early Bajocian), Neuquén basin, Argentina. *J. Paleontol.* **1999**, *73*, 677–681. [CrossRef]
132. Fernández, M. Ophthalmosauria (Ichthyosauria) forefin from the Aalenian-Bajocian boundary of Mendozo Province, Argentina. *J. Vertebr. Paleontol.* **2003**, *23*, 691–694. [CrossRef] [PubMed]
133. Cau, A.; Fanti, F. The oldest known metriorhynchid crocodylian from the Middle Jurassic of North-eastern Italy: Neptunidraco ammoniticus gen. et sp. nov. *Gondwana Res.* **2011**, *19*, 550–565. [CrossRef]
134. Maxwell, E.E.; Fernández, M.S.; Schoch, R.R. First diagnostic marine reptile remains from the Aalenian (Middle Jurassic): A new ichthyosaur from southwestern Germany. *PLoS ONE* **2012**, *7*, e41692. [CrossRef] [PubMed]
135. Bardet, N.; Falconnet, J.; Fischer, V.; Houssaye, A.; Jouve, S.; Pereda Suberbiola, X.; Pérez-García, A.; Rage, J.-C.; Vincent, P. Mesozoic marine reptile palaeobiogeography in response to drifting plates. *Gondwana Res.* **2014**, *26*, 869–887. [CrossRef]
136. Sachs, S.; Hornung, J. Marine Tetrapoden aus dem Mittleren Jura (Aalenium und Bajocium) von Bielefeld (Nordwestdeutschland) *Berichte Naturwiss. Ver. Für Bielef. Und Umgeg.* **2015**, *53*, 52–73.
137. Arkell, W.J. *The Jurassic System in Great Britain*; Clarendon Press: Oxford, UK, 1933.
138. Araujo, R.; Smith, A.S.; Liston, J. The Alfred Leeds Fossil Vertebrate Collection of the National Museum of Ireland—Natural History. *Ir. J. Earth Sci.* **2008**, *26*, 17–32.
139. Martill, D.M. The stratigraphic distribution of fossil vertebrates in the Oxford Clay of England. *Mercian Geol.* **1986**, *10*, 161–186.
140. Martill, D.M.; Hudson, J.D. (Eds.) *Fossils of the Oxford Clay, Palaeontological Association Field Guide to Fossils*; Palaeontological Association: London, UK, 1991.
141. Hudson, J.D.; Martill, D.M. The Peterborough Member (Callovian, Middle Jurassic) of the Oxford Clay Formation at Peterborough, UK. *J. Geol. Soc.* **1994**, *151*, 113–124. [CrossRef]
142. Delair, J.B. The Mesozoic reptiles of Dorset, part two. *Proc. Dorset Nat. Hist. Archaeol. Soc.* **1960**, *80*, 52–75.
143. Foffa, D.; Young, M.; Brusatte, S.L.; Graham, M.R.; Steel, L. A new metriorhynchid crocodylomorph from the Oxford Clay Formation (Middle Jurassic) of England, with implications for the origin and diversification of Geosaurini. *J. Syst. Palaeontol.* **2018**, *16*, 1123–1143. [CrossRef]
144. Rogov, M.; Zakharov, V.; Kiselev, D. Molluscan immigrations via biogeographical ecotone of the Middle Russian Sea during the Jurassic. *Vol. Jurass.* **2008**, *6*, 143–152.
145. Foffa, D.; Young, M.; Brusatte, S. Filling the Corallian gap: New information on Late Jurassic marine reptile faunas from England. *Acta Palaeontol. Pol.* **2018**, *63*, 287–313. [CrossRef]

diversity

MDPI

Article

Exceptional In Situ Preservation of Chondrocranial Elements in a Coniacian Mosasaurid from Colombia

María Eurídice Páramo-Fonseca [1,2,*], José Alejandro Narváez-Rincón [1,2], Cristian David Benavides-Cabra [1,2] and Christian Felipe Yanez-Leaño [3]

1 Departamento de Geociencias, Universidad Nacional de Colombia, Bogotá 111321, Colombia; janarvaezr@unal.edu.co (J.A.N.-R.); cdbenavidesc@unal.edu.co (C.D.B.-C.)
2 Museo Geológico Nacional José Royo y Gómez, Servicio Geológico Colombiano, Bogotá 111321, Colombia
3 Grupo de Investigación Clínica SHAIO, Fundación Clínica SHAIO, Bogotá 111121, Colombia; christian.yanez@shaio.org
* Correspondence: meparamof@unal.edu.co

Abstract: The first record of well-preserved chondrocranial elements in mosasaurids is here described. These elements are preserved in situ in a Coniacian skull found in north-central Colombia, inside a calcareous concretion. Based on a 3D model generated from computed tomography scans, we identified elements of the nasal and orbitotemporal regions. Our descriptions show that in this specimen, the chondrocranium was reduced, more so than in most lacertilians (including their closest recent relatives, the varanids), but not as severely as in snakes or amphisbaenians (which have an extremely reduced chondrocranium and limbs). The new evidence suggests that the reduction in the chondrocranium in mosasaurids could be related to modification of their limbs when adapting to aquatic environments, but also that in mosasaurids, the olfactory tract was reduced, and the optic muscle insertions occurred mainly in the interorbital septum. The exceptional preservation of the chondrocranial elements in the specimen is facilitated by a gray mineralization covering them. XRD analysis and thin section observations indicated that this mineralization is composed of microcrystalline quartz and calcite. We infer that this material was produced by a partial silicification process promoted by lower pH microenvironments associated with bacterial breakdown of non-biomineralized tissues during early diagenesis.

Keywords: mosasaurid; chondrocranium; Coniacian; Colombia

1. Introduction

The chondrocranium is the cartilaginous portion of the vertebrate braincase [1–3]. Among gnathostomes, the chondrocranium has six recognizable components during development [2,3]: the nasal capsules, which support the nasal apparatus and may form the ethmoid plate; the orbital cartilages, located medial to the eyes; the otic capsules, which contain the inner ear; the parachordals, which form the posterior base of the braincase; a pair of rod-like trabeculae cranni that sit between the parachordals and the nasal capsules beneath the orbital cartilage and the interorbital septum; and the occipital and preoccipital arches, which enclose the posterior part of the brain. The trabeculae cranii eventually meet in the midline anteriorly to form the internasal septum [2,3], which ossifies to constitute the ethmoid and sphenethmoid bones [4]. The parachordals give rise to the basal plate, which ossifies to constitute the basioccipital-basisphenoid region of the skull [4]. The otic capsules ossify to form the prootic, opisthotic, and epiotic, and the occipital arch ossifies to form the supraoccipital and the exoccipitals [4] (Figure 1A). The nasal and orbital cartilages remain cartilaginous in adulthood [4]. The chondrocranium varies greatly among taxa regarding its frame-like structure, mineralization, and when and to what extent it is replaced by bone during ontogeny [1,3]. In reptiles, the grade of ossification and chondrification varies greatly [5]. In this group of vertebrates, the adult chondrocranium is divided into three regions: the nasal, the orbitotemporal, and the otic-occipital region [2,6–9]. Anteriorly, the nasal region is formed by the nasal capsules, which are separated by the nasal septum, and posteriorly, it is formed by the planum antorbitale [9]. The orbitotemporal region is formed by the interorbital septum, the planum supraseptale, and the taeniae and pilae chondrified in diverse degrees [9]. The otic and occipital regions are formed by the otic capsules and the basal plate [9].

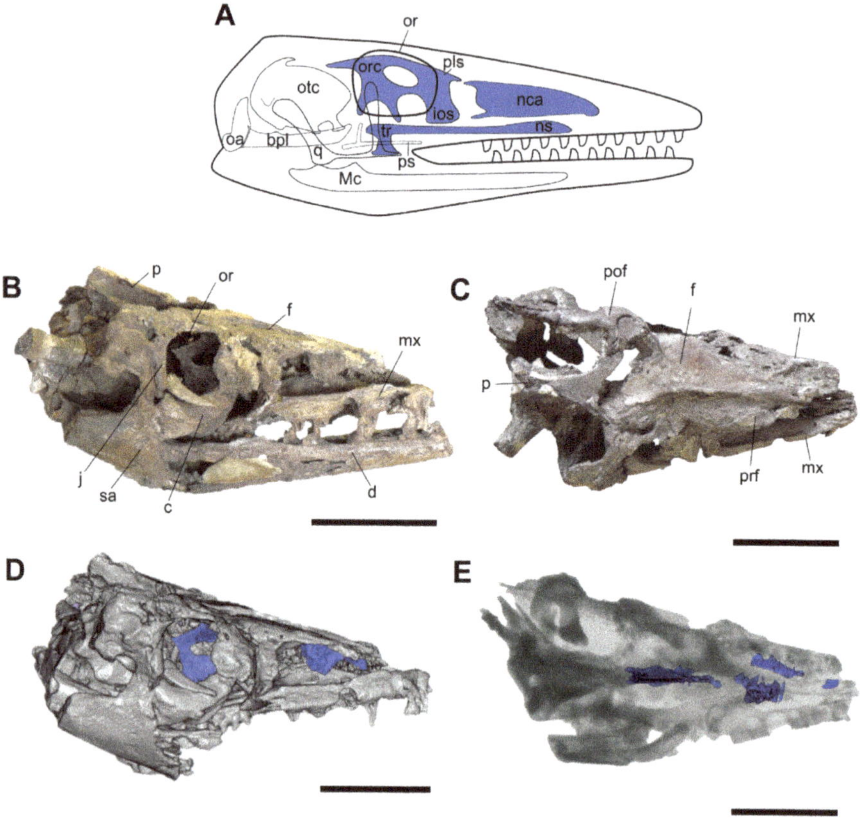

Figure 1. (**A**) Schematic diagram of a lizard-like chondrocranium in right lateral view (redrawn and modified from [2,3]) adapted to a mosasaur-like skull silhouette. (**B–E**) IGMp879524 skull showing the preserved in situ chondrocranial elements. (**B**,**C**) Photographs of the skull in (**B**), right lateral and (**C**) dorsal views. (**D**) Opaque 3D model of the skull in right lateral view (3D Slicer option CT-AAA). (**E**) Transparent 3D model of the skull in dorsal view (3D Slicer option CT-X-ray). The preserved chondrocranial elements are highlighted in blue. Part of the mandible and the braincase are not included in the 3D model. **Abbreviations**: **bpl**, basal plate; **c**, coronoid; **d**, dentary; **f**, frontal; **ios**, interorbital septum; **j**, jugal; **Mc**, Meckel's cartilage; **mx**, maxilla; **nca**, nasal capsules; **ns**, nasal septum; **oa**, occipital arch; **or**, orbit; **orc**, orbital cartilage; **otc**, otic capsule; **p**, parietal; **pof**, postorbitofrontal; **prf**, prefrontal; **pls**, planum supraseptale; **ps**, parasphenoid; **q**, quadrate; **sa**, surangular; **tr**, trabeculae cranii. Scale bars: 100 mm.

In lepidosaurs, a significant portion of the chondrocranium is generally retained into adulthood [3]. In most adult lepidosaurs, the nasal capsules, a nasal septum, an interorbital septum, and a central framework of slender bars (taeniae and pilae) are retained [3]. In some lizards, parts of the taenia medialis and pila metoptica ossify, forming the orbitosphenoid [2].

In mosasaurs (Late Cretaceous marine lepidosaurs), the chondrocranium is almost unknown. Camp [10] described an orbitosphenoid in a *Plotosaurus* (*Kolposaurus*) skull from the Maastrichtian (Upper Cretaceous) of California. According to this author, the orbitosphenoid encloses the optic chiasma and provides support to the ventrolateral wall of the cerebrum. Additionally, Camp [10] described a brownish stain accompanied by bone granules and identified it as a partially preserved interorbital septum anterior to the orbitosphenoid. Bellairs [5], based on descriptions of the chondrocranium development in sauropsids, including varanids, stated that the partially interorbital septum identified by [10] might be called septosphenoid. In some specimens of *Platecarpus* and one of *Mosasaurus*, a bone with a similar form of the orbitosphenoid of *Plotosaurus* has been interpreted as orbitosphenoid [11].

Another mosasaurid specimen preserving chondrocranial elements was reported by [12]. It consists of a skull (IGMp879524), reported by [12] under the catalog number IPN-2, preserving in situ an almost complete chondrocranium. The specimen was found in Coniacian beds from north-central Colombia and constitutes the first record of well-preserved chondrocranial elements in a mosasaurid. The osteo-anatomic description and the taxonomic determination of the specimen is currently under publication. In this contribution, we fully describe and identify the preserved chondrocranial elements of this specimen, discuss the probable functional implications of their anatomy, and comment on their preservation features.

2. Materials and Methods

The specimen (IGMp879524) is housed in the paleontological collection of the Museo Geológico Nacional José Royo y Gómez at the Servicio Geológico Colombiano (SGC). It consists of a good portion of an articulated three-dimensionally preserved skull, lacking its anterior end, and maintaining some delicate intracranial structures, such as chondrocranial elements (Figure 1A,B). It was found in a calcareous concretion from Coniacian beds [12] of the Galembo Formation [13], a geologic unit formerly known as the Galembo Member of the La Luna Formation [14]. The specimen was collected in the surroundings of the Lebrija municipality, Santander department, northern Colombia [12].

The preparation of IGMp879524 was mainly performed chemically, with an initial mechanical treatment. The chemical preparation revealed the presence of a gray mineralization, indissoluble in acid, which was found coating a large part of the bone surfaces and forming some laminae in the concretion matrix. This material was also found firmly adhered to intracranial thin and delicate bones and cartilages, as the chondrocranium components. In these cases, we preserve the coverage without subjecting the specimen to mechanical preparation to avoid compromising the integrity of the fossilized bones or cartilages. The position of some bones of the skull, as the quadrates, which are displaced and rotated dorsally into the temporal fenestrae, suggests that the skull was deposited upside down. This is important because a few delicate detached fragments found on the ventral surface of the frontal could correspond to separated elements from the chondrocranium that fell upon the internal surface of the frontal.

To describe and identify the chondrocranial elements of the skull (IGMp879524), we built a 3D model from computed X-ray tomography (CT) scans of the specimen (Figure 1C,D). Most of the mandibles and some fragments of the occipital region, which are separated from the main skull piece, were not joined for scanning given the fragility of these elements. The CT scan was performed at the Clinica SHAIO, Bogotá, Colombia, using a Canon Aquilon One with a precision of 0.5 mm, KVP: 120, head protocol, and bone filter. The final voxel size was 0.25 mm. A 3D model was made using the open-source software 3D Slicer 5.2.2. The identification of the chondrocranial elements follows [5–7]. We adopted the terminology used by these authors in their description of the development of the chondrocranium in squamate reptiles.

To evaluate the composition of the gray mineralization, we performed an X-ray diffraction (XRD) analysis of a sample of this material recovered from the concretion. Also, to contrast the matrix and the gray mineralization composition, we made observations on three thin sections: one of the calcareous matrix with bone, one of the gray mineralization, and one including both the mineralization and the calcareous matrix. The XRD analysis was performed at the Lithogeochemical Characterization Laboratory of the Departamento de Geociencias of the Universidad Nacional de Colombia. The sample was prepared for the XRD analysis following the parameters established by [15,16]. The measurements were taken using a Bruker Co. Karlsruhe, Germany D2 PHASER with a copper lamp, and the results were interpreted using the software Difract.EVA.V4.2.1. The thin sections were made at the Petrographic Techniques Laboratory of the Departamento de Geociencias of the Universidad Nacional de Colombia. All the figures in this contribution were assembled using the open-source software Inkscape 1.3.2.

3. Anatomical Description

Based on direct observations on the fossil, as well as on the CT scan slices and 3D model, we identified the following chondrocranial elements (see Figures 2 and 3): some poorly preserved nasal elements, which probably represent remains of the nasal capsules and fragments of the nasal septum; orbitotemporal elements, including well-preserved planum supraseptale and interorbital septum with the posterior septal fenestra, the taenia medialis, the cartilago hypochiasmatica, and the trabecula communis. All these elements were preserved covered by a gray mineralization. Nevertheless, the surfaces of these elements, mainly the interorbital septum, show a porous texture, indicating a cartilaginous origin (Figure 2F).

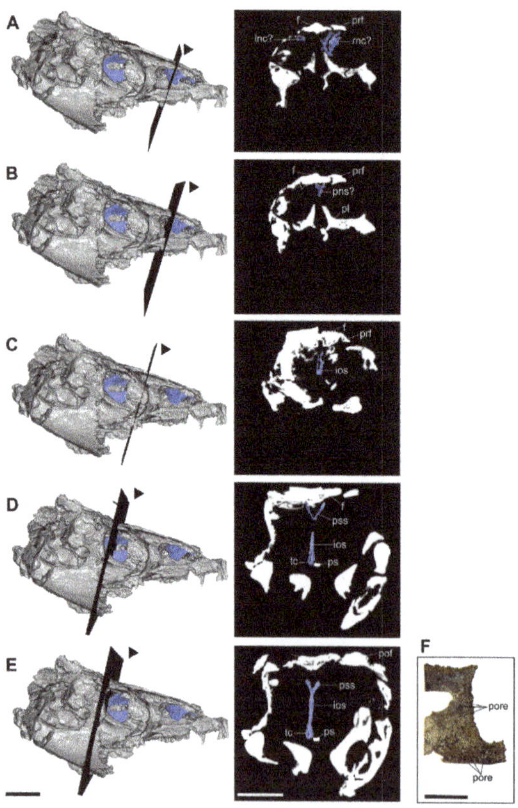

Figure 2. Location of the chondrocranial elements in IGMp879524 skull. (**A–E**) Three-dimensional model (left) and cross-section scans (right) of the skull without the complete mandible. The chondrocranial

elements are highlighted in blue. The black planes and arrows in the 3D models indicate the position and orientation of the cross-sections shown on the right side. (F) Detail of the interorbital septum in its left side showing its porous texture. **Abbreviations**: **f**, frontal; **ios**, interorbital septum; **lnc?**, probable element of the left nasal capsule; **pl**, palatine; **pns?**, probable element of the posterior region of the nasal septum; **pof**, postorbitofrontal; **prf**, prefrontal; **ps**, parasphenoid rostrum; **pss**, planum supraseptale; **rnc?**, probable element of the right nasal capsule; **tc**, trabeculae comunis. Scale bars: (**A–E**) 50 mm; (**F**) 30 mm.

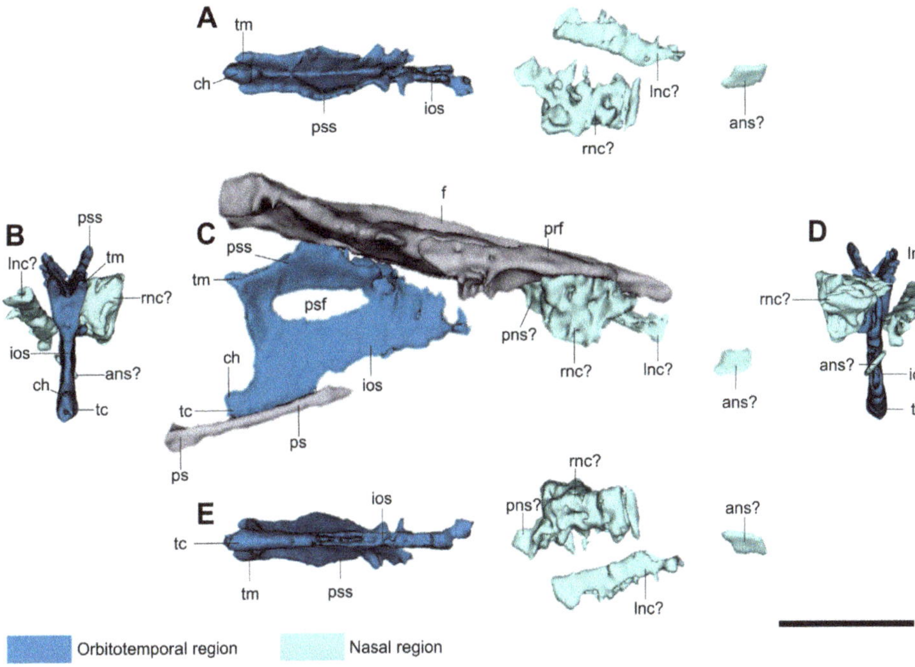

Figure 3. Three-dimensional model of the chondrocranial elements and contacting bones of IGMp879524 in (**A**) dorsal, (**B**) posterior, (**C**) right lateral, (**D**) anterior, and (**E**) ventral views. **Abbreviations**: **ans?**, probable element of the anterior region of the nasal septum; **ch**, cartilago hypochiasmatica; **f**, frontal; **ios**, interorbital septum; **lnc?**, probable element of the left nasal capsule; **pns?**, probable element of the posterior region of the nasal septum; **prf**, prefrontal; **ps**, parasphenoid rostrum; **psf**, posterior septal fenestra; **psr**, parasphenoid rostrum; **pss**, planum supraseptale; **rnc?**, probable element of the right nasal capsule; **tc**, trabeculae comunis; **tm**, taenia medialis. Scale bar: 50 mm.

3.1. Nasal Region

We recognize three chondrocranial structures preserved in the nasal region. Anteriorly, there are two irregular bulbous masses, one on each side of the sagittal plane, contacting the prefrontals (Figures 2A and 3C). The right one, the best preserved, increases its size anteroventrally reaching the palatine. It has an internal narrow canal, almost vertical in cross-section (Figure 2A). The left mass seems incomplete; it is dorsoventrally compressed and slightly inclined anteroventrally and shows an internal space, but in this case, it is a narrow subhorizontal canal (Figure 2A). According to the *Varanus* cross-sections presented by [17], the canals of these two structures in IGMp879524 could represent the olfactory chambers for the nasal sacs, since they are located between frontal and palatine. However, in this same anatomical position, Konishi [18] proposes the presence of salt-glands in a halisaurine mosasaur. The third structure, a probable element of the nasal septum, is located posterior to the described masses and is a mid-sagittal structure that dorsally contacts the anterior end of the frontal internarial bar (Figure 2B). This structure is posteriorly short and subreniform in cross-section, and anteriorly it increases in height and becomes subtriangular in cross-section, pointing ventrally and with a small canal located in the dorsal midline (Figure 2B).

The preserved nasal elements, as well as the presence of a narrow canal in the lateral masses, are reminiscent of the morphology of the nasal septum and nasal capsules described by [7,9,17]. However, the poor preservation of these structures prevents us from confidently affirming their identification as the nasal septum and capsules.

3.2. Orbitotemporal Region

In the orbitotemporal region, the interobital septum, the planum supraseptale, the taeniae medialis, the trabecula communis, and the cartilago hypochiasmatica are preserved in situ. The taenia marginalis is completely absent on both sides of the skull, and no pilae are preserved in situ. However, since the skull was deposited in an upside-down position, a few small fragments of flattened rod-like structures laying on the ventral surface of the frontal, at the level of the anterior interorbital septum and near both displaced epipterygoids, could be interpreted as broken fragments from both pilae metoptica that have become detached by the thrust of the falling epipterygoids. No orbitosphenoids were identified among these fragments. Considering the symmetric absence of the taeniae marginalis, and the presence of the gray mineralized coverage in all other anterior and posterior intracranial elements, we consider that the absence of these elements is not due to preservation effects; instead, it is because they were not originally present.

The interorbital septum is the main element preserved in the orbitotemporal region. It is a thin mid-sagittal plate dividing both orbits (Figure 2D,E). It is anteriorly reduced and does not contact the nasal elements of the chondrocranium (Figure 3). The interorbital septum, together with the planum supraseptale, form a Y-shaped structure in vertical cross-section (Figures 2E and 3B). Ventrally, the interorbital septum thickens in the trabecula communis (Figures 2E and 3B,D) and reaches the slender anterior rostrum (cultriform process) of the parasphenoid. Due to taphonomic processes, the interorbital septum does not coincide exactly with the parasphenoid rostrum in the sagittal plane (Figure 2C,D). The interorbital septum is perforated on its upper half by a large oval foramen (Figure 3C), which is recognized as the posterior septal fenestra (following [7]).

The planum supraseptale is formed by two short latero-dorsally directed ala-like plates, which form a dorsal angled concavity (Figure 2D,E). It extends anteroposteriorly in two different planes. Anteriorly, the planum supraseptale is horizontal or parallel to the ventral surface of the frontal, whereas posteriorly it is inclined, separating it from the skull roof (Figure 3C). The planum supraseptale ventrally connects to the interorbital septum and forms a portion of the dorsal margin of the posterior septal fenestra (Figures 2D and 3C). Dorsally, in its horizontal portion, the planum supraseptale reaches the frontal, just where the anterior end of the cerebral hemispheres and the olfactory canal are located on the ventral surface of the frontal.

The trabecula communis forms a wedge-shaped thickening of the ventral border of the interorbital septum (Figure 2D,E). It shortly projects posteriorly from the posteroventral margin of the interorbital septum (Figure 3). Above this short projection of the trabecula communis there is a small bulge directed dorsally that could be identified as the cartilago hypochiasmatica (Figure 3C). The taeniae medialis are represented by two short bars projected posterodorsally from the posterior ends of the planum supraseptale (Figure 3). The posterior margin of the interorbital septum, the ventral margin of the taenia medialis, and the preserved dorsal margin of the cartilago hypochiasmatica should have formed the anterior and part of the dorsal and ventral margins of the optic fenestra, as has been illustrated for most lacertilians [2,5,7].

4. Functional Implications

The chondrocranium in IGMp879524 is comparable to that of nearly all described extant lepidosaurs [2,5,9,19–21]; it has similar components arranged in the nasal and orbitotemporal regions. This condition, previously unknown in mosasaurids, allows us to project some of the morphofunctional studies carried out on the chondrocranium of recent lepidosaurs, towards the functionality of the chondrocranium in mosasaurids.

In the studied specimen (IGMp879524), the nasal region, although poorly preserved, allows us to affirm that the chondrification of the internasal septum was very reduced. The orbitotemporal region is also reduced; the planum supraseptale is narrow even though the skull is wide; it is not laterally expanded, differing from most other lizards [5,22]; the pilae are completely absent, except for probable detached fragments of the pila metóptica, and from the taeniae, only a reduced taenia medialis is present. Following the parameters used by [22] to evaluate the reduction degree of the chondrocranium in lepidosaurs, this condition shows that the orbitotemporal region of the chondrocranium of IGMp879524 is reduced, although not extremely reduced. According to these features, some functional implications can be analyzed.

Following the anatomical description of the chondrocranium of some lizards presented by [2,22], the planum supraseptale supports the olfactory tract dorsally, and posterodorsally it supports parts of the telencephalon and diencephalon. The reduced planum supraseptale contacting the frontal in IGMp879524 forms a narrow cavity for the olfactory tract, suggesting this tract was slender in this specimen. In addition, Bellairs [2] and Jones et al. [20] show that in most lizards, behind the posteroventral border of the interorbital septum, the optic nerves (nerve II) enter the brain (optic chiasma), delimited posteriorly by the pila metoptica on each side, and dorsally by the taenia medialis on each side of the interorbital septum. In IGMp879524, each of the taenia medialis and pila metoptica, forming the foramen for the optic nerve, seems greatly reduced, suggesting a narrow foramen for these nerves.

Although it has been proposed that the structural variation of the chondrocranium in squamates could have functional implications in skull mechanics, its biomechanical role in vertebrates remains poorly understood [20]. Jones et al. [3] found no evidence to support a vertical strut role for the chondrocranium in adult lizards. In contrast, according to [3], the role of the chondrocranium is more evident in an adequate growth of the skull during the embryologic development. Jones et al. [20], modeling the responses of strains on the chondrocranium of a lizard (*Salvator merianae*), shows that the chondrocranium only helps to dampen the load on the cranial bones from the stresses generated by the tensions and compressions when biting. These observations could indicate that in mosasaurids, as in most vertebrates, the chondrocranium played an important role during the embryologic development, to obtain an undeformed cranial morphology in the adult. They also indicate that, in the adult stage, the chondrocranium helped to dampen the loads on the cranial bones when biting, as well as to protect the olfactory tract and support muscles and other soft tissues.

According to [21], in the embryos of *Tuatara punctatus*, there are some muscles for moving the eye that are inserted mainly into the interorbital septum, but also into the planum supraseptale and the pila metoptica. The latter is highly fused with the pila antotica and pila accesoria, forming a continuous plate in *T. punctatus*. This shows the great importance of these parts of the chondrocranium in the proper functioning of the eye. The well-preserved interorbital septum and the absence of a fused pilae plate in IGMp879524 suggest that the optic muscle insertions in this mosasaur occurred mainly in the interorbital septum.

Yaryhin et al. [22] suggested that the reduction in the chondrocranial elements could be related to modifications on the appendicular skeleton. These authors show how in lepidosaurs with reduced limbs, there is a tendency for the chondrocranium to shrink. Mosasaurs had not been analyzed in this aspect because of the lack of information on their chondrocranium. Our description provides for the first-time reliable information on the chondrocranium of mosasaurs and shows that, in these aquatic lizards, the chondrocranium was reduced, more than in most lacertilians, including their closest recent relatives, the varanids, but not as much as in snakes or amphisbaenians, which have an extreme reduction in their chondrocranium and their limbs [22]. The reduced chondrocranium of IGMp879524 constitutes a new datum for future analyses. Although much information is still lacking, the reduction in the chondrocranium in our specimen allows us to suggest that a reduction in the chondrocranium in mosasaurs could be related to the modification of their limbs for adaptation to an aquatic life.

5. Preservation

The specimen IGMp879524 is preserved in a concretion (nodule) found in a limestone sequence of the Galembo Formation. The skull has a gray mineralization that is indissoluble in formic acid, firmly adhered, and covering a large part of the bone surfaces and all preserved chondrocranium elements (Figure 4A,B). This mineralization is also present in the concretion matrix as irregular laminae (Figure 4A,B). The XDR analysis performed on this gray mineralization shows that it is composed of Quartz (66.2%) and Calcite (33.8%) (Figure 4C).

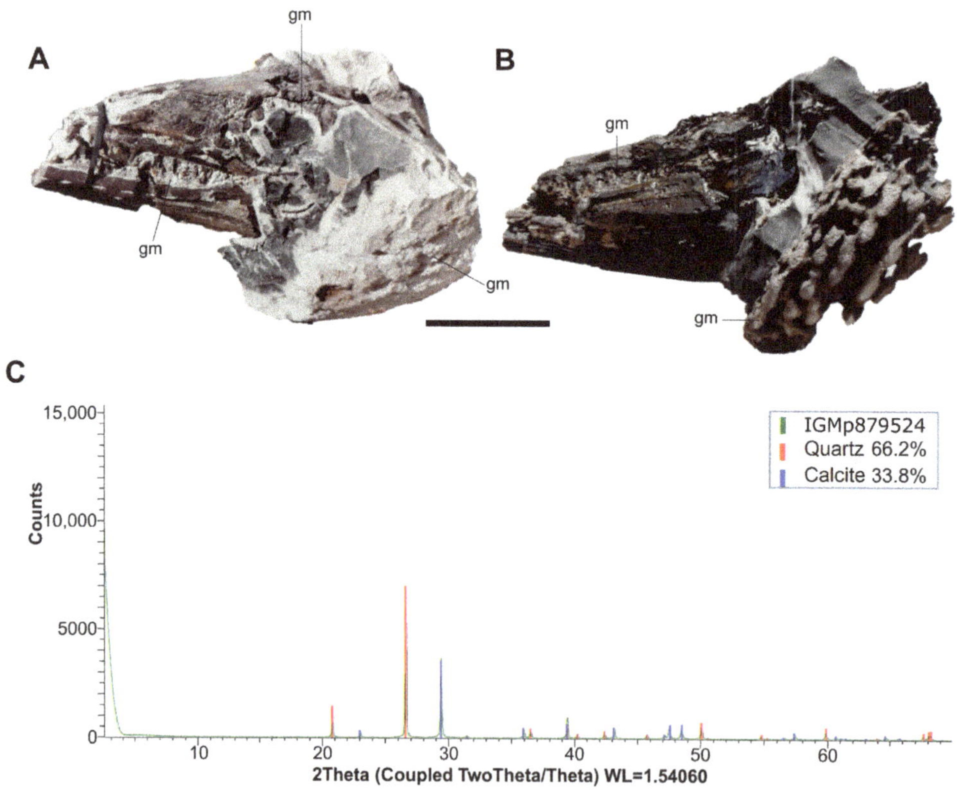

Figure 4. (**A**,**B**) Gray mineralization in the IGMp879524 skull. (**A**) Left lateral view of the skull within the concretion before acid preparation. (**B**) Left lateroventral view of the skull and concretion during chemical preparation. (**C**) Diffractogram of the gray mineralization; **Abbreviations**: **gm**, gray mineralization. Scale bar in (**A**,**B**): 10 cm.

Our petrographic observations of the thin sections indicated that the gray mineralization is mineralogically composed mainly of microcrystalline quartz (15 μm approximately) and, in a smaller proportion, of microcrystalline calcite (micrite) randomly distributed

(Figure 5A). Some poorly preserved round structures resembling those of calcareous dinoflagellate cysts (following [23] (Figure 9C)) were identified in the gray mineralization (Figure 5D). The calcareous matrix of the concretion is mainly composed of micrite with a small proportion of pseudosparite and scattered clusters of microcrystalline quartz (Figure 5B). A few chloritized pseudosparite calcite crystals were found throughout both the gray mineralization and concretion. Thus, the main difference between the gray mineralization and the calcareous matrix is the amount of quartz (Figure 5C). The bone is composed of phosphate, and its pores are filled with sparitic calcite crystals (Figure 5E). Near the bone, there are small veins filled with microcrystalline quartz (Figure 5F).

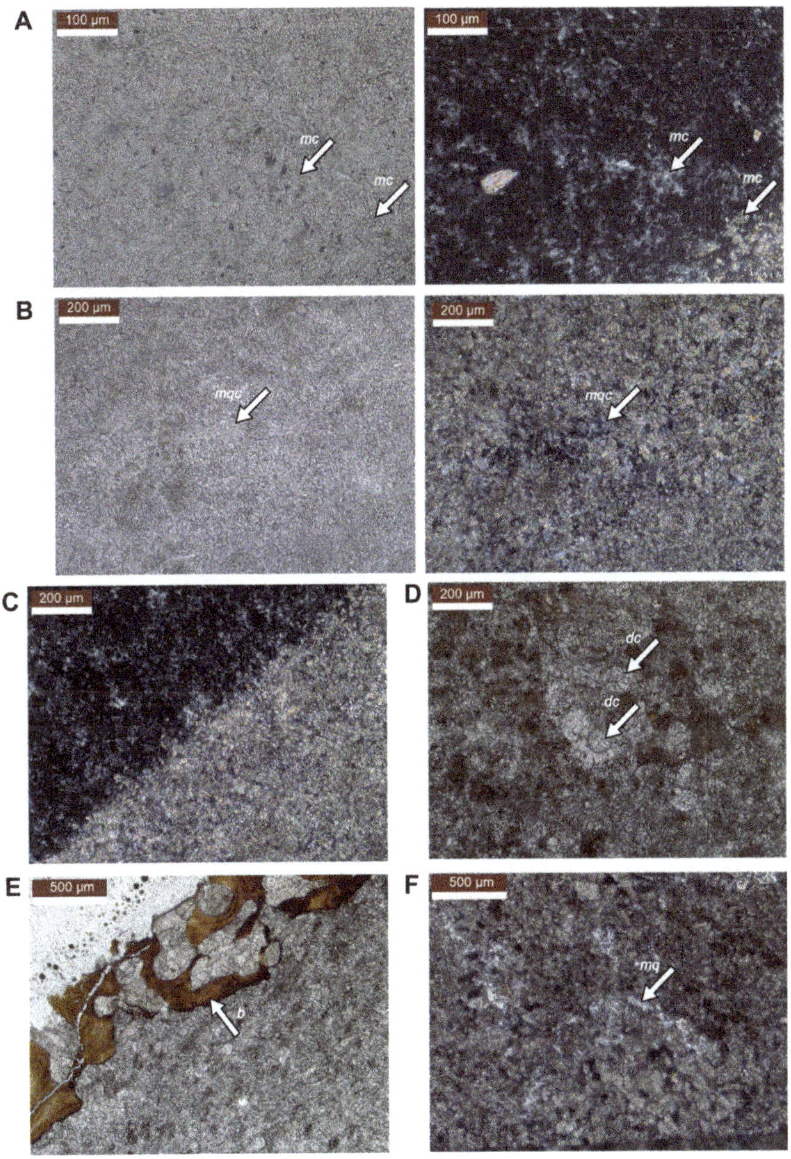

Figure 5. Thin section microphotographs. (**A**,**B**) With plane polarized light PPL (left) and crossed polarized light XPL (right). (**A**) Gray mineralization showing microcrystalline quartz with a smaller

proportion of microcrystalline calcite; (**B**) calcareous matrix showing microcrystalline calcite with scattered clusters of microcrystalline quartz. (**C**) Contact between the gray mineralization rich in quartz (left) and the matrix rich in **calcite** (right) (XPL). (**D**) Dinoflagellate cysts in the gray mineralization (PPL); (**E**) phosphatic bone with sparitic calcite crystals filling its pores (PPL). (**F**) Microcrystalline quartz filling veins in the calcareous matrix (XPL). **Abbreviations**: **b**, bone; **dc**, dinoflagellate cyst; **mc**, microcrystalline calcite; **mq**, microcrystalline quartz; **mqc**, microcrystalline quartz cluster.

In the Galembo Formation, where IGMp879524 comes from, the presence of beds and lenses of chert is frequent [13]. According to [24], silicification is found as a replacement feature in limestone sequences, usually as nodules and occasionally as beds. Trewin and Fayers [24] and Knauth [25] established that authigenic quartz is found abundantly in sedimentary rocks in the form of chert, commonly as granular microcrystalline quartz or microquartz (diameter generally less than 5–20 μm) that has replaced (silicified) pre-existing sediments such as carbonate, opal, or evaporite minerals. The presence of chert in the Galembo Formation and microcrystalline quartz and micrite in the gray mineralization and the concretion matrix of IGMp879524 suggests a partial silicification in the specimen IGMp879524 and an authigenic quartz replacing part of the calcite.

In the gray mineralization of the specimen, the silicification is around 60–70%, and in the concretion matrix, it is around 10–30%, showing a higher silicification in the gray mineralization than in the matrix surrounding the fossil. According to [24], the silicification process is promoted by a high concentration of carbonate ions, organic matter, and reduced pH; the silica is soluble in alkaline waters with a pH above 9, whereas calcite solubility increases with decreasing pH. Diagenetic waters undergoing a decrease in pH (acid) can be expected to dissolve calcite while silica precipitates, promoting the silicification of carbonates [24,25]. According to [25], the earliest stages of silicification in carbonate sequences occur in shell material, in which localized silica precipitation/carbonate dissolution are promoted by the bacterial breakdown of organic matter. This occurs because the decay of the organic matter (decaying tissues) contributes to the development of microenvironments, which geochemically differ in various ways from their immediate surroundings [26,27]. These microenvironments may form internally or externally of carcasses, and their geochemical conditions might promote the precipitation of minerals, such as silica [26,27]. Thus, selective silicification may result in scattered silicified fossils or fossil parts within limestone [24], as we found in our specimen. The scattered silicification in IGMp879524 indicates that there were spots of dissolution of calcite and precipitation of silica promoted by the formation of microenvironments with lower pH, probably associated with the bacterial breakdown of the non-biomineralized tissues. Nonetheless, microenvironment development only leads to mineral formation if the chemical species that precipitate in response to microbial metabolisms are present in sufficient supply [27]. Therefore, we can deduce that for silicification to occur, the concentration of dissolved silica in the diagenetic waters must be enough so it can precipitate and replace carbonates when the geochemical conditions are appropriate. The availability of silica in the environment during the deposition of the Galembo Formation is evidenced by the presence of beds of chert. The dinoflagellate cysts found in the gray mineralization suggests that the dissolution of dinoflagellate skeletons in the water could have been a source of silica.

Silicification in limestone sequences appears to be a relatively early diagenetic process, taking place during shallow burial [24,25]. The presence of silicification in the concretion containing IGMp879524 probably comes from the silicification of carbonates during early diagenesis. That is supported by the presence of calcite within the skull cavities, the gravity-displaced position of some bones in the skull, and the preservation of some chondrocranial cartilages, which indicate that the more labile cranial soft tissues were decaying during the diagenesis of the concretion. This decay process generated the acid microenvironments optimal for the dissolution of calcite and silicification.

6. Conclusions

In this study, we describe the exceptionally preserved chondrocranial elements found in a skull of a mosasaurid collected in north-central Colombia (specimen IGMp879524). We identified the interorbital septum, the planum supraseptale, parts of the trabeculae comunis, and remains in the position of the nasal septum and nasal capsules. The chondrocranium in IGMp879524 is greatly reduced, with the taeniae and pilae very poorly represented.

The chondrocranium of IGMp879524 has a narrow planum supraseptale that contacts the frontal, creating a narrow cavity for the olfactory tract. The well-preserved interorbital septum and absence of a fused pilae plate in IGMp879524 indicate that the optic muscle insertions were primarily inserted in the interorbital septum in this mosasaurid. Our results suggest that the chondrocranium in mosasaurids was reduced, more than in most lacertilians, including their closest recent relatives, the varanids. Nonetheless, this reduction was not as severe as that seen in snakes or amphisbaenians, which have extremely reduced chondrocranium and limbs. Although much information is still lacking, the new evidence presented in this contribution suggests that the reduction in chondrocranium in mosasaurids could be related to the modification of their limbs by adaptation to aquatic life.

We established that the concretion containing IGMp879524 suffered a partial and scattered silicification during early diagenesis, which allows the exceptional preservation of the chondrocranium. This silicification was possible by the supply of sufficient concentration of dissolved silica in the diagenetic waters during the formation of the concretion. We interpret that the silicification process was promoted by the formation of microenvironments with lower pH, associated with the bacterial breakdown of the non-biomineralized tissues of the skull.

Author Contributions: Conceptualization, M.E.P.-F., C.D.B.-C. and J.A.N.-R.; methodology, M.E.P.-F., C.D.B.-C. and J.A.N.-R.; software, C.F.Y.-L. and J.A.N.-R.; investigation, M.E.P.-F., C.D.B.-C. and J.A.N.-R.; writing—original draft preparation, M.E.P.-F., C.D.B.-C. and J.A.N.-R.; writing—review and editing, M.E.P.-F., C.D.B.-C. and J.A.N.-R.; visualization, C.F.Y.-L. and J.A.N.-R.; supervision, M.E.P.-F. All authors have read and agreed to the published version of the manuscript.

Funding: This research received no external funding.

Institutional Review Board Statement: Not applicable.

Data Availability Statement: The data that support the findings of this study are available from the corresponding author upon reasonable request.

Acknowledgments: We are grateful to the Museo Geológico Nacional José Royo y Gómez for granting access to the specimen under their care. The SHAIO CLINIC research group provided the CT scan. We want to thank Ricardo Buitrago, Marco Chala, and Juliana Morantes, members of the SHAIO research group, for all their assistance during the scanning of the specimen. We would like to thank CT scan technicians Martin Meneses and Julio Quiroga from the Clinica SHAIO for their collaboration in the scanning process. Furthermore, we thank the Lithogeochemical Characterization Laboratory of the Departamento de Geociencias of the Universidad Nacional de Colombia for allowing us to run the XRD analysis. We also want to thank Carlos Rodriguez Esquivel for his assistance in the XRD analysis. Likewise, we would like to thank Hernán Ortiz from the Vertebrate Paleontology Laboratory and Armando Sanchez from the Petrographic Techniques Laboratory of the Departamento de Geociencias of the Universidad Nacional de Colombia for their assistance in the elaboration of the thin sections. We would like to thank Diana Montoya for their useful comments on the elements seen in the thin sections. We are indebted to Ingmar Werneburg for sharing essential bibliography. Likewise, we are grateful to the two anonymous reviewers whose comments help to improve the final version of this manuscript.

Conflicts of Interest: The authors declare no conflicts of interest.

References

1. De Beer, G.R. The early development of the chondrocranium of the lizard. *Q. J. Microsc. Sci.* **1930**, *73*, 707–739.
2. Bellairs, A.D.; Kamal, A.M. The chondrocranium and the development of the skull in recent reptiles. In *Biology of the Reptilia, Volume 11, Morphology F*; Gans, C., Parsons, T.S., Eds.; Academic Press: New York, NY, USA, 1981; Volume 11, pp. 1–283.

3. Jones, M.E.; Gröning, F.; Aspden, R.M.; Dutel, H.; Sharp, A.; Moazen, M.; Fagan, M.J.; Evans, S.E. The biomechanical role of the chondrocranium and the material properties of cartilage. *Vertebr. Zool.* **2020**, *70*, 699–715. [CrossRef]
4. Kardong, K. *Comparative Anatomy, Function, Evolution*, 8th ed.; Mc Graw Hill: New York, NY, USA, 2019; pp. 1–814.
5. Bellairs, A.D. The anterior brain-case and interorbital septum of Sauropsida, with a consideration of the origin of snakes. *Zool. J. Linn. Soc.* **1949**, *41*, 485–512. [CrossRef]
6. Shrivastava, R.K. The structure and development of the chondrocranium of Varanus. Part I. The development of the ethmoidal region. *Okajima Folia Anat. Jpn.* **1963**, *39*, 55–83. [CrossRef] [PubMed]
7. Shrivastava, R.K. The structure and development of the chondrocranium of Varanus. Part II. The development of the orbito-temporal region. *J. Morphol.* **1964**, *115*, 97–108. [CrossRef]
8. Shrivastava, R.K. The structure and development of the chondrocranium of Varanus. Part III. The otic and occipital regions, basal plate, viscerocranium, and certain features of the osteocranium. *Morphol. Jahrb.* **1964**, *106*, 147–187.
9. Yaryhin, O.; Werneburg, I. The origin of orbitotemporal diversity in lepidosaurs: Insights from tuatara chondrocranial anatomy. *Vertebr. Zool.* **2020**, *69*, 169–181. [CrossRef]
10. Camp, C.L. *California Mosasaurs*; Memoirs of the University of California; University of California Press: Berkeley, CA, USA, 1942; Volume 13, pp. 1–68.
11. Russel, D. Systematics and morphology of American mosasaurs (Reptilia, Sauria). *Bull. Peabody Mus. Nat. Hist.* **1967**, *23*, 1–241.
12. Páramo-Fonseca, M.E. Mosasauroids from Colombia. *Bull. Soc. Geol. Fr.* **2012**, *183*, 103–109. [CrossRef]
13. Terraza-Melo, R. "Formación La Luna": Expresión espuria en la geología colombiana. In *Estudios Geológicos y Paleontológicos Sobre el Cretácico en la Región del Embalse del río Sogamoso, Valle Medio del Magdalena*, 1st ed.; Etayo-Serna, F., Ed.; Servicio Geológico Colombiano: Bogotá, Colombia, 2019; Compilación de los Estudios Geológicos Oficiales en Colombia Volume XXIII; pp. 303–362.
14. Morales, L.G.; Podesta, D.J.; Hatfield, W.C.; Tanner, H.; Jones, S.H.; Barker, M.H.S.; O'Donoghue, D.J.; Mohler, C.E.; Dubois, E.P.; Jacobs, C.; et al. General Geology and oil occurrence of Middle Magdalena Valley, Colombia. In *Habitat of Oil*; American Association of Petroleum Geologist: Tulsa, OK, USA, 1958.
15. Thorez, J. *Practical Identification of Clay Minerals*; Institute of Mineralogy—Liege State University: Liege, Belgium, 1976; pp. 1–90.
16. Bonilla, G.; Sarmiento, G.A.; Gaviria, S. Proveniencia y transformación diagenética de minerales arcillosos del Maastrichtiano—Paleoceno al norte de Bogotá, Cordillera Oriental de Colombia. *Geol. Column* **2011**, *36*, 179–195.
17. Bellairs, A.D. Observations on the snout of Varanus, and a comparison with that of other lizards and snakes. *J. Anat.* **1949**, *83*, 116–146. [PubMed]
18. Konishi, T. A mosasaur (Squamata: Mosasauridae) sneeze: A hypothesis concerning salt excretion in the top predators of the Cretaceous Seas. In Proceedings of the 75th Annual Meeting of the Society of Vertebrate Paloentology, Dallas, TX, USA, 14–17 October 2015; MacKenzie, A., Maxwell, E., Miller-Camp, J., Eds.; 2015.
19. Yaryhin, O.; Werneburg, I. Tracing the developmental origin of a lizard skull: Chondrocranial architecture, heterochrony, and variation in lacertids. *J. Morphol.* **2018**, *279*, 1058–1087. [CrossRef] [PubMed]
20. Jones, M.E.; Gröning, F.; Dutel, H.; Sharp, A.; Fagan, M.J.; Evans, S.E. The biomechanical role of the chondrocranium and sutures in a lizard cranium. *J. R. Soc. Interface* **2017**, *14*, 20170637. [CrossRef] [PubMed]
21. Zhang, Z.; Yaryhin, O.; Koyabu, D.; Werneburg, I. Morphological associtaion betwwen muscle attachments and ossification sites in the late cartilaginous skull of tuatara embryos. *J. Morphol.* **2022**, *283*, 908–931. [CrossRef] [PubMed]
22. Yaryhin, O.; Klembara, J.; Pichugin, Y.; Kaucka, M.; Werneburg, I. Limb reduction insquamate reptiles correlates with the reduction of the chondrocranium: A case study on serpentiform anguids. *Dev. Dyn.* **2021**, *250*, 1300–1317. [CrossRef] [PubMed]
23. Montoya Arenas, D.M. Formación La Paja: Descripción de la sección tipo. Influencia de los tapices microbiales en su génesis. In *Estudios Geológicos y Paleontológicos Sobre el Cretácico en la Región del Embalse del Río Sogamoso, Valle Medio del Magdalena*, 1st ed.; Etayo-Serna, F., Ed.; Servicio Geológico Colombiano: Bogotá, Colombia, 2019; Compilación de los Estudios Geológicos Oficiales en Colombia Volume XXIII; pp. 55–156.
24. Trewin, N.H.; Fayers, S.R. Sedimentary Rocks | Chert. In *Encyclopedia of Geology*; Richard, C., Selley, L., Cocks, M., Ian, R., Palmer, R., Eds.; Elsevier: Amsterdam, The Netherlands, 2007; Volume 3, pp. 154–196. [CrossRef]
25. Knauth, L.P. Chapter 7. Petrogenesis of Chert. In *Silica: Physical Behavior, Geochemistry, and Materials Applications*, 1st ed.; Heaney, P., Prewit, C., Gibbs, G., Eds.; De Gruyter: Berlin, Germany, 1994; pp. 233–258. [CrossRef]
26. McNamara, M.E.; Orr, P.J.; Kearns, S.L.; Alcala, L.; Anadon, P.; Penalver Molla, E. Soft tissue preservation in Miocene frogs from Libros, Spain: Insights into the genesis of decay microenvironments. *PALAIOS* **2009**, *24*, 107–117. [CrossRef]
27. Muscente, A.D.; Schiffbauer, J.D.; Broce, J.; Laflamme, M.; O'Donnell, K.; Boag, T.H.; Meyer, M.; Hawkins, A.D.; Huntley, J.W.; McNamara, M.; et al. Exceptionally preserved fossil assemblages through geologic time and space. *Gondwana Res.* **2017**, *48*, 164–188. [CrossRef]

diversity

MDPI

Article

Bone Connectivity and the Evolution of Ichthyosaur Fins

Marta S. Fernández [1], Lisandro Campos [1,*], Agustina Manzo [1] and Evangelos Vlachos [2]

1 CONICET—División Paleontología Vertebrados, Museo de La Plata, Facultad de Ciencias Naturales y Museo, Universidad Nacional de La Plata, La Plata B1900AVW, Buenos Aires, Argentina; martafer@fcnym.unlp.edu.ar (M.S.F.); agustinamanz@gmail.com (A.M.)

2 CONICET—Museo Paleontológico Egidio Feruglio, Av. Fontana 140, Trelew U9100GYO, Chubut, Argentina; evlacho@mef.org.ar

* Correspondence: lcampos@fcnym.unlp.edu.ar

Abstract: After the end-Triassic extinction, parvipelvian ichthyosaurs diversified and became dominant elements of marine ecosystems worldwide. By the Early Jurassic, they achieved a thunniform body plan that persisted for the last 100 m.y.a of their evolution. Diversification and extinctions of thunniform ichthyosaurs, and their swimming performance, have been studied from different perspectives. The transformation of limbs into hydrofoil-like structures for better control and stability during swimming predates thunniform locomotion. Despite their importance as control surfaces, fin evolution among thunnosaurs remains poorly understood. We explore ichthyosaur fin diversity using anatomical networks. Our results indicate that, under a common hydrofoil controller fin, the bone arrangement diversity of the ichthyosaur fin was greater than traditionally assumed. Changes in the connectivity pattern occurred stepwise throughout the Mesozoic. Coupled with other lines of evidence, such as the presence of a ball-and-socket joint at the leading edge of some derived Platypterygiinae, we hypothesize that fin network disparity also mirrored functional disparity likely associated with different capabilities of refined maneuvering. The ball-and-socket articulation indicates that this local point could be acting like a multiaxial intrafin joint changing the angle of attack and thus affecting the maneuverability, similar to the alula of flying birds. Further studies on large samples and quantitative experimental approaches would be worthy to test this hypothesis.

Keywords: anatomical networks; ichthyosaur fins; evolution

1. Introduction

Ichthyosauromorphs diversified in the aftermath of the Permo-Triassic mass extinction [1,2]. The macromorphological evolutionary changes in their body plan provide canonical examples of convergence among tetrapods secondarily adapted to the marine environment (SECAD from hereon) [3]. As early as the Anisian (Middle Triassic), some ichthyosauromorphs evolved fusiform bodies with dorsal and well-developed caudal fins [4]. Since then, and throughout the Jurassic and much of the Cretaceous, Ichthyosauria Ichthyosauriomorphs (ichthyosaurs from hereon) have been dominant elements in marine ecosystems worldwide. Within this clade, thunnosaurian ichthyosaurs are easily recognizable by their streamlined body deepest at the pectoral region and tapering posteriorly to the peduncle of the lunate caudal fin [5,6] (Figure 1). Alongside Neoceti cetaceans, ichthyosaurs were the only tetrapods to evolve a thunniform body plan suitable for long-distance cruising [7–9] and the first vertebrates to achieve thunniform bodies [10].

As required, throughout the wide arc of SECAD lineages, the shift from continental to marine lifestyle was coupled with the transformation of the columnar and weight-bearing limbs of continental forms into paddles or fins, both for propulsion and/or steering during swimming [11–13]. Both functional categories of modified limbs (paddle-shaped limb or hydrofoil-shaped) imply the enclosing of limb bones into soft-tissue envelopes and the lengthening of the distal region by the addition of bones [14]. As a result, all SECAD have better-integrated limbs in comparison with their terrestrial ancestors. However, among them, the evolutionary strategy and adaptation path followed by ichthyosaurs were unique. Network analysis by Fernández et al. [15] showed that the most widespread evolutionary strategy among SECAD was the enclosing of limb bones in soft-tissue "envelopes" (like "baby mittens"), without drastically impacting the underlying connectivity pattern of the bones. In contrast, the strategy depicted by ichthyosaurs involved "zipping up" their fingers so that digital bones (transformed into carpal-like elements) were connected not only proximodistally with the surrounding bones but also laterally. This strategy resulted in highly integrated and homogeneous forefins in ichthyosaurs, allowing them to explore new regions of the morphospace [15].

In the last decades, the knowledge of the speed and mode of ichthyosaur evolution and extinction increased significantly. Integrative analyses of disparity and evolutionary rates indicate that the evolution of the lineage was characterized by a Triassic early burst followed by an evolutionary bottleneck leading to a long-term reduction of the evolutionary rates and disparity throughout the Jurassic and Cretaceous [2]. On the other hand, disparity and diversity data of Cretaceous forms show that the extinction of ichthyosaurs was characterized by a two-phase pathway: an early Cenomanian extinction that radically reduced their ecological diversity, and a final extinction event at the end of the Cenomanian [16]. However, within this general framework, two key episodes of ichthyosaur evolution are particularly significant due to their impact on the diversity and morphological innovation of the group, and both had ophthalmosaurian parvipelvians as their main protagonists: the Early/Middle Jurassic and the Jurassic/Cretaceous transitions. This clade of parvipelvians accounts for more than half of the entire evolutionary history of ichthyosaurs and is known for drastic transformations in their forefins, including the emergence of pre-radial and post-ulnar zeugopodial elements and numerous accessory digits. The Early/Middle Jurassic transition, although poorly documented [17–19], witnessed the emergence of the ophthalmosaurians. In contrast, the Jurassic/Cretaceous transition marks a profound drop in the diversity (and probably disparity) of the clade [16,20].

Understanding the evolutionary transformation of ichthyosaur fins is crucial for taking the first steps in comprehending the role of forefins during swimming in these marine reptiles, particularly as they evolved into efficient thunnosaurian cruisers. Here we analyze the morphological disparity of ichthyosaurs by exploring how the underlying connectivity pattern of fins transformed during ichthyosaurs' evolutionary history. We increased the taxon sample of anatomical networks of fins from 3 [15] to 16 including forefins of *Mixosaurus cornalianus* and 14 parvipelvians. Finally, framed against the phylogeny, we track the changes in the connectivity pattern of ichthyosaur forefins over 147 million years (from the Annisian up to the Albian) comprising most of the evolutionary history of the ichthyosauromorphs.

The results of analyses of the fin networks highlighted that, within a clear trend towards better integrated and modular forefins, ichthyosaurs depicted a broad array of connectivity patterns. The overall similarity of the fin morphology (i.e., hydrofoil design) hides a striking underlying disparity of bone arrangements. We also found that major evolutionary changes in fin networks occur stepwise. Given the significance of forefins as control surfaces during swimming we proposed that the forefin disparity mirrored functional disparity as well, likely associated with disparity of the refined maneuverability principally among derived thunniform swimmers.

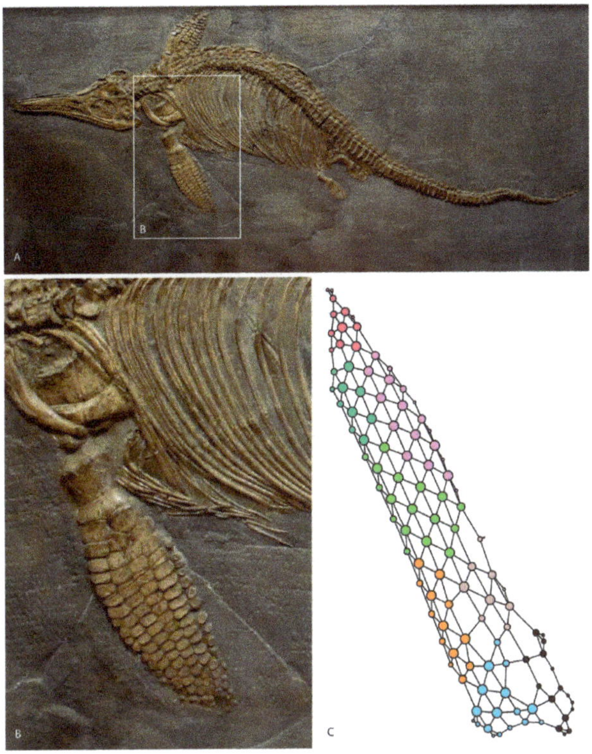

Figure 1. *Ichthyosaurus somersetensis* holotype from the Hettangian of England modified from [21] (**A**). Left forefin on dorsal view (**B**). Anatomical network model of the forefin (**C**).

2. Materials and Methods

We built undirected and unweighted anatomical network models of the forefin for a total of 16 Ichthyosaurian taxa (Supplementary Material, Table S1), in addition to the SECAD dataset of [15]. For the selection of taxa and specimens, we chose complete fins in their anatomical position without any deformation. In cases where this was not possible, we reconstructed the missing parts using all available information, ensuring that at least the minimum number of fin elements were positioned in their most conservative configuration. Anatomical network analysis seeks to describe and analyze the underlying connectivity pattern of the bone elements and their connections, being sutures, contacts, and articulations. This kind of analysis adapts concepts of network analysis to anatomy, where network metrics are interpreted as metrics of anatomical complexity, integration, heterogeneity, and modularity (following [22] and references therein). Each element of the forefin is represented as a node, and contacts among them are depicted as links connecting the nodes. Osteological information is based on personal examination (MF, LC, AM) and published specimens. Network models were created in the open-source software Gephi v.0.10.0 [23], which was implemented for calculation of the network's descriptors, including those descriptors developed specifically for anatomical networks (heterogeneity and parcellation based on [22]). These metrics are anatomically interpreted as measures of the complexity of connections (density, number of connections divided by the maximum possible number of connections), anatomical integration both locally (average clustering coefficient, number of connections between the neighbors of a node divided by the maximum possible number of connections in the neighborhood, on average) as well as along the entire length of the structure (average path length, average of the path length between any pair of nodes), the variability of connections (heterogeneity, standard deviation of connections divided by the mean number of connections), and anatomical modularity (parcellation, based on the number of modules and the number of nodes in each module). For a detailed description of the network metrics and how they are calculated see [22] and references therein. Data from ichthyosaur limbs was subjected to two PCA analyses: one with the complete SECAD dataset from [15] adding new network models obtained herein (Figure 2) and a second considering solely the ichthyosaur information to gain detailed observations (Figure 3). A major change compared to the [15] analysis, is that we now include the average path length metric as well, under normalized variance–covariance correlation because the av. path length is measured in different units compared to the other metrics. Finally, based on the phylogenetic hypothesis presented in [24], a reconstruction of the ancestral states was made in TNT v. 1.6 [25] by mapping the network metrics as continuous characters using the built-in optimization.

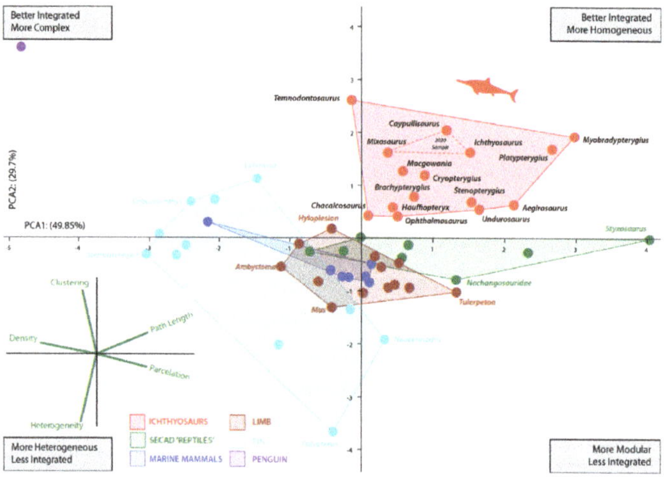

Figure 2. Principal component analysis (PCA) scatter diagram showing morphospace occupation defined by the first two PCAs explaining 77.288% of the variation. Red dashed lines represent the convex hull morphospace occupied by the three ichthyosaurs previously analyzed [15]. See Table S2 for details on network properties of analyzed taxa.

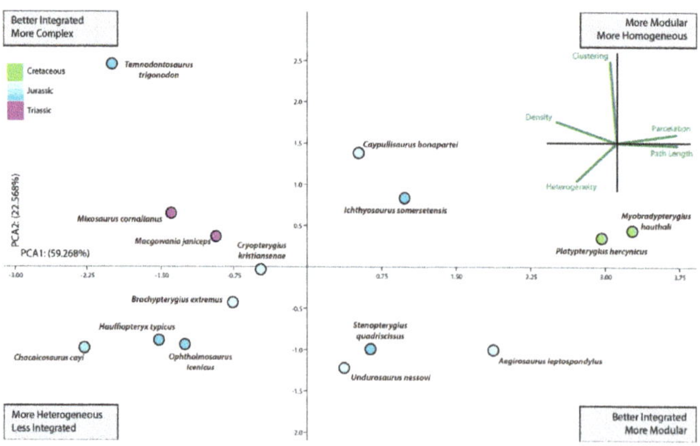

Figure 3. Ichthyosauria forefin morphospace plotted separately to aid comparison of morphospace occupancy through time. This is a second PCA using only data from ichthyosaurs. If we focus on the

Ichthyosauria forefin morphospace occupancy over time derived from the second PCA (Figure 3), from Late Triassic represented by *Mixosaurus* up to the Albian (Late Cretaceous) represented by *Platypterygius hercynicus*, there were no major shifts in the fin morphospace occupation but an overall trend toward better integrated and more modular fins. Thus, this long-term tendency spanned for approximately 137 million years comprising most of the evolutionary history of the Ichthyosauriomorpha.

3. Results

3.1. Morphospace Analyses

The increased taxon sampling and the inclusion of another descriptor in the analyses (average path length) complemented previous results. As in the former analysis including only three ichthyosaurs, the increased sample shows that the limb-to-fin transition of ichthyosaurs followed a unique strategy among SECAD. After the initial shift between the pattern of the basal ichthyosauromorh (e.g., *Nanchangosaurus* and *Hupehsuchus*) and ichthyosaur fins, ichthyosaurs explore new regions of the morphospace. As depicted in Figure 2, the morphospace occupied by ichthyosaurs does not overlap with that of any other SECAD, a difference that is also confirmed statistically with a PERMANOVA analysis (Supplementary Material File S1). Nonetheless, this finding should be interpreted with caution until more taxa from other lineages of marine reptiles can be incorporated into the study. Within a general path to homogeneous reintegration (sensu [15]), the pattern of connectivity changes depicted by their networks indicates that the disparity among ichthyosaur fins was greater than previously assumed. Thus, the morphospace is expanded in all directions.

After the Triassic/Jurassic crisis, ichthyosaurs occupied a large morphospace (in blue color on Figure 3) spreading alongside positive values on PCA1 and PCA2 except for the outlying *Temnodontosaurus* with low negative values on PC1. Within the common path to complex reintegration of their fore appendages, Jurassic forms spread across the empty morphospace. Some of them, like *Chacaicosaurus*, *Hauffiopteryx*, and *Ophthalmosaurus*, have the proximal elements better connected than phalanges resulting in relatively more heterogeneous networks. On the other hand, the connections of *Ichthyosaurus* and *Caypullisaurus* fins are distributed almost evenly across the networks resulting in a relatively more homogeneous fin. *Temnodontosaurus trigonodon* is the only parvipelvian with a diverging pattern (less homogeneous connectivity across the fin). The disparate location of this taxon is not surprising as this taxon reduced the number of primary digits to three. Cretaceous *Myobradypterygius hauthali* and *Platypterygius hercynicus* are clustered together and separated from the Jurassic thunniforms; this is due to their distinctive fin morphology characterized by the increased number of tightly packed phalanges resulting in extremely homogeneous and better-integrated fins. However, the low sampling of Late Cretaceous taxa could underestimate the morphospace occupation during the last episodes of the evolutionary history of the lineage.

3.2. Connectivity Changes in the Forefins across Phylogeny

The analysis of the anatomical networks of SECAD fins [15] indicated that as early as the Middle Triassic, the evolutionary strategy in ichthyosaurs of "zipping-up" their fingers was established and that, through the Jurassic, thunniform ichthyosaurs followed an adaptation path to homogeneous reintegration of their forefins. The analysis of an expanded sample (Table S2), mapped across phylogeny under maximum parsimony, indicates four points where major changes happen and that these evolutionary changes occurred stepwise (Figures 4 and S1). The first step is noted at the Ichthyosauria node, denoted the early and drastic changes in the underlying connectivity pattern of limb elements, promoted by the "re-integration" of the fingers, that clearly impacted the network parameters. The whole fin integration increases but without losing much of its modularity. While nodes, edges, and average clustering coefficient increase, heterogeneity and parcellations decrease. At the Parvipelvian node, no major changes occurred except for the ongoing trends toward more homogeneous fins expressed by a decrease in heterogeneity (H) values. The second step likely occurred in the Early Jurassic. At this point, the fins became larger, but slightly less integrated and modular. After relatively long stability, two successive steps took place during the Middle and Late Jurassic. The last step is the one that registers the most abrupt change in the values of the network descriptors, marking a notable increase toward even more integrated and homogeneous networks.

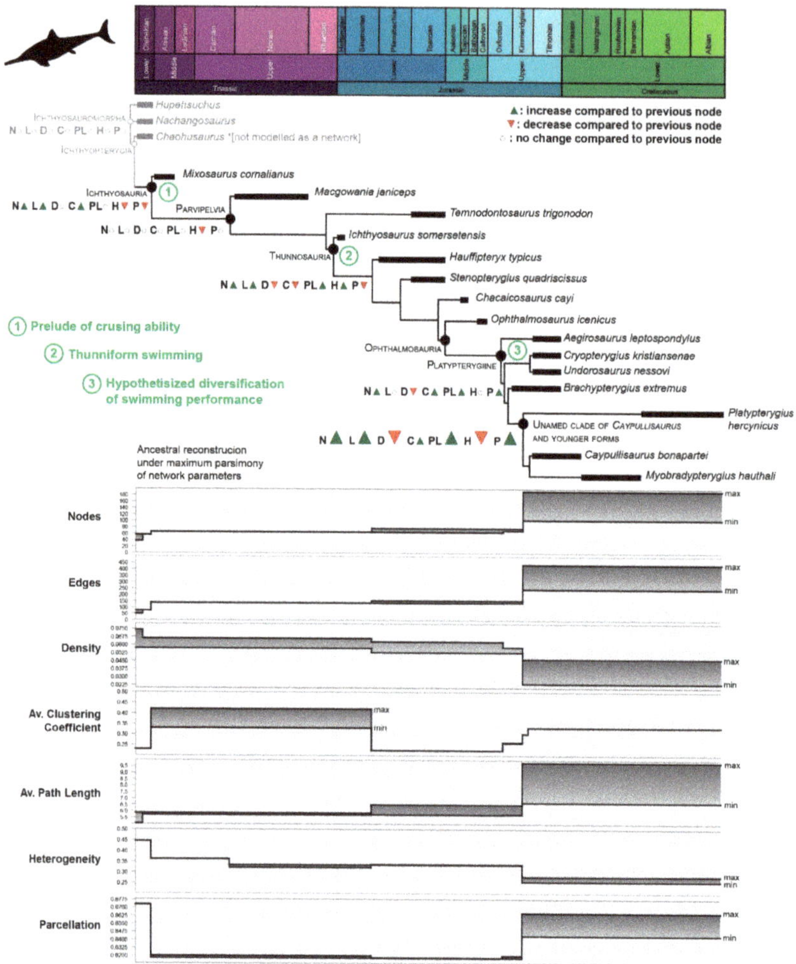

Figure 4. Fin evolution of Ichthyosauria. Changes in the connectivity pattern through phylogeny. On cladogram, in gray, non-Ichthyosauria ichthyosauromorphs were added for comparison; in green, 1–3, major evolutionary events related to swimming. Bottom: stepwise pattern of connectivity changes, each network property is illustrated separately to aid visualization. Abbreviations of network properties as for Table 1.

Table 1. Network properties of analyzed ichthyosauromorphs. C, average clustering coefficient; D, density; E, edges; H, heterogeneity; N, nodes; P, parcellation; PL, average path length. Complete dataset in Supplementary Materials Table S2.

Taxa	Nodes	Edges	Density	Clustering	Path Length	Heterogeneity	Parcellation	Group
Aegirosaurus leptospondylus	109	199	0.034	0.141	8.079	0.253	0.854	ICHTHYOSAURS
Brachypterygius extremus	70	139	0.058	0.262	5.761	0.337	0.825	
Chacaicosaurus cayi	48	84	0.074	0.227	5.213	0.381	0.795	
Cryopterygius kristiansenae	77	155	0.053	0.307	6.504	0.322	0.817	
Hauffiopteryx typicus	59	108	0.063	0.224	5.745	0.350	0.798	
Macgowania janiceps	65	138	0.066	0.331	5.726	0.322	0.825	
Myobradypterygius hauthali	198	453	0.023	0.332	11.364	0.255	0.865	
Ophthalmosaurus icenicus	61	113	0.062	0.207	5.198	0.346	0.816	
Platypterygius hercynicus	193	436	0.024	0.303	9.803	0.242	0.869	
Stenopterygius quadriscissus	89	163	0.042	0.167	7.484	0.279	0.823	
Temnodontosaurus trigonodon	50	114	0.093	0.494	5.442	0.263	0.799	
Undurosaurus nessovi	97	187	0.040	0.259	8.419	0.406	0.840	
Ichthyosaurus somersetensis	113	252	0.040	0.420	6.965	0.334	0.871	
Caypullisaurus bonapartei	103	245	0.047	0.433	6.685	0.280	0.844	
Mixosaurus cornalianus	78	171	0.057	0.425	5.861	0.362	0.804	
Hupehsuchus	37	49	0.074	0.230	5.041	0.446	0.873	REPTILE SECAD
Nanchangosaurus	56	76	0.049	0.160	6.097	0.495	0.884	

4. Discussion

4.1. Morphospace Occupation

Analyses of ichthyosaur disparity, based on phylogenetic data sets [2,26], identified clear differences in morphospace occupation between Triassic and post-Triassic forms. These contributions proposed that ichthyosaurs passed through an evolutionary bottleneck close to the Triassic–Jurassic boundary and that after this key period, ichthyosaur evolution showed a long-term reduction in evolutionary rates and disparity. Other approaches integrating ecomorphological metrics and functional disparity for ecospace modeling [27–29] agreed with these general results. Particularly, [28] found that, after the Triassic–Jurassic crisis, ichthyosaurs again achieved relatively high diversity in the Early Jurassic but throughout the Middle and Late Jurassic, the proportional disparity of ichthyosaurs becomes increasingly diminished. However, these general outcomes do not match with the disparity of the connectivity pattern of the forefins found here (Figure 3). The analysis of the forefin networks showed no evidence of disparity retraction after the Early Jurassic as depicted by the morphospace occupation of the Middle Jurassic and younger thunnosaurs. Noteworthy, within a general tendency towards more integrated and modular fins, the thunnosauria morphospace is expanded in all directions. Similar results have been obtained through the analysis of humerus and zeugopodium morphology among ophthalmosaurids [20]. Other lines of evidence, like those provided by bone microanatomy, e.g., [30,31], also suggest that thunnosaurs, and particularly ophthalmosaurids, were ecologically diverse throughout the Jurassic.

4.2. Fin Connectivity and Functional Disparity

The exploration of functional disparity focuses on morphological diversity (and its innovations) with a recognized impact on the way of life of animals [28,32]. In the particular case of the Mesozoic SECAD, since the pioneering contributions of Massare [5,33], most of the ecomorphological approaches have been focused on the feeding apparatus [34–36] and paleohistology [37,38]. However, swimming performance is a key factor for the SECAD not only for dispersal during steady swimming but also for foraging. Thus, the skeletal thunniform body plan has been linked to the ecological abilities for the capture of fast pelagic prey such as fast swimming belemnite cephalopods [10]. The evolution of the thunniform body plan of ichthyosaurus has also been explored in terms of energetic performance. Assuming that all post-Triassic ichthyosaurs were thunniform swimmers, it has been proposed that body size was a key factor in the evolution of swimming [39]. These contributions deal mainly with the steady locomotion of ichthyosaurs; however, different maneuverability performances are crucial for surviving escaping from predators and/or capturing elusive prey. Although belemnites were important items of thunnosaur diets, the gut content of Cretaceous Platypterygiinae, as well as tooth and skull morphology [16,40], indicate that they probably fed on a wide range of prey, including other vertebrates.

The role of the pectoral appendages of vertebrate swimmers as control surfaces is well known. Changes in the orientation of the control surfaces with respect to the body axis, as well as small changes in orientation at the leading and trailing edges, have an impact on stability and maneuverability. This is true for flexible pectoral fins of fishes [41] but also, although to a lesser extent, in relatively stiff flippers like those of sharks and odontocetes. Among odontocetes, the lack of maneuverability is compensated by changing small turn radii of flexible forms for higher turning rates and they depict different turning performances [42]. In sharks, the majority of the pectoral fin area is internally supported by collagenous ceratotrichia, which cannot be actively moved [43]. Most of the stabilization relies on changes in the angle of attack or asynchronous pectoral fin movement [44,45]. Despite ichthyosauromorphs being axial swimmers through their evolution and having paired fins that must have acted on stability and maneuverability, the disparity of hydrofoils across thunnosaur clades has not been explored other than as an eventual source of phylogenetic or taxonomic information [24,46]. Given the functional relevance of fins as control surfaces, features such as density, clustering, or path length of their bone arrangements could be considered not only as expressions of morphological disparity but also as functional disparity among thunnosaurs and, thus, suggests different ecological niches.

In addition to the observed disparity of connectivity patterns of the forefin of ichthyosaurs, an eloquent feature that still remains undescribed must be addressed. This is the presence of a ball-and-socket joint between the distal end of the humerus and extra-zeugopial accessory elements on the leading edge of the forefin of some Late Jurassic-Cretaceous ichthyosaurs. Thus in *Platypterygius australis* QM F3348 [47] and in the Late Jurassic Platypterygiine MLP 85-I-15-1 [48] (Figure S2) the proximal surface of the extra zeugopodial element anterior to radius is short (antero-posteriorly) and notably convex and articulates with a strongly concave and small distal articular surface of the humerus. A similar condition occurred in the forefin of the Late Jurassic Platypterygiine *Sumpalla* [20] although in this taxon the articular facet on the distal humerus is not so well demarcated. This peculiar ball-and-socket joint between the humerus and the pre-radial accessory fin elements indicates that this local point could be acting like a multiaxial joint. If so, then subtle, intrafin movements at this point would indicate considerable changes on the leading edge. That is as a vortex generator that increases the lift force and enhances maneuverability during locomotion analogous to the function of the alula in flying birds [49]. Noteworthy, the forefin of *Platypterygius americanus* (UW 2421, Figure S2) [50], shows another very interesting condition: a ball-and-socket joint occurs on the trailing edge between the humerus and a pisiform. This condition suggests that the diversity of maneuvering abilities among derived ichthyosaurus may have been even greater. Quantitative experimental approaches would be worthy to test this hypothesis.

Unfortunately, the fins of QM F3348, MLP 85-I-15-1, and UW 2421, which are eloquent examples of ball-and-socket joints, could not be modeled for this study because they are very incomplete. It is expected that the exploration of deposits such as those of the Cretaceous Zapata Formation in Southern Chile [51,52] may provide more complete specimens in the near future.

4.3. Stepwise Evolution of Ichthyosaur Hydrofoils

Along the phylogeny is a clear trend, expressed across the succession of major steps of connectivity changes, towards better integrated, more modular, and more homogeneous fins in ichthyosaurs (Figure 4). These major changes could be interpreted as steps of a stepwise evolutionary pattern of limb-to-controller hydrofoil transition within ichthyosaurs. It is known that, on very broad scales, morphological iteration (and convergence) occurs frequently [53,54]. Whether this stepwise pattern denotes, at the lowest scale, a morphological iteration in the evolution of more efficient controller flipper-hydrofoils is a worthy question to be empirically tested in the future.

The results of network analysis framed against phylogeny show that the underlying connectivity patterns changed as ichthyosaurs evolved thunniform body plan very early in phylogeny. Thus, the first step of the connectivity changes coincides with the emergence of Ichthyosauria soon after the emergence of Ichthyosauromorhs at the Olenekian [1]. Some noteworthy modifications on the forefin, as the lack of centralia [55] pre-dated these changes. The ongoing fin evolution throughout the Early Jurassic indicates that morphological changes that accompanied the emergence of parvipelvians and thunnosaurs, such as mesopodialization and the development of the thunniform body plan, respectively, predate the next steps of important changes in bone connectivity. The paucity of Aalenian–Bathonian records [56] obscures understanding of the fin transition between the Early and Middle Jurassic and the sudden appearance of ophthalmosaurid ichthyosaurs. Unfortunately, the most complete specimens of early ophthalmosaurids (i.e., *Mollesaurus* and *Argovisaurus*) lack their fins [19,57]. However, the comparison between *Chacaicosaurus* and *Ophthalmosaurus icenicus*, as well as the ancestral reconstruction using parsimony analysis of the network parameters (Figure 4), suggest that the complexity of the propodeal–epipodial joint (as was the morphological innovation of the appearance of the pae) did not produce drastic changes in the connectivity pattern. In the same way, the rise of Platypterygiine by the Late Jurassic is not mirrored by changes in the fin networks. It is likely that along evolution, the morphological innovations of the forefins (associated with the emergency of major clades) provided the structural framework that allowed the subsequent diversification of the bone connectivity that ultimately triggered an ecological diversity (e.g., diversity of refined maneuverability among thunniform swimming).

Noteworthy, as also indicated by other ecological and diversity parameters [16,20], the Jurassic–Cretaceous transition seems to reduce the disparity of the forefin. The only survival lineage shows the most extreme pattern of homogeneous integration but also a restricted occupation of the morphospace.

5. Conclusions and Future Directions

The generalized hydrofoil design of ichthyosaur fins hides a great diversity of bone arrangements. The occupation of the morphospace through time shows a clear evolutionary trend towards better integrated and modular forefins. Within this common path, the disparity of thunnosaurs (as mirrored by the large occupation of morphospace areas) persisted throughout the Jurassic. A key period occurred at the Jurassic–Cretaceous boundary. Late Cretaceous-derived Platypterygiine explores a vacant restricted new area of the available space.

The connectivity pattern diversity (i.e., variations of density, clustering, path length, and nodes and edge values) may also represent functional diversity. Based on the role of the forefin as the control surface of swimming, we argue that the morphospace occupation can be interpreted in ecological–functional terms. The controller hydrofoils of ichthyosaurs are assumed to be relatively stiff and with restricted mobility [37]. However, the number of nodes, density, clustering, and path length of their bony arrangement indicate that not all fins should have had the same performance in terms of partial surface deformation and/or in terms of relative stiffness. Noteworthily, some derived Platypterygiines had a ball-and-socket joint point on the leading edge of their fins that could have facilities for localized bending of the leading edge substantially affecting the angle of attack during swimming. Based on the integration of the outcomes of network analysis and gross anatomy of the leading edge we propose diverse maneuverability capacities among members of the large clade Platypterygiine. Further studies on large samples and quantitative experimental approaches would be worthy to test this hypothesis. The mapping of the bone arrangements of the forefin on phylogeny shows that evolutionary changes occurred stepwise along the Mesozoic.

Supplementary Materials: The following supporting information can be downloaded at: https://www.mdpi.com/article/10.3390/d16060349/s1, Figure S1: Mapped network properties of analyzed taxa; Figure S2: Forefins showing ball-and-socket joints; File S1: Results of PERMANOVA; Table S1: List of the specimens used for the construction of the anatomical networks of the forefin; Table S2: Network properties of analyzed taxa.

Author Contributions: Conceptualization, M.S.F., E.V. and L.C.; methodology, E.V., L.C. and A.M.; investigation, M.S.F., L.C., A.M. and E.V.; writing—original draft preparation, M.S.F.; writing—review and editing, L.C. and A.M. All authors contributed to drafting the manuscript. All authors have read and agreed to the published version of the manuscript.

Funding: This research was partially funded by the Universidad Nacional de La Plata, programa de incentivos docentes UNLP-N981 (M.S.F.) and the Agencia Nacional de Investigaciones científicas de Argentina PICT 2020-2067 (M.S.F.), PICT 2022-0963 (L.C.) and PICT 2019-0327 (M.S.F., E.V.).

Data Availability Statement: The authors confirm that the data supporting the findings of this study are available within the article and its Supplementary Materials.

Acknowledgments: Yanina Herrera (MLP), Martín Ezcurra (MACN), Alberto Garrido, and Belén Bollini (MOZ) are thanked for allowing access to the specimens under their care. The authors also wish to extend their thanks to the reviewers for their comments, which substantially improved the quality of the manuscript. Finally, we want to thank Nathalie Bardet, the academic editor, for her invitation to contribute to this special volume dedicated to Mesozoic marine reptile faunas.

Conflicts of Interest: The authors declare no conflicts of interest.

References

1. Motani, R.; Jiang, D.; Tintori, A.; Ji, C.; Huang, J. Pre- versus Post-Mass Extinction Divergence of Mesozoic Marine Reptiles Dictated by Time-Scale Dependence of Evolutionary Rates. *Proc. R. Soc. B Biol. Sci.* **2017**, *284*, 20170241. [CrossRef] [PubMed]
2. Moon, B.C.; Stubbs, T.L. Early High Rates and Disparity in the Evolution of Ichthyosaurs. *Commun. Biol.* **2020**, *3*, 68. [CrossRef]
3. Kelley, N.P.; Pyenson, N.D. Evolutionary Innovation and Ecology in Marine Tetrapods from the Triassic to the Anthropocene. *Science* **2015**, *348*, aaa3716. [CrossRef] [PubMed]
4. Renesto, S.; Dal Sasso, C.; Fogliazza, F.; Ragni, C. New Findings Reveal That the Middle Triassic Ichthyosaur *Mixosaurus cornalianus* Is the Oldest Amniote with a Dorsal Fin. *Acta Palaeontol. Pol.* **2020**, *65*, 511–522. [CrossRef]
5. Massare, J.A. Swimming Capabilities of Mesozoic Marine Reptiles: Implications for Method of Predation. *Paleobiology* **1988**, *14*, 187–205. [CrossRef]
6. Massare, J.A. Faunas, Behavior, and Evolution. In *Ancient Marine Reptiles*, 1st ed.; Callaway, J.M., Nicholls, E.L., Eds.; Elsevier: Amsterdam, The Netherlands, 1997; pp. 401–421, ISBN 978-0-12-155210-7. [CrossRef]
7. Webb, P.W. Body Form, Locomotion and Foraging in Aquatic Vertebrates. *Am. Zool.* **1984**, *24*, 107–120. [CrossRef]
8. Webb, P.W. Simple Physical Principles and Vertebrate Aquatic Locomotion. *Am. Zool.* **1988**, *28*, 709–725. [CrossRef]
9. Motani, R. Scaling Effects in Caudal Fin Propulsion and the Speed of Ichthyosaurs. *Nature* **2002**, *415*, 309–312. [CrossRef]
10. Motani, R.; Shimada, K. Skeletal Convergence in Thunniform Sharks, Ichthyosaurs, Whales, and Tunas, and Its Possible Ecological Links through the Marine Ecosystem Evolution. *Sci. Rep.* **2023**, *13*, 16664. [CrossRef]
11. Caldwell, M.W. Modified Perichondral Ossification and the Evolution of Paddle-Like Limbs in Ichthyosaurs and Plesiosaurs. *J. Vertebr. Paleontol.* **1997**, *17*, 534–547. [CrossRef]
12. Caldwell, M.W. From Fins to Limbs to Fins: Limb Evolution in Fossil Marine Reptiles. *Am. J. Med. Genet.* **2002**, *112*, 236–249. [CrossRef] [PubMed]
13. Maxwell, E.E. Unraveling the Influences of Soft-Tissue Flipper Development on Skeletal Variation Using an Extinct Taxon. *J. Exp. Zoolog. B Mol. Dev. Evol.* **2012**, *318*, 545–554. [CrossRef]
14. DeBlois, M.C.; Motani, R. Flipper Bone Distribution Reveals Flexible Trailing Edge in Underwater Flying Marine Tetrapods. *J. Morphol.* **2019**, *280*, 908–924. [CrossRef] [PubMed]
15. Fernández, M.S.; Vlachos, E.; Buono, M.R.; Alzugaray, L.; Campos, L.; Sterli, J.; Herrera, Y.; Paolucci, F. Fingers Zipped up or Baby Mittens? Two Main Tetrapod Strategies to Return to the Sea. *Biol. Lett.* **2020**, *16*, 20200281. [CrossRef]
16. Fischer, V.; Bardet, N.; Benson, R.B.J.; Arkhangelsky, M.S.; Friedman, M. Extinction of Fish-Shaped Marine Reptiles Associated with Reduced Evolutionary Rates and Global Environmental Volatility. *Nat. Commun.* **2016**, *7*, 10825. [CrossRef] [PubMed]
17. Gasparini, Z. A New Pliosaur from the Bajocian of the Neuquen Basin, Argentina. *Palaeontology* **1997**, *40*, 135–147.
18. Fernández, M.S.; Talevi, M. Ophthalmosaurian (Ichthyosauria) Records from the Aalenian–Bajocian of Patagonia (Argentina): An Overview. *Geol. Mag.* **2014**, *151*, 49–59. [CrossRef]
19. Miedema, F.; Bastiaans, D.; Scheyer, T.M.; Klug, C.; Maxwell, E.E. A Large New Middle Jurassic Ichthyosaur Shows the Importance of Body Size Evolution in the Origin of the Ophthalmosauria. *BMC Ecol. Evol.* **2024**, *24*, 34. [CrossRef]

20. Campos, L.; Fernández, M.S.; Herrera, Y.; Garrido, A. Morphological Disparity in the Evolution of the Ophthalmosaurid Forefin: New Clues from the Upper Jurassic of Argentina. *Pap. Palaeontol.* **2021**, *7*, 1995–2020. [CrossRef]
21. Lomax, D.R.; Massare, J.A. Two New Species of *Ichthyosaurus* from the Lowermost Jurassic (Hettangian) of Somerset, England. *Pap. Palaeontol.* **2017**, *3*, 1–20. [CrossRef]
22. Esteve-Altava, B.; Pierce, S.E.; Molnar, J.L.; Johnston, P.; Diogo, R.; Hutchinson, J.R. Evolutionary Parallelisms of Pectoral and Pelvic Network-Anatomy from Fins to Limbs. *Sci. Adv.* **2019**, *5*, eaau7459. [CrossRef] [PubMed]
23. Bastian, M.; Heymann, S.; Jacomy, M. Gephi: An Open Source Software for Exploring and Manipulating Networks. *Proc. Int. AAAI Conf. Web Soc. Media* **2009**, *3*, 361–362. [CrossRef]
24. Campos, L.; Fernández, M.S.; Bosio, V.; Herrera, Y.; Manzo, A. Revalidation of *Myobradypterygius hauthali* Huene, 1927 and the Phylogenetic Signal within the Ophthalmosaurid (Ichthyosauria) Forefins. *Cretac. Res.* **2024**, *157*, 105818. [CrossRef]
25. Goloboff, P.A.; Morales, M.E. TNT version 1.6, with a graphical interface for MacOS and Linux, including new routines in parallel. *Cladistics* **2023**, *39*, 144–153. [CrossRef]
26. Thorne, P.M.; Ruta, M.; Benton, M.J. Resetting the Evolution of Marine Reptiles at the Triassic-Jurassic Boundary. *Proc. Natl. Acad. Sci. USA* **2011**, *108*, 8339–8344. [CrossRef] [PubMed]
27. Dick, D.G.; Maxwell, E.E. The Evolution and Extinction of the Ichthyosaurs from the Perspective of Quantitative Ecospace Modelling. *Biol. Lett.* **2015**, *11*, 20150339. [CrossRef]
28. Stubbs, T.L.; Benton, M.J. Ecomorphological Diversifications of Mesozoic Marine Reptiles: The Roles of Ecological Opportunity and Extinction. *Paleobiology* **2016**, *42*, 547–573. [CrossRef]
29. Reeves, J.C.; Moon, B.C.; Benton, M.J.; Stubbs, T.L. Evolution of Ecospace Occupancy by Mesozoic Marine Tetrapods. *Palaeontology* **2021**, *64*, 31–49. [CrossRef]
30. Houssaye, A.; Martin Sander, P.; Klein, N. Adaptive Patterns in Aquatic Amniote Bone Microanatomy—More Complex than Previously Thought. *Integr. Comp. Biol.* **2016**, *56*, 1349–1369. [CrossRef]
31. Talevi, M.; Fernández, M.S. Unexpected Skeletal Histology of an Ichthyosaur from the Middle Jurassic of Patagonia: Implications for Evolution of Bone Microstructure among Secondary Aquatic Tetrapods. *Naturwissenschaften* **2012**, *99*, 241–244. [CrossRef]
32. Anderson, P.S.L.; Friedman, M.; Brazeau, M.D.; Rayfield, E.J. Initial Radiation of Jaws Demonstrated Stability despite Faunal and Environmental Change. *Nature* **2011**, *476*, 206–209. [CrossRef] [PubMed]
33. Massare, J.A. Tooth Morphology and Prey Preference of Mesozoic Marine Reptiles. *J. Vertebr. Paleontol.* **1987**, *7*, 121–137. [CrossRef]
34. Kelley, N.P.; Motani, R. Trophic Convergence Drives Morphological Convergence in Marine Tetrapods. *Biol. Lett.* **2015**, *11*, 20140709. [CrossRef] [PubMed]
35. Foffa, D.; Young, M.T.; Stubbs, T.L.; Dexter, K.G.; Brusatte, S.L. The Long-Term Ecology and Evolution of Marine Reptiles in a Jurassic Seaway. *Nat. Ecol. Evol.* **2018**, *2*, 1548–1555. [CrossRef]
36. Delsett, L.L.; Pyenson, N.; Miedema, F.; Hammer, Ø. Is the Hyoid a Constraint on Innovation? A Study in Convergence Driving Feeding in Fish-Shaped Marine Tetrapods. *Paleobiology* **2023**, *49*, 684–699. [CrossRef]
37. Houssaye, A.; De Buffrénil, V. Bone Histology and the Adaptation to Aquatic Life in Tetrapods. In *Vertebrate Skeletal Histology and Paleohistology*; CRC Press: Boca Raton, FL, USA, 2021; pp. 744–756, ISBN 978-1-351-18959-0.
38. Houssaye, A.; Fish, F.E. Functional (Secondary) Adaptation to an Aquatic Life in Vertebrates: An Introduction to the Symposium. *Integr. Comp. Biol.* **2016**, *56*, 1266–1270. [CrossRef] [PubMed]
39. Gutarra, S.; Moon, B.C.; Rahman, I.A.; Palmer, C.; Lautenschlager, S.; Brimacombe, A.J.; Benton, M.J. Effects of Body Plan Evolution on the Hydrodynamic Drag and Energy Requirements of Swimming in Ichthyosaurs. *Proc. R. Soc. B Biol. Sci.* **2019**, *286*, 20182786. [CrossRef] [PubMed]
40. Kear, B.P.; Boles, W.E.; Smith, E.T. Unusual Gut Contents in a Cretaceous Ichthyosaur. *Proc. R. Soc. Lond. B Biol. Sci.* **2003**, *270*, S206–S208. [CrossRef]
41. Gerstner, C.L. Maneuverability of Four Species of Coral-Reef Fish That Differ in Body and Pectoral-Fin Morphology. *Can. J. Zool.* **1999**, *77*, 1102–1110. [CrossRef]
42. Fish, F.E.; Lauder, G.V. Control Surfaces of Aquatic Vertebrates: Active and Passive Design and Function. *J. Exp. Biol.* **2017**, *220*, 4351–4363. [CrossRef]
43. Fish, F.E. Aquatic locomotion: Environmental constraints that drive convergent evolution. In *Convergent Evolution: Animal Form and Function*; Springer International Publishing: Cham, Switzerland, 2023; pp. 477–522.
44. Fish, F.E.; Shannahan, L.D. The role of the pectoral fins in body trim of sharks. *J. Fish Biol.* **2000**, *56*, 1062–1073. [CrossRef]
45. Hoffmann, S.L.; Porter, M.E. Body and pectoral fin kinematics during routine yaw turning in bonnethead sharks (*Sphyrna tiburo*). *Integr. Org. Biol.* **2019**, *1*, obz014. [CrossRef] [PubMed]
46. Motani, R. Phylogeny of the Ichthyopterygia. *J. Vertebr. Paleontol.* **1999**, *19*, 473–496. [CrossRef]
47. Wade, M. *Platypterygius australis*, an Australian Cretaceous Ichthyosaur. *Lethaia* **1984**, *17*, 99–113. [CrossRef]
48. Fernández, M.S. Nuevo Material de *Caypullisaurus bonapartei* Fernández (Reptilia: Ichthyosauridae) Del Jurásico Superior de La Cuenca Neuquina, Argentina. *Ameghiniana* **1998**, *35*, 21–24.
49. Lee, S.; Kim, J.; Park, H.; Jabłoński, P.G.; Choi, H. The Function of the Alula in Avian Flight. *Sci. Rep.* **2015**, *5*, 9914. [CrossRef] [PubMed]

50. Maxwell, E.E.; Kear, B.P. Postcranial Anatomy of *Platypterygius americanus* (Reptilia: Ichthyosauria) from the Cretaceous of Wyoming. *J. Vertebr. Paleontol.* **2010**, *30*, 1059–1068. [CrossRef]
51. Pardo-Pérez, J.; Frey, E.; Stinnesbeck, W.; Fernández, M.S.; Rivas, L.; Salazar, C.; Leppe, M. An Ichthyosaurian Forefin from the Lower Cretaceous Zapata Formation of Southern Chile: Implications for Morphological Variability within *Platypterygius*. *Palaeobiodivers. Palaeoenviron.* **2012**, *92*, 287–294. [CrossRef]
52. Stinnesbeck, W.; Frey, E.; Rivas, L.; Perez, J.P.; Cartes, M.L.; Soto, C.S.; Lobos, P.Z. A Lower Cretaceous Ichthyosaur Graveyard in Deep Marine Slope Channel Deposits at Torres Del Paine National Park, Southern Chile. *Geol. Soc. Am. Bull.* **2014**, *126*, 1317–1339. [CrossRef]
53. Vermeij, G.J. Historical Contingency and the Purported Uniqueness of Evolutionary Innovations. *Proc. Natl. Acad. Sci. USA* **2006**, *103*, 1804–1809. [CrossRef]
54. Jablonski, D. Scale and Hierarchy in Macroevolution. *Palaeontology* **2007**, *50*, 87–109. [CrossRef]
55. Motani, R.; Jiang, D.-Y.; Tintori, A.; Rieppel, O.; Chen, G.-B.; You, H. First Evidence of Centralia in Ichthyopterygia Reiterating Bias from Paedomorphic Characters on Marine Reptile Phylogenetic Reconstruction. *J. Vertebr. Paleontol.* **2015**, *35*, e948547. [CrossRef]
56. Fischer, V.; Weis, R.; Thuy, B. Refining the Marine Reptile Turnover at the Early–Middle Jurassic Transition. *PeerJ* **2021**, *9*, e10647. [CrossRef] [PubMed]
57. Fernández, M.S. A New Ichthyosaur from the Los Molles Formation (Early Bajocian), Neuquen Basin, Argentina. *J. Paleontol.* **1999**, *73*, 677–681. [CrossRef]

diversity

MDPI

Article

Skin Anatomy, Bone Histology and Taphonomy of a Toarcian (Lower Jurassic) Ichthyosaur (Reptilia: Ichthyopterygia) from Luxembourg, with Implications for Paleobiology

Ida Bonnevier Wallstedt [1,2,*], Peter Sjövall [3], Ben Thuy [4], Randolph G. De La Garza [1], Mats E. Eriksson [1] and Johan Lindgren [1]

1 Department of Geology, Lund University, 223 62 Lund, Sweden; randolph.de_la_garza@geol.lu.se (R.G.D.L.G.); mats.eriksson@geol.lu.se (M.E.E.); johan.lindgren@geol.lu.se (J.L.)
2 Natural History Museum of Denmark, 1350 Copenhagen, Denmark
3 Materials and Production, RISE Research Institutes of Sweden, 501 15 Borås, Sweden; peter.sjovall@ri.se
4 National Museum of Natural History Luxembourg, L-2160 Luxembourg, Luxembourg; ben.thuy@mnhn.lu
* Correspondence: idawallstedt@gmail.com

Abstract: A partial ichthyosaur skeleton from the Toarcian (Lower Jurassic) bituminous shales of the 'Schistes Carton' unit of southern Luxembourg is described and illustrated. In addition, associated remnant soft tissues are analyzed using a combination of imaging and molecular techniques. The fossil (MNHNL TV344) comprises scattered appendicular elements, together with a consecutive series of semi-articulated vertebrae surrounded by extensive soft-tissue remains. We conclude that TV344 represents a skeletally immature individual (possibly of the genus *Stenopterygius*) and that the soft parts primarily consist of fossilized skin, including the epidermis (with embedded melanophore pigment cells and melanosome organelles) and dermis. Ground sections of dorsal ribs display cortical microstructures reminiscent of lines of arrested growth (LAGs), providing an opportunity for a tentative age determination of the animal at the time of death (>3 years). It is further inferred that the exceptional preservation of TV344 was facilitated by seafloor dysoxia/anoxia with periodical intervals of oxygenation, which triggered phosphatization and the subsequent formation of a carbonate concretion.

Keywords: bone histology; ichthyosaur; molecular paleontology; paleocolour; phosphatization; Schistes Carton; soft tissues

1. Introduction

Since the original description in 1699 [1], ichthyosaurs (marine 'fish lizards' of the Mesozoic) have been the subject of extensive research. Owing to numerous fossil findings, many of which are preserved in a pristine condition, our understanding of ichthyosaur ecology and phylogeny is relatively comprehensive [2]. In addition, recent structural and molecular analyses of ichthyosaur soft tissues have enabled inferences to be made about certain aspects of their biology [3], thereby strengthening hypotheses regarding similarities between derived ichthyosaurs and extant toothed whales [4].

Ichthyosaur soft-tissue remains were first described in 1836 by William Buckland, in the form of presumed skin associated with a Lower Jurassic specimen from Lyme Regis, southern England. Since then, numerous finds of ichthyosaur soft parts have been made, mainly from Jurassic Konservat-Lagerstätten in Germany and the United Kingdom [2]. These findings comprise several different types of organs and tissues, including skin [3,5–7], connective tissue [8] and internal organs (e.g., liver, intestines and stomach) [3,7,9]. Regarding skin, all three layers of the amniote integument, i.e., the epidermis, dermis and hypodermis, have been documented in ichthyosaur fossils [2,3]. Furthermore, remnant cells and endogenous biomolecular traces, including melanophores (pigment cells) and melanosomes (melanin-bearing cellular organelles), have been retrieved from some specimens [3,10]. The distribution of melanophores and melanosomes in ichthyosaur fossils additionally has shown that at least some species were countershaded in life [3]. This color scheme likely served a variety of functions, including camouflage, thermoregulation and protection against harmful ultraviolet (UV) radiation [3]. Blubber is another type of peripheral tissue that has recently been documented in ichthyosaurs [3], emphasizing the evolutionary convergence between these ancient marine reptiles and modern odontocetes (toothed whales).

In this contribution, an ichthyosaur fossil (hereafter referred to by its accession number, MNHNL TV344), represented by both incomplete postcranial elements and extensive soft tissues, from Toarcian (Lower Jurassic) strata of southern Luxembourg is described and illustrated.

Institutional abbreviations. MH, Urweltmuseum Hauff, Holzmaden, Germany; MNHNL, National Museum of Natural History Luxembourg.

2. Geological Setting

TV344 was collected from a temporary outcrop exposing Lower Toarcian sediments near the town of Dudelange in southern Luxembourg (Figure 1). In this area, the Jurassic marine strata are sub-divided into the *Dactylioceras tenuicostatum*, *Harpoceras serpentinum* and *Hildoceras bifrons* ammonite zones.

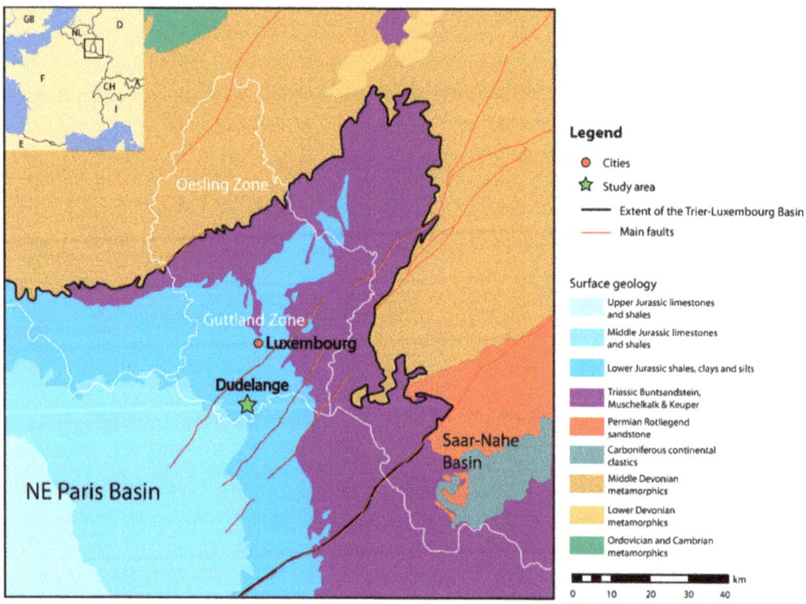

Figure 1. Geological map of Luxembourg and its surroundings, with the study area indicated by a green star (modified from Figure 1 in Ref. [11]). White lines indicate national borders.

Global oceanic anoxia occurred periodically during the Early Jurassic, promoting the formation of organic-rich black shales across large parts of what is now Europe [12]. The Toarcian Oceanic Anoxic Event (hereafter referred to as the T-OAE) is thought to be one of many environmental perturbations caused by the Karoo–Ferrar Large Igneous Province in present-day Southern Africa and Antarctica [13].

The Posidonienschiefer Formation of Germany comprises an extensive accumulation of bituminous and highly fossiliferous black shales with intercalated limestone horizons, formed during the T-OAE [14]. Roughly contemporaneous and lithostratigraphically equivalent to the Posidonienschiefer Formation is the 'Schistes Carton' unit of southern Luxembourg [15,16]. The strata of the 'Schistes Carton' unit have been subjected to extensive compression and are in at least three levels rich in carbonate concretions, which have yielded numerous vertebrate and invertebrate remains [15,17,18]. The fossil-producing nodules are generally grey, ovoid, of decimeter-size or more, and have a finely laminated internal texture. The lithological similarity between these nodules and the surrounding host rock implies that the formation of the former took place at about the same time as the main sediments were deposited. This, in turn, suggests that fossilization within the nodules occurred synsedimentary [18,19]. It has been argued that the formation of the nodules was initiated as a result of sulfate-reducing bacteria locally increasing the pH of the pore waters, causing intense local carbonate precipitation [20].

The 'Schistes Carton' unit contains a diverse fossil macrofauna. In addition to numerous ichthyosaur specimens, plesiosaurs and metriorhynchoids have also been recorded from these strata [20,21]. Aside from marine reptiles, fish, squids, belemnites, ammonites, gastropods, bivalves, and insects have further been documented [20]. The presence of abundant insect fossils supports an interpretation that the sediments of the 'Schistes Carton' unit accumulated in a relatively nearshore environment [20,21].

3. Materials and Methods

TV344 was acquired from an amateur paleontologist who collected the fossil from a temporary exposure of marine strata of the 'Schistes Carton' unit at a construction site near the town of Dudelange, Luxembourg, in the early 1990s. No data concerning the exact position or orientation of the nodule in the stratigraphic succession is available. However, lithological comparisons with similarly-looking nodules suggests that it originates from a bed within to the *Harpoceras serpentinum* ammonite Zone (*Harpoceras exaratum* ammonite Subzone), immediately below a level interpreted to be stratigraphic equivalent to 'Unterer Stein' in south-western Germany and roughly corresponding to the base of the Jet Rock series of Yorkshire, England [15,17,22–24]. The specimen was initially prepared by the collector; however, the methods employed were not documented. Nonetheless, it can be inferred that mechanical preparation beyond initial cracking of the nodule was minimal, and that the exposed surfaces were treated with an organic preservative.

We photographed the ichthyosaur fossil under both polarized and UV light using a Nikon D3500 camera equipped with a 18–55 mm zoom lens and a Laowa 60 mm macro lens. Images were produced using the methods outlined in [25] for polarized light photography and [26] for UV photography.

To gain a better understanding of the structural layering of the various tissues, two petrographic sections were produced from an area with both soft tissues and skeletal elements (Figure 2A). A sample was removed from the fossil and embedded in epoxy. Two slices were then produced using a slow-speed diamond saw and ground using a Struers® (Copenhagen, Denmark) RotoPol-25 with a 600 grit diamond plate. The rock chips were mounted on glass slides using epoxy resin and ground to adequate thickness (70–100 μm) using the Struers RotoPol-25 and alternating between 600 and 1200 grit diamond plates. Finally, the ground sections were polished using 1 μm and then 0.25 μm diamond paste.

Soft-tissue samples for ultrastructural and molecular analyses were collected by removing some of the surrounding sediment, thereby revealing a previously unexposed part of the fossil. From there, two soft-tissue samples were extracted; these were demineralized using 0.5 M ethylenediaminetetraacetic acid (EDTA) and then rinsed with Milli-Q water. Ultrastructural and molecular analyses were conducted using a combination of scanning electron microscopy and mass spectrometry, as described below.

Time-of-flight secondary ion mass spectrometry (ToF-SIMS) was employed to assess the molecular composition of the soft-tissue samples. The sample which provided the most distinct molecular signal was then selected for a subsequent scanning electron microscopy investigation.

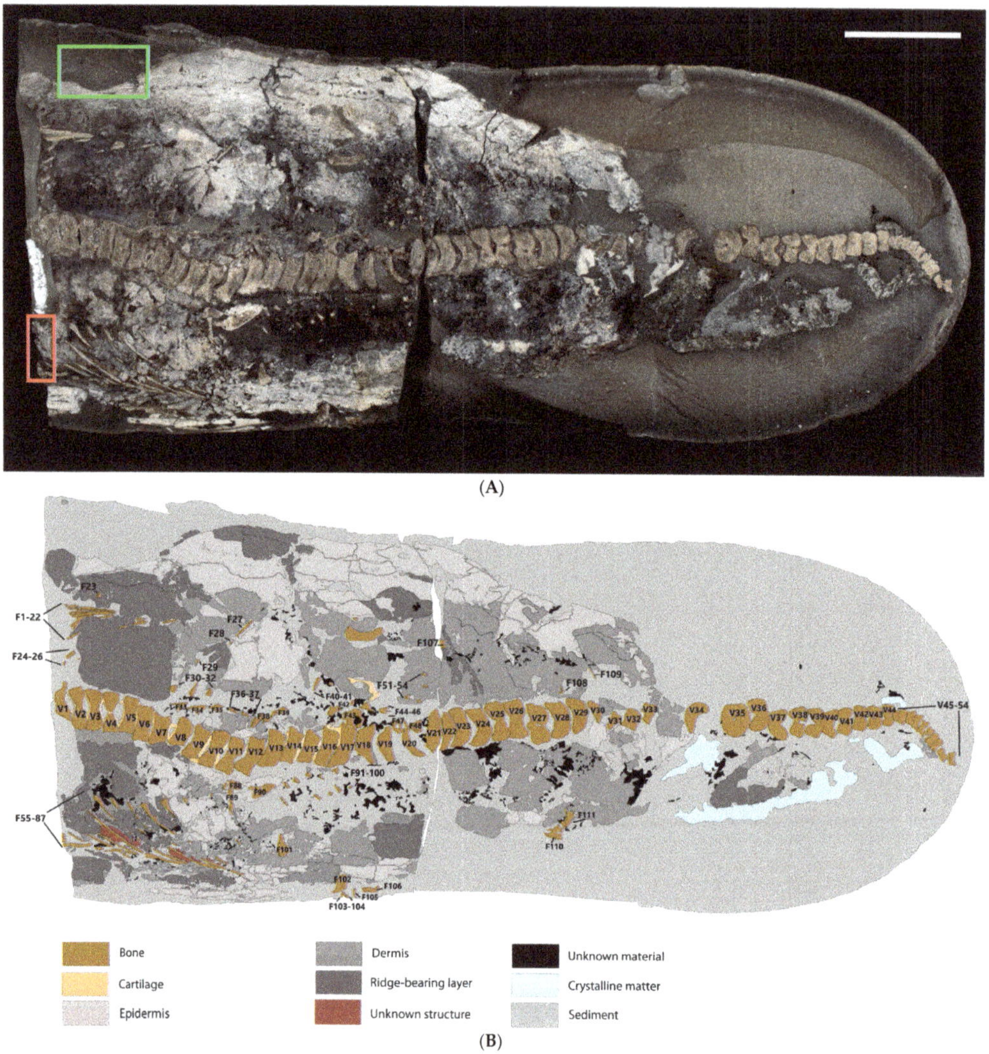

Figure 2. TV344, a partial ichthyosaur from the 'Schistes Carton' unit of Luxembourg. (**A**): Photograph taken under polarized light. The sample site for the petrographic sections is marked with a red rectangle, whereas the samples used in our molecular analyses are demarcated by a green rectangle. The scale bar represents 5 cm. (**B**): Sketch map outlining the different components of the fossil and its surrounding matrix.

Field Emission Gun Scanning Electron Microscopy (FEG-SEM) was employed in two ways: firstly, it was used to analyze one of the samples that had previously undergone ToF-SIMS analysis. This was done using a Zeiss® (Jena, Germany) Supra 40VP FEG-SEM at RISE in Borås, Sweden. Before being placed in the microscope, the sample was coated with 15 nm chromium. Secondly, FEG-SEM coupled with EDX was employed to investigate the petrographic sections and to obtain an overview of the chemical composition of the various skin layers and bones. A Tescan® (Brno, Czechia) Mira3 High Resolution Schottky FEG-SEM at the Department of Geology, Lund University, was used for producing micrographs, and a Oxford Instruments® X-MaxN 80, linked to this instrument, was used for the EDX analysis. Before being placed in the microscope, the petrographic sections were coated with 5 nm of a platinum/palladium mixture.

4. Systematic Paleontology

ICHTHYOSAURIA de Blainville, 1835 [27].
NEOICHTHYOSAURIA Sander, 2000 [28].

5. Results

5.1. General Description

The carbonate nodule that houses TV344 measures approximately 38 cm in total length and 16 cm in maximum width. The nodule has been cracked into part and counterpart sections, revealing a semi-articulated vertebral column that comprises 54 consecutive vertebral centra (Figure 2). The largest of these measure ~1.9 cm in height and ~0.6 cm in length, whereas the smallest ones measure ~0.5 cm in height and ~0.3 cm in length. Towards the posterior end of the specimen, the vertebral centra rapidly decrease in size. Vertebrae and associated bone fragments are hereafter referred to by the following numbers: V1–V54 (referring to vertebrae) and F1–F111 (other bone fragments), respectively (Figure 2B). V45–V54 bend sharply downward relative to V1–V44.

Owing to the presence of one set of ribs on either side of the vertebral column, and that the backbone is mostly articulated, the animal is interpreted to have settled on the seafloor on its back. Therefore, the area below the vertebral column (when the anterior end is facing towards the left; Figure 2) is hereafter referred to as the left side of the animal, whereas the area above the vertebral column is considered to be the right side. It should be noted, however, that this is only a preliminary assessment as there are no accompanying field data indicating which direction is stratigraphically up in the nodule.

5.2. Vertebral Column

As a result of the nodule being broken into part and counterpart sections, the vertebral centra have been split along their longitudinal axis. The bony tissues have a spongious inner texture. Moreover, the centra have an amphicoelous shape characteristic of ichthyosaurs, with a notochordal foramen visible in eight of these skeletal elements (Figure 2). F38–39 and F42–48 are located directly above and between the vertebral centra (Figure 2B). Based on their shape and distribution, they likely represent remains of neural arches. Between V1–10 and V13–17, as well as above F44–46, ossified matter is preserved that resembles bone in color and luminescence but is not readily identifiable as vertebrae (Figure 2, A1). This material is tentatively interpreted as cartilage (Figure 2B).

The ribs of TV344 transition from longer to shorter between V11 and V17. Based on this, the trunk–tail transition likely occurs in the area between V11 and V17. Consequently, F88–F90, and possibly also F101, likely constitute parts of the pelvic girdle. This assessment is corroborated by the number of caudal vertebrae present in the fossil. Vertebral columns of the common Early Jurassic genera *Ichthyosaurus* and *Stenopterygius* generally have 25 to 35 vertebrae in the region between the pelvis and onset of the tail bend [29]. TV344 has 33 vertebral centra between the inferred pelvic girdle and tail bend, which is consistent with these numbers. This inference would mean that V1–V11 are all dorsal vertebrae, whereas V12–V54 represent caudal ones.

5.3. Ribs

Two sets of ribs are present in TV344, one on either side of the vertebral column. The ribs positioned on the left side of the body (F55–87) are exposed at lower angles than those on the right side (F1–26). F1–26 and F55–87 presumably represent dorsal ribs, owing to their elongate appearance and distance to the vertebrae, whereas the shorter F30–37 and F91–100 likely are the remains of sacral and caudal ribs.

F33–37 and F30–32 are oriented in the same direction as the dorsal ribs, and positioned immediately to the right of the vertebral column. F33 and F34 both have a dichocephalous shape at the end facing the centra, to suggest that they—alongside the fragments surrounding them—represent smaller rib fragments. F91–100 are located directly to the left of the vertebral column. They are of sub-centimeter size, and mostly oriented at the same angle as the rib section F55–87; thus, they likely also represent remnant ribs. F88–90 are positioned next to the left set of ribs, and directly to the left of the vertebral column, and the individual elements measure between 1 and 2 cm in length. F101 is located near these fragments and is of similar size as the former elements.

Twelve dorsal ribs are visible in cross-section in the petrographic slides; two occur near the surface of the sections and ten in a gently bent row underneath (Figure A3). The two surficial ribs are sectioned more proximally than are those in the underlying row. While the two upper ribs are entirely cancellous and sub-circular in cross-section (Figure 3B), the ribs of the underlying row display circular cortical bone, an inner ring of cancellous bone, and a medullary cavity (Figure 3A,C). In the lower rib sections, at least three, but possibly four, regularly spaced, concentric lines can be seen in the peripheral cortex (Figure 3C). Osteocyte lacunae additionally occur throughout the peripheral bony tissues (Figure 3C).

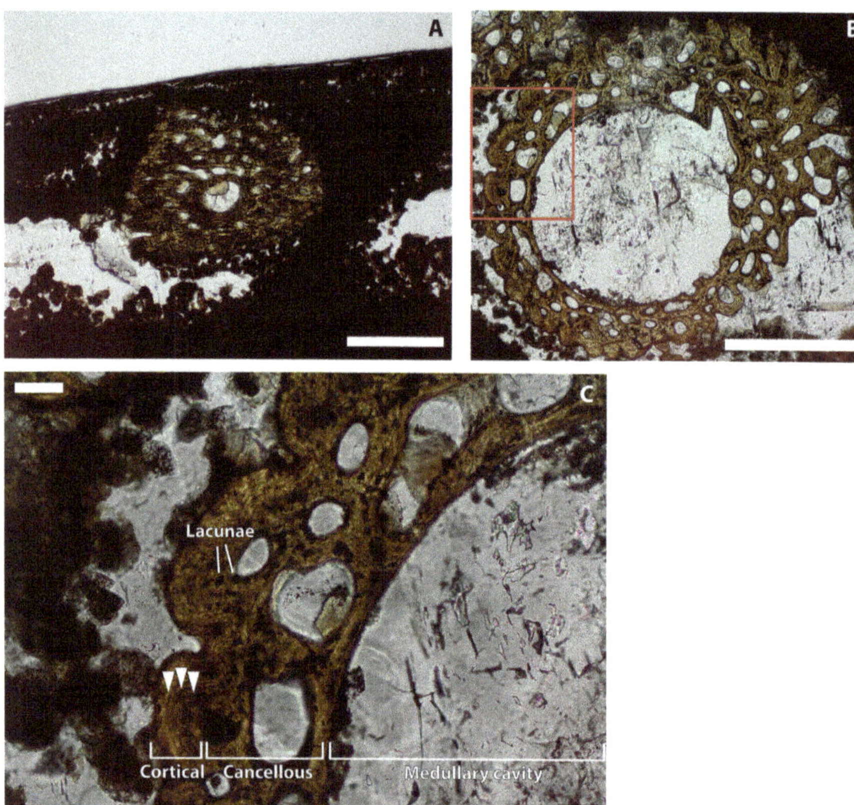

Figure 3. Petrographic section produced from a set of ribs on the left-hand side of TV344. (**A**): Upper part of the section, showing a more proximally located section through a rib. Scale bar represents 1 mm. (**B**): Lower part of the petrographic slide, showing a distal cross-section of a rib. Scale bar represents 500 µm. (**C**): Magnification of (**B**), as indicated by red frame. Cortical bone, cancellous bone and the medullary cavity are all indicated. Circumferential lines in the cortex are highlighted by arrowheads. Lacunae are indicated by white lines. Scale bar represents 50 µm.

5.4. Soft Tissues

Most of the soft-tissue structures preserved in TV344 are identifiable as traces of the animal's integument (see also Section 6.1), and are therefore treated as such in the following description. Assuming that the animal came to rest on its back (see above), the integumentary tissues surrounding the skeleton all likely derive from the dorsal side. The externalmost layer (facing the sediment) is light in color (Figure 4B), highly luminescent under UV-light (Figure A1), and is herein interpreted as the epidermis. The inner termination of the epidermis interfingers with an underlying material, a boundary that is evident in the elemental maps where the epidermis can be seen as a phosphorus-enriched layer into which calcium- and magnesium-rich ridges (~50 μm in diameter) protrude (Figure 5). These ridges have localized concentrations of iron and sulfur, likely stemming from pyrite, and correlate with dense, angular grains visible in the backscatter electron images (Figure 5). The undulating surface at the bottom of the epidermis corresponds to a layer of ridges that is observable under light microscopy (Figure 4A,B). These ridges extend parallel to the longitudinal axis of the fossil, thereby corresponding in orientation to dermal ridges of odontocetes [30].

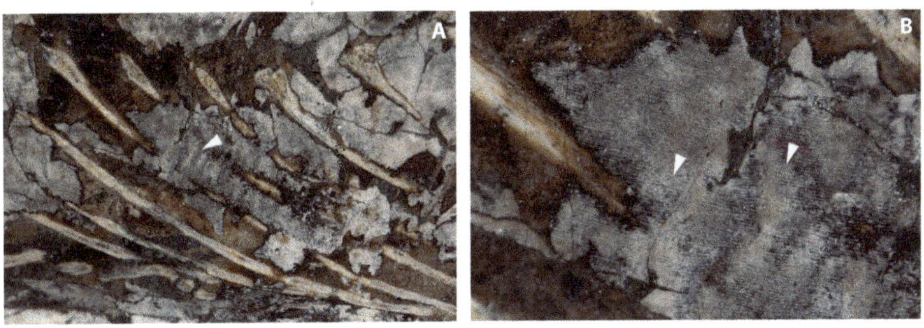

Figure 4. Detailed photographs of TV344. (**A**): Ribs from the left side of the fossil covered by rippled skin (indicated by an arrowhead). (**B**): Close-up view of skin covering the ribs, showing the transition between the epidermis and underlying ridged layer (indicated by arrowheads).

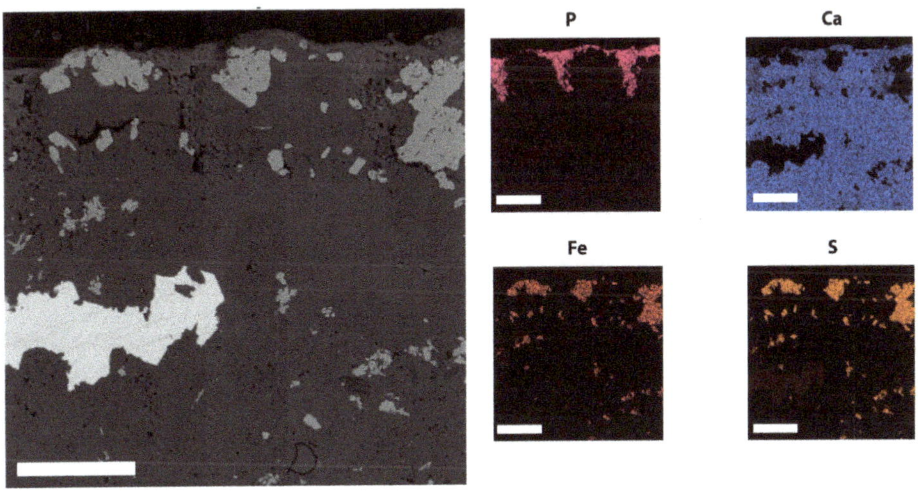

Figure 5. FEG-SEM and EDX micrographs of TV344 integument. Back-scattered electron image to the left and EDX elemental maps to the right. Note dermal ridges protruding into the phosphatized epidermis, and localized concentrations of iron and sulfur in the ridges. Scale bars represent 50 μm.

At higher magnification, dark, branching bodies about 10–20 μm in diameter can be seen (Figure 6A). These also occur in SEM micrographs as clusters of small (~1 μm), oblong bodies at and near the boundary between the epidermis and dermis (Figure A4). ToF-SIMS analysis of TV344 did reveal a chemical signature broadly comparable to that of eumelanin at two sites in one of our samples (Figure A2); however, the mass spectra also revealed peaks characteristic of phthalates and other components that can be assigned to epoxy resins, indicating contamination by the preservative used to consolidate the fossil. Under FEG-SEM, microbodies were identified at the sites from where the eumelanin-like signal originated (Figure 6C,D). The similarity in shape and size between these microbodies and previously identified remnant melanosomes [3,31] suggests that these microstructures likewise are fossilized pigment organelles. Wherever melanosomes were found, they were always bunched together into dense clusters. However, these aggregations were sparsely distributed, which might explain the ambiguity of the 'melanin' signal.

Figure 6. Pigmentation in ichthyosaur fossils. (**A**): Epidermis of TV344 displaying dark, branching bodies (highlighted by arrowheads) with a diameter of ~10–20 μm. Scale bar represents 100 μm. (**B**): Epidermis of *Stenopterygius* specimen MH 432 similarly displaying dark, branching melanophores (see [3]). Melanophores highlighted by arrowheads. Scale bar represents 200 μm. (**C**,**D**): FEG-SEM micrographs of demineralized TV344 integument showing remnant melanosomes embedded in amorphous organic matter. Scale bars represent 1 μm.

Visible in Figure 4A,B are ripple-like structures interpreted as skin folds, similar to what have previously been described from preserved ichthyosaur integument [3,32]. Such ripples might form when the integument loses its structural integrity during the initial stages of decomposition [3].

6. Discussion

6.1. Integument of TV344

Based on both ultrastructural and compositional similarities with previously documented soft-tissue ichthyosaurs [3], the topographically outermost layer in TV344 most likely represents the epidermis. Furthermore, in the lower part of this stratum, along its undulating inner termination, clusters of small, oblong bodies occur (Figure A4). These are consistent with melanosomes in both size and shape [3,31].

Immediately underlying the epidermis is a layer of ridge-like structures oriented longitudinally to the main axis of the body (Figures 3A,B and 6A,C). Similar structures have been observed in other ichthyosaur specimens and interpreted as integumental fibers [33–35]. In the aforementioned studies, the authors also reported additional fiber layers extending in opposing directions, collectively forming a cross-layered structure. Alternatively, ridges in ichthyosaur skin have been interpreted as representing the interface between the epidermis and dermis [3]. Owing to the position of this layer in TV344 directly underneath the epidermis and the lack of structures running in opposing directions, we here favor the second interpretation (which is also consistent with the architecture of odontocete integument [30]).

TV344 exhibits signs of what could be a dark skin pigmentation, similar to what has previously been documented in other ichthyosaur fossils [3,6,10]. Under light microscopy, dark, branching bodies that bear superficial resemblance to melanophores are observed (Figure 6A,B). Melanosomes were subsequently identified using FEG-SEM, occurring as small clusters (Figure 6C,D). Based on these observations, the integumental layers have been reconstructed in Figure 7.

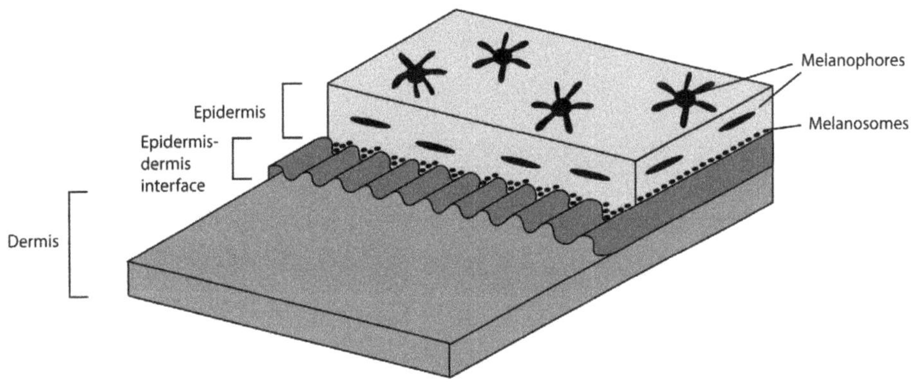

Figure 7. Principal sketch of the inferred layering of the integument in TV344.

6.2. Ontogenetic Stage

When considering the age of the animal at the time of death, there is plenty of evidence that TV344 represents a juvenile individual, including the appearance of the vertebral centra. In juvenile ichthyosaurs, the spongiosa tend to grow continuously denser during ontogeny [36]. The osseous trabeculae of the vertebral centra in TV344 are set widely apart relative to larger and thus presumably skeletally more mature ichthyosaurs. Furthermore, the open notochordal foramen and absence of secondary osteons are additional indicators of skeletal immaturity in TV344.

A more precise age could be determined by analyzing the peripheral lines visible in one of the rib sections (Figure 3C). If these markings represent annual lines of arrested growth (LAGs), then the animal was at least three years old at the time of death. Notably, LAGs have been previously reported in ichthyosaur cortical bone, e.g., in *Mixosaurus* [37]. Annuli have also been observed in a number of ichthyosaurs, e.g., in *Ichthyosaurus* and *Stenopterygius* [38–40].

The notion that the animal which ultimately gave rise to TV344 could have been at least three years old when it died would imply that this individual exhibited a long-lasting juvenile stage and slow growth rates for at least the first few years of its life. Growth rates of derived Jurassic ichthyosaurs are generally considered to have been rapid, as indicated by their highly vascularized fibrolamellar bone [41] and purported absence of LAGs. Nonetheless, slow growth in juveniles followed by more rapid growth as adulthood approaches have been previously documented in many vertebrates, both extant and extinct ones [42].

6.3. Implications of Rib Anatomy

In contrast to previously described Early–Middle Jurassic ichthyosaurs, such as *Stenopterygius* and *Mollesaurus* [40,43], the ribs of TV344 have an open medullary cavity. This cavity is rather small at the proximal end but considerably larger towards the distal termination. An open medullary cavity has been previously observed in the Triassic ichthyosaur *Omphalosaurus* [41], as well as in both juvenile and adult specimens of the Late Jurassic to Early Cretaceous genus *Caypullisaurus* [44]. However, this condition has hitherto not been reported in any Early Jurassic ichthyosaur. Instead, published histological studies of Early Jurassic ichthyosaurs have noted an absence of a medullary cavity (for instance, the medullary region in the ribs of a *Stenopterygius quadriscissus* has a rounded intertrabecular region [40]). Since TV344 represents an immature individual, this could imply that an open medullary cavity is a juvenile feature in Early Jurassic ichthyosaurs. Secondary bone would then deposit in the medullary region as the animal aged.

In *S. quadriscissus*, it has also been noted that the internal morphology of the rib varies, with thicker cortical bone and fewer vascular canals near the mid-shaft than at the proximal and distal ends [40]. This parallels with the rib morphology of TV344, which shows that the mid-distal portion is less spongy than the proximal part.

6.4. Taphonomy

In cases where the integument survives degradative processes, fossilization usually is facilitated by rapid mineral precipitation [31]. Phosphatization, i.e., replacement of organic matter with phosphatic minerals, is a common mode of fossilization of soft tissues [45]. For phosphatization to occur, the depositional environment must be supersaturated with calcium and phosphate ions and a pH below 6.4 must prevail to prevent the formation of carbonates [46]. Microbes, specifically sulfate-oxidizing bacteria, are presumed to have a significant function in this process [47]. As a carcass decomposes, the bacteria lower the pH of the surrounding environment, thereby contributing to the release of phosphate from the decaying body—a process that is enhanced under dysoxic and anoxic conditions [48,49]. The combined effect of dysoxia/anoxia and a lowered pH is enhanced precipitation of calcium phosphate. Furthermore, when these processes take place during early taphonomic stages, they can even facilitate the preservation of endogenous biomolecules [31].

In the 'Schistes Carton' unit, the first phase of fossilization presumably took place during a period of seawater oxygenation in the otherwise dysoxic/anoxic Tethys Ocean. Reactive iron oxides can cycle phosphate ions into the bottom waters by reacting with ions in the upper water column, subsequently releasing phosphorous into the bottom waters in a process known as the 'iron pumping cycle' [31]. This cycling of phosphates was enhanced by seasonal contributions of oxygenated waters. These caused the redox chemocline to move from the water column into the sediment, which enabled sulfate-oxidizing bacteria to absorb phosphates [31]. The build-up of phosphate ions below the chemocline was further increased by downfall of organic detritus materials, such as plankton, from the surface. Sulfate-oxidizing bacteria in the soupy substrate locally lowered the pH to enable precipitation of calcium phosphate, which then rapidly encrusted and entombed the carcass, thereby facilitating the preservation of soft parts. Phosphate precipitation may have been further facilitated by an overall lowered seawater pH during the T-OAE [50]. The phosphate and calcium ions in the seawater and/or sediment stemmed both from the decaying carcass itself and, in the case of phosphate ions, from phosphates participating in the iron pumping cycle (Figure 8). This specific mode of preservation, where seasonal input of oxygenated waters facilitates mineralization, has been described in several Toarcian formations across the Tethys and Panthalassa oceans [31,51]. Carbonate ions might also have been present, originating from decaying organic matter and calcium carbonate shells [52], but were not favored in the precipitation process because of the relatively low pH. Precipitation of calcium phosphate likely occurred during a relatively brief time span [53] and during one of the oxygenation intervals that led to phosphatization also in other Early Toarcian strata.

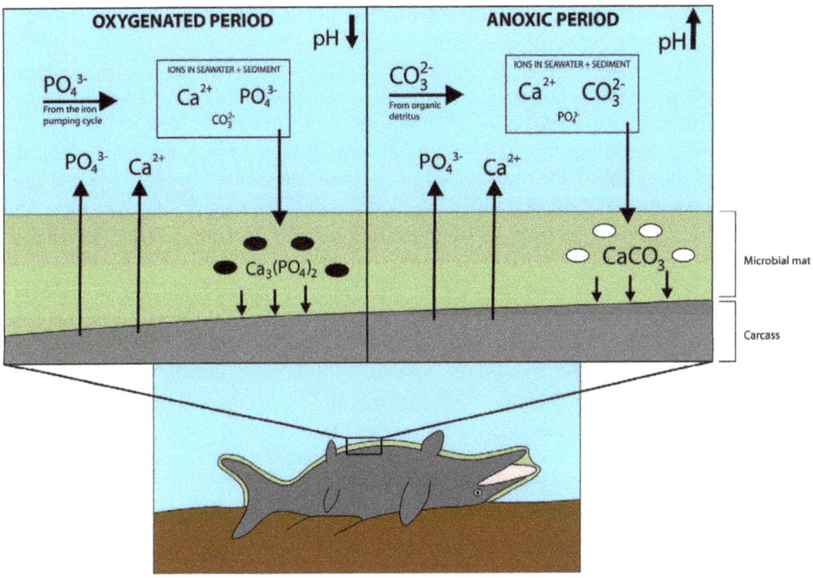

Figure 8. Model for the taphonomic conditions that enabled the fossilization of TV344. The carcass is covered by a microbial mat containing both sulphate-oxidizing and sulphate-reducing bacteria. During brief oxygenated periods (upper left half of figure), sulphate-oxidizing bacteria (black ovals) locally lower the pH, thereby enabling ions in the seawater and sediment to form a sheet of calcium phosphate on top of the carcass. During dysoxic/anoxic periods (upper right half of figure), sulphate-reducing bacteria (white ovals) locally elevated the pH, thereby facilitating the formation of a calcium carbonate nodule from ions in the seawater and sediment.

The second fossilization phase followed in between periods of oxygenation, when the bottom waters were again dysoxic/anoxic. During this phase, sulfate-reducing bacteria locally elevated the pH, enabling precipitation of calcium carbonate (Figure 8). Phosphate, calcium and carbonate ions were still available, but carbonate ions were favored over phosphate ions due to the elevated pH. The concentration of phosphate ions might further have been lower, as the redox chemocline would have been lifted out of the sediment and into the water column, thereby reducing the amount of trapped phosphates. Hence, the combined effects of an elevated pH and a relatively higher concentration of carbonate ions caused the precipitation of calcium carbonate [23]. Furthermore, it has been suggested that nodule beds in shales formed during the T-OAE correspond to a general rise in oceanic pH caused by higher fluxes of organic matter [51]. The formation of carbonate concretions was by all likelihood rapid. In addition to altering the pH of the seawater surrounding a carcass, microorganisms might also have shielded the decomposing organic matter through the formation of microbial mats.

As a result of these taphonomic processes, the integument of TV344 was preserved as an outer phosphatized layer (the epidermis), underlain by the epidermis-dermis interface and the dermis, which are diagenetically altered to varying extent. During decomposition and subsequent fossilization, sediments likely infiltrated the carcass; this is inferred from the extensive fragmentation of the integument. This intrusion likely led to large portions of the lower epidermis being consumed by microbes, in turn leading to a concentration of melanosomes. Some epidermal melanophores remained, however, as evidenced in Figure 6A. Similarly, the dermis was extensively infiltrated by sediments, leading to large-scale replacement of the organic matter, as shown by the ridging remaining as impressions in the overlying epidermis. While several integumental layers are observable to the naked eye (Figure 4), they are not readily apparent in the petrographic sections. This is presumably because these layers are unevenly distributed across the body due to patchy preservation. In some parts, the dermis presumably was lost altogether.

The preservation of skeletal elements in TV344 is incomplete; the vertebral column is intact, but many of the appendicular elements have been lost, presumably due to post-depositional disturbance of the bones. Furthermore, a significant portion of the animal's body is missing from the fossil. This can be attributed to the nodule being partial; the appearance of the nodule (one rounded edge and one edge which appears to have been broken off) suggests that there originally was another part present. The anterior continuation of the body likely was preserved in this remaining nodule; however, the MNHNL is not in possession of any additional nodule material.

7. Conclusions

This study presents the first thorough description of a soft-tissue ichthyosaur (TV344) from the Toarcian 'Schistes Carton' unit of southern Luxembourg. We conclude the following:

- TV344 represents an incomplete juvenile individual belonging to the clade Neoichthyosauria, and possibly the genus *Stenopterygius*;
- The fossil comprises a consecutive series of 54 vertebral centra, associated ribs, 20 bone fragments, as well as extensive soft-tissue remains;
- The soft tissues surrounding the vertebral column are largely integumentary in nature. In addition, there is a distinct layering of the skin, where the epidermis, dermis and epidermis–dermis interface are preserved. Furthermore, the epidermis contains melanophores and melanosomes;
- Concentric ring-like structures observed in sectioned ribs of TV344 may represent annual LAGs, to suggest that the animal was at least three years old at the time of death;
- The taphonomic conditions that enabled the exceptional preservation of TV344 included a combination of seafloor dysoxia/anoxia with periodical oxygenation, thus facilitating rapid phosphatization of the soft tissues during oxygenated periods and the formation of the carbonate nodule during dysoxic/anoxic intervals. Microbial activity aided by lowering the pH during the oxygenated phases, which further enhanced phosphate precipitation.

Author Contributions: Conceptualization, I.B.W., J.L., M.E.E. and B.T.; methodology, I.B.W. and J.L.; formal analysis, P.S. and I.B.W.; investigation, P.S., I.B.W., J.L. and R.G.D.L.G.; resources, B.T., P.S. and J.L.; writing—original draft preparation, I.B.W.; writing—review and editing, I.B.W., J.L., M.E.E., B.T., P.S. and R.G.D.L.G.; visualization, I.B.W.; supervision, J.L., M.E.E. and B.T.; project administration, J.L. and I.B.W.; funding acquisition, J.L. and M.E.E. All authors have read and agreed to the published version of the manuscript.

Funding: Financial support was provided by project grants (#2020-03542), (#2019-03516) and (#2019-03731) from the Swedish Research Council to J.L., M.E.E. and P.S., respectively, and a research grant (#20220563) from the Crafoord Foundation to J.L.

Institutional Review Board Statement: Not applicable.

Data Availability Statement: The original contributions presented in the study are included in the article. Further inquiries can be directed to the corresponding authors.

Acknowledgments: The authors thank Martin Jarenmark (LU) for assistance preparing samples and Carl Alwmark (LU) for assistance with SEM/EDX analysis. The authors also wish to thank three anonymous reviewers for their rigorous review, which served to improve the finished manuscript.

Conflicts of Interest: The authors declare no conflicts of interest.

Appendix A

Figure A1. TV344 photographed under ultraviolet light.

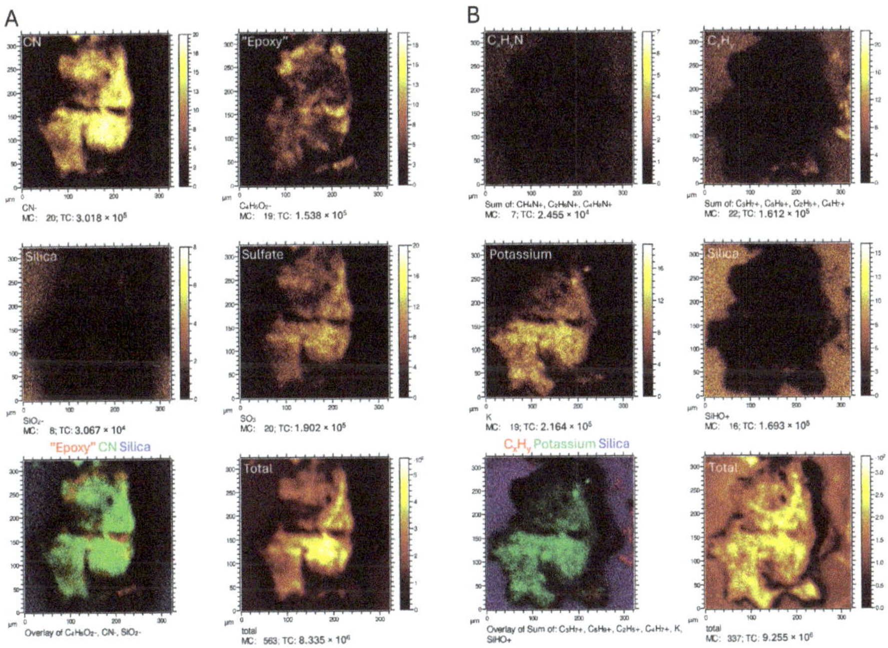

Figure A2. (**A**): ToF-SIMS images of negative ions taken at Sample Point #5 on the right-hand side of the fossil (see Figure 2A), representative of nitrogen-containing organics (CN^-), epoxy ($C_4H_5O_2^-$), silica (SiO_2^-) and sulfate (SO_3^-), respectively. The two panels in the bottom row depict an overlay of epoxy, CN^- and silica together with the total signal intensity distribution. (**B**): ToF-SIMS images of positive ions acquired at Sample Point #5 on the right-hand side of the fossil (see Figure 2A), representative of amino acid-containing organics ($C_xH_yN^+$), aliphatic hydrocarbons ($C_xH_y^+$), potassium (K^+) and silica ($SiOH^+$). The two panels in the bottom row include an overlay of aliphatic hydrocarbons, potassium and silica, and the total signal intensity distribution.

Figure A3. Overview of a complete petrographic section obtained from TV344. Scale bar represents 5 mm.

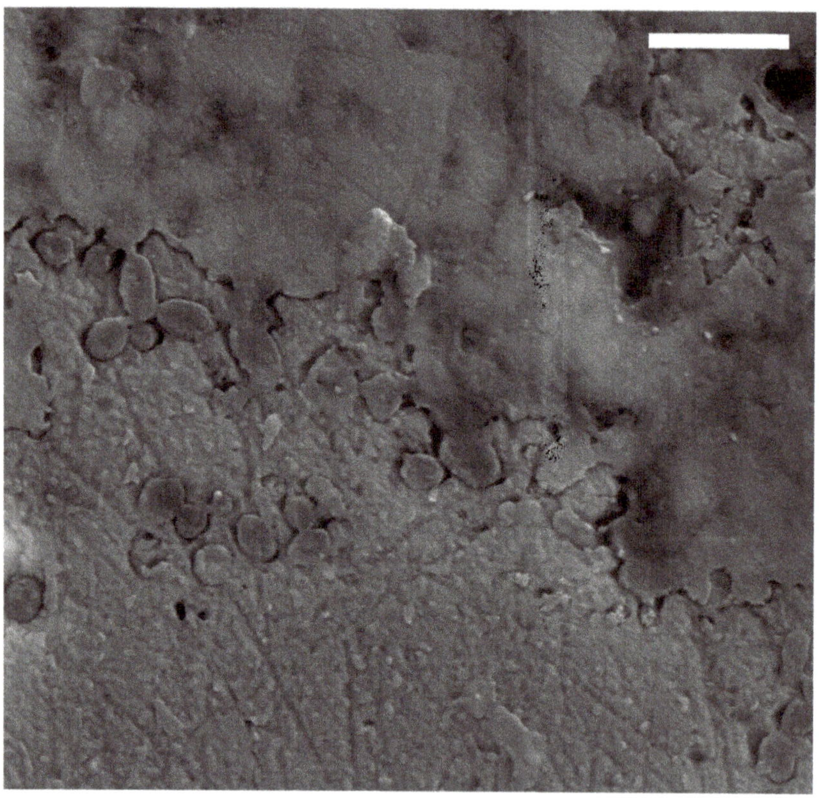

Figure A4. FEG-SEM micrograph of TV344 integument, showing oblong microbodies clustered near the inner termination of the epidermis. Scale bar represents 2 μm.

References

1. Lhuyd, E. *Lithophylacii Brittannici Ichnographia, Sive Lapidum Aliorumque Fossilium Brittanicorum Singulari Figura Insignium;* Gleditsch & Weidmann: London, UK, 1699.
2. Eriksson, M.E.; De La Garza, R.; Horn, E.; Lindgren, J. A review of ichthyosaur (Reptilia, Ichthyopterygia) soft tissues with implications for life reconstructions. *Earth-Sci. Rev* **2022**, *226*, 103965. [CrossRef]

3. Lindgren, J.; Sjövall, P.; Thiel, V.; Zheng, W.; Ito, S.; Wakamatsu, K.; Hauff, R.; Kear, B.P.; Engdahl, A.; Alwmark, C.; et al. Soft-tissue evidence for homeothermy and crypsis in a Jurassic ichthyosaur. *Nature* **2018**, *564*, 359–365. [CrossRef] [PubMed]
4. Motani, R. Evolution of fish-shaped reptiles (Reptilia: Ichthyopterygia) in their physical environments and constraints. *Annu. Rev. Earth Planet. Sci.* **2005**, *33*, 395–420. [CrossRef]
5. Lingham-Soliar, T.; Plodowski, G. Taphonomic evidence for high-speed adapted fins in thunniform ichthyosaurs. *Sci. Nat.* **2007**, *94*, 65–70. [CrossRef] [PubMed]
6. Lindgren, J.; Sjövall, P.; Carney, R.M.; Uvdal, P.; Gren, J.A.; Dyke, G.; Schultz, B.P.; Shawkey, M.D.; Barnes, K.R.; Polcyn, M.J. Skin pigmentation provides evidence of convergent melanism in extinct marine reptiles. *Nature* **2014**, *506*, 484–488. [CrossRef] [PubMed]
7. Renesto, S.; Dal Sasso, C.; Fogliazza, F.; Ragni, C. New findings reveal that the Middle Triassic ichthyosaur Mixosaurus cornalianus is the oldest amniote with a dorsal fin. *Acta Palaeontol. Pol.* **2020**, *65*, 511–522. [CrossRef]
8. Martill, D.M. An ichthyosaur with preserved soft tissue from the Sinemurian of southern England. *Paleontology* **1995**, *38*, 897–903.
9. Jacobs, M.L.; Martill, D.M. A new ophthalmosaurid ichthyosaur from the Upper Jurassic (Early Tithonian) Kimmeridge Clay of Dorset, UK, with implications for Late Jurassic ichthyosaur diversity. *PLoS ONE* **2020**, *15*, e0241700. [CrossRef]
10. Whitear, M. XCVII.—On the colour of an ichthyosaur. *Ann. Mag. Nat. Hist.* **1956**, *9*, 742–744. [CrossRef]
11. Schintgen, T.; Förster, A. Geology and basin structure of the Trier-Luxembourg Basin—Implications for the existence of a buried Rotliegend graben. *ZDGG* **2013**, *164*, 615–637. [CrossRef]
12. Muscente, A.D.; Vinnes, O.; Sinha, S.; Schiffbauer, J.D.; Maxwell, E.; Schweigert, G.; Martindale, R.C. What role does anoxia play in exceptional fossil preservation? Lessons from the taphonomy of the Posidonia Shale (Germany). *Earth-Sci. Rev.* **2023**, *238*, 104323. [CrossRef]
13. Fantasia, A.; Föllmi, K.B.; Adatte, T.; Spangenberg, J.E.; Montero-Serrano, J.-C. The Early Toarcian oceanic anoxic event: Paleoenvironmental and paleoclimatic change across the Alpine Tethys (Switzerland). *Glob. Planet. Chang.* **2018**, *162*, 53–68. [CrossRef]
14. Ruebsam, W.; Schmid-Röhl, W.; Al-Husseini, M. Astronomical timescale for the early Toarcian (Early Jurassic) Posidonia Shale and global environmental changes. *Palaeogeogr. Palaeoclimatol. Palaeoecol.* **2023**, *623*, 111619. [CrossRef]
15. Hermoso, M.; Delsate, D.; Baudin, F.; Le Callonnec, L.; Minoletti, F.; Renard, M.; Faber, A. Record of Early Toarcian carbon cycle perturbations in a nearshore environment: The Bascharage section (easternmost Paris Basin). *Solid Earth* **2014**, *5*, 793–804. [CrossRef]
16. Bechly, G. First record and a new species of the fossil dragonfly genus Proinogomphus (Odonata: Liassogomphidae) from the Early Jurassic of Bascharage in the Grand Duchy of Luxembourg. *Zootaxa* **2018**, *4450*, 108–114. [CrossRef] [PubMed]
17. Heads, S.W.; Thuy, B.; Tamarri, J. Two new species of Archelcana (Sharov (Orthoptera: Elcanidae) from the Lower Jurassic of Luxembourg. *Kentiana* **2022**, *1*, 1–13. [CrossRef]
18. Hanzo, M. A propos de nodules carbonatés du Toarcien inférieur de la région de Bettembourg (Grand-Duché de Luxembourg). In Proceedings of the 103e Congrès National de Sociétés Savants, Nancy-Metz, Nancy, France, 10-15 April 1978.
19. Hanzo, M. Milieu de dépôt et évolution diagénétique des argilites toarciennes d'après l'etude nodules carbonatés des "Schistescarton" de Bettembourg (Grand-Duché de Luxembourg). *Mém. Sci. Terre Nancy* **1979**, *23*, 45–49.
20. Henrotay, M.; Marques, D.; Paicheler, J.-C.; Gall, J.-C.; Nel, A. Le Toarcien inferieur des régions de Bascharage et de Bettembourg (Grand-Duché du Luxembourg): Évidences paléontologiques et sédimentologiques d'environnements restreints proches de l'émersion. *Geodiversitas* **1998**, *20*, 263–284.
21. Johnson, M.M.; Young, M.T.; Brusatte, S.L.; Thuy, B.; Weis, R. A catalogue of teleosauroids (Crocodylomorpha: Thalattosuchia) from the Toarcian and Bajocian (Jurassic) of southern Luxembourg. *Hist. Biol.* **2019**, *31*, 1179–1194. [CrossRef]
22. Riegraf, W.; Werner, G.; Lörcher, F. *Der Posidonienschiefer. Biostratigraphie, Fauna und Fazies des südwestdeutschen Untertoarciums (Lias epsilon)*; Enke Verlag: Stuttgart, Germany, 1984; 195p.
23. Frimmel, A. Hochauflösende Untersuchungen von Biomarkern an epikontinentalen Schwarzschiefern des Unteren Toarciums (Posidonienschiefer, Lias ε) von SW-Deutschland. Ph.D. Thesis, Eberhard-Karls-Universität Tübingen, Metzingen, Germany, 2003.
24. Thibault, N.; Ruhl, M.; Ullmann, C.V.; Korte, C.; Kemp, D.B.; Gröcke, D.R.; Hesselbo, S.P. The wider context of the Lower Jurassic Toarcian oceanic anoxic event in Yorkshire coastal outcrops, UK. *Proc. Geol. Assoc.* **2018**, *129*, 372–391. [CrossRef]
25. Crabb, P. The use of Polarised light in photography of macrofossils. *Palaeontology* **2001**, *44*, 659–664. [CrossRef]
26. Tischlinger, H.; Arratia, G. Ultraviolet light as a tool of investigating Mesozoic fishes with a focus on the ichthyofauna of the Solnhofen Limestone. In *Mesozoic Fishes 5. Global Diversity and Evolution, Proceedings of the International Meeting, Saltillo, Mexico, 2010*; Arratia, G., Schultze, H.-P., Wilson, M.V.H., Eds.; Verlag Dr. Friedrich Pfeil: München, Germany, 2010; pp. 549–560.
27. de Blainville, H.M.D. Description of some species of reptiles of California, preceded by the analysis of a general system of Herpetology and amphibiology. *Ann. Mag. Nat. Hist.* **1835**, *4*, 236–296.
28. Sander, P.M. Ichthyosauria: Their diversity, distribution, and phylogeny. *Palaontol. Z.* **2000**, *74*, 1–35. [CrossRef]
29. Romer, A.S. *Osteology of the Reptiles*; University of Chicago Press: Chicago, IL, USA, 1956; p. 772.
30. Cozzi, B.; Huggenberger, S.; Oelschläger, H. *Anatomy of Dolphins: Insights into Body Structure and Function*; Academic: Amsterdam, The Netherlands, 2017.
31. De La Garza, R.; Sjövall, P.; Hauff, R.; Lindgren, J. Preservational modes of some ichthyosaur soft tissues (Reptilia, Ichthyopterygia) from the Jurassic Posidonia Shale of Germany. *Palaeontology* **2023**, *66*, e12668. [CrossRef]

32. Smithwick, F. A Taphonomic and Paleoecological Approach to the Study of Paleocolour. Ph.D. Thesis, University of Bristol, Bristol, UK, 2019.
33. Delair, J.B. Unusual preservation of fibrous elements in an ichthyosaur skull. *Nature* **1966**, *212*, 575–576. [CrossRef]
34. Lingham-Soliar, T. Rare soft tissue preservation showing fibrous structures in an ichthyosaur from the Lower Lias (Jurassic) of England. *Proc. Biol. Sci.* **1999**, *266*, 2367–2373. [CrossRef]
35. Lingham-Soliar, T. The ichthyosaur integument: Skin fibers, a means for a strong, flexible and smooth skin. *Lethaia* **2001**, *34*, 287–302. [CrossRef]
36. Houssaye, A.; Nakajima, Y.; Sander, P.M. Structural, functional, and physiological signals in ichthyosaur vertebral centrum microanatomy and histology. *Geodiversitas* **2018**, *40*, 161–170. [CrossRef]
37. Kolb, C.; Sánchez-Villagra, M.R.; Scheyer, T.M. The palaeohistology of the basal ichthyosaur Mixosaurus Baur, 1887 (Ichthyopterygia, Mixosauridae) from the Middle Triassic: Palaeobiological implications. *Comptes Rendus Palevol* **2011**, *10*, 403–411. [CrossRef]
38. Seitz, A.L.L. *Vergleichende Studien über den Mikroskopischen Knochenbau Fossiler und Rezenter Reptilien und Dessen Bedeutung für das Wachstum und Umbildung des Knochengewebes im Allgemeinen*; Druck von E. Karras: Halle, Germany, 1907; p. 172.
39. Gross, W. Die Typen des mikroskopischen Knochenbaues bei fossilen Stegoccphalen und Reptilien. *Z. Anat. Entwickl.* **1934**, *103*, 731–764. [CrossRef]
40. Anderson, K.L.; Druckenmiller, P.S.; Erickson, G.M.; Maxwell, E.E. Skeletal microstructure of Stenopterygius quadriscissus (Reptilia, Ichthyosauria) from the Posidonienschiefer (Posidonia Shale, Lower Jurassic) of Germany. *Palaeontology* **2019**, *62*, 433–449. [CrossRef]
41. Houssaye, A. Bone histology of aquatic reptiles: What does it tell us about secondary adaptation to an aquatic life? *Biol. J. Linn. Soc.* **2012**, *108*, 3–21. [CrossRef]
42. Horner, J.R.; Padian, K. Age and growth dynamics of Tyrannosaurus rex. *Proc. Biol. Sci.* **2004**, *271*, 1875–1880. [CrossRef] [PubMed]
43. Talevi, M.; Fernandez, M. Unexpected skeletal histology of an ichthyosaur from the Middle Jurassic of Patagonia: Implications for evolution of bone microstructure among secondary aquatic tetrapods. *Naturwissenschaften* **2012**, *99*, 241–244. [CrossRef] [PubMed]
44. Talevi, M.; Fernandez, M.; Salgado, L. Variación Ontogenética en la Histología Ósea de Caypullisaurus Bonapartei Fernández, 1997 (Ichthyosauria: Ophthalmosauridae). *Ameghiniana* **2012**, *49*, 38–46. [CrossRef]
45. Briggs, D.E.G. The Role of Decay and Mineralization in the Preservation of Soft-Bodied Fossils. *Annu. Rev. Earth Planet. Sci.* **2003**, *31*, 275–301. [CrossRef]
46. Briggs, D.E.G.; Wilby, P.R. The role of the calcium carbonate-calcium phosphate switch in the mineralization of soft-bodied fossils. *J. Geol. Soc.* **1996**, *153*, 665–668. [CrossRef]
47. Iniesto, M.; Buscalioni, Á.D.; Carmen Guerrero, M.; Benzerara, K.; Moreira, D.; López-Archilla, A.I. Involvement of microbial mats in early fossilization by decay delay and formation of impressions and replicas of vertebrates and invertebrates. *Sci. Rep.* **2016**, *6*, 25716. [CrossRef]
48. Peterson, J.E.; Lenczewski, M.E.; Scherer, R.P. Influence of microbial biofilms on the preservation of primary soft tissue in fossil and extant archosaurs. *PLoS ONE* **2010**, *5*, e13334. [CrossRef]
49. Brock, J.; Schulz-Vogt, H.N. Sulfide induces phosphate release from polyphosphate in cultures of a marine Beggiatoa strain. *ISME J.* **2011**, *5*, 497–506. [CrossRef]
50. Bomou, B.; Suan, G.; Schlögl, J.; Grosjean, A.-S.; Suchéras-Marx, B.; Adatte, T.; Spangenberg, J.E.; Fouché, S.; Zacaï, A.; Gibert, C.; et al. The paleoenvironmental context of Toarcian vertebrate-yielding shales of southern France (Hérault). *Geol. Soc. Spec. Publ.* **2021**, *514*, 121–152. [CrossRef]
51. Muscente, A.D.; Martindale, R.C.; Schiffbauer, J.D.; Creighton, A.L.; Bogan, B.A. Taphonomy of the Lower Jurassic Konservat-Lagerstätte at Ya Ha Tinda (Alberta, Canada) and its significance for exceptional fossil preservation during oceanic anoxic events. *Palaios* **2019**, *34*, 515–541. [CrossRef]
52. Berner, R.A. Calcium carbonate concretions formed by the decomposition of organic matter. *Science* **1968**, *159*, 195–197. [CrossRef]
53. Briggs, D.E.G.; Kear, A.J. Fossilization of Soft Tissue in the Laboratory. *Science* **1993**, *259*, 1439–1442. [CrossRef]

diversity MDPI

Article

First Virtual Reconstruction of a Mosasaurid Brain Endocast: Description and Comparison of the Endocast of *Tethysaurus nopcsai* with Those of Extant Squamates

Rémi Allemand [1,*], Michael J. Polcyn [2,3], Alexandra Houssaye [4], Peggy Vincent [1], Camilo López-Aguirre [5] and Nathalie Bardet [1]

1 CR2P Centre de Recherche en Paléontologie de Paris, UMR 7207 CNRS-MNHN-Sorbonne Université, Muséum National d'Histoire Naturelle, 57 Rue Cuvier CP-38, 75005 Paris, France; peggy.vincent@mnhn.fr (P.V.); nathalie.bardet@mnhn.fr (N.B.)
2 Faculty of Geosciences, Utrecht University, Princetonlaan 8a, 3584 CB Utrecht, The Netherlands; m.j.polcyn@uu.nl
3 Huffington Department of Earth Sciences, Southern Methodist University, 3225 Daniel Ave., Dallas, TX 75275, USA; mpolcyn@smu.edu
4 MECADEV Mécanismes Adaptatifs et Évolution, UMR 7179 CNRS-MNHN-Sorbonne Université, Muséum National d'Histoire Naturelle, 57 Rue Cuvier CP-55, 75005 Paris, France; alexandra.houssaye@mnhn.fr
5 Department of Anthropology, University of Toronto Scarborough, 1265 Military Trail, Toronto, ON M1C 1A4, Canada; c.lopezaguirre@utoronto.ca
* Correspondence: remi.allemand1@mnhn.fr

Abstract: Paleoneurological studies of mosasaurids are few and limited to old partial reconstructions made from latex casts on *Platecarpus* and *Clidastes*. Here, the brain endocasts of three specimens of the early mosasaurid *Tethysaurus nopcsai* from the Turonian of Morocco are reconstructed for the first time by using micro-computed tomography. Comparisons between *Tethysaurus* and the later *Platecarpus* and *Clidastes* show that distinct endocranial organizations have occurred within the clade through time, including differences in the flexure of the endocast and the size of the parietal eye. The physiological consequences of such variability remain unclear and further investigations are required to better interpret these variations. In addition, the endocast of *Tethysaurus* was compared to those of extant anguimorphs, iguanians, and snakes, using landmark-based geometric morphometrics. The results revealed that *Tethysaurus* exhibits a unique combination of endocranial features compared to extant toxicoferans. Contrary to previous statements, we find no strong resemblance in endocast morphology between *Tethysaurus* and varanids. Rather, the endocast of *Tethysaurus* shows some morphological similarities with each of the clades of anguimorphs, iguanians, and snakes. In this context, while a notable phylogenetic signal is observed in the variability of squamate endocasts, it is premature to establish any phylogenetic affinities between mosasaurids and extant squamates based solely on endocast morphologies.

Keywords: mosasaurids; squamates; brain endocast; landmarks; geometric morphometrics

1. Introduction

Mosasauridae is a clade of squamates that became secondarily adapted for marine life in the early Late Cretaceous and went extinct at the end of the Maastrichtian, during the K/Pg biological event [1,2]. During their existence, mosasaurids showed increasing adaptations to the marine environment through time [1]. They became increasingly efficient swimmers capable of deep prolonged repetitive diving (e.g., [3,4]) and thrived in many marine habitats from rocky shores to open oceans, including freshwater environments [2]. Mosasaurids were very diversified and occupied a wide range of ecological niches, showing a wide range of body sizes, locomotor styles, diets, and feeding strategies (e.g., [5–11]). By the end of the Cretaceous, they were the apex predators in many marine ecosystems and attained a cosmopolitan distribution (e.g., [1,2,12–14]). Fossil remains of mosasaurids have been recovered from all continents and from latitudes near the Arctic [15] to the Antarctic [16–18]. Mosasauridae comprises at least four subfamilies: Halisaurinae, Mosasaurinae, Tylosaurinae, and Plioplatecarpinae [19], although the details of mosasaurid relationships may vary depending on authors and phylogenetic analyses (e.g., [20–22]). Mosasaurids include "early diverging" taxa that exhibit plesiomorphic morphological characteristics (i.e., *Tethysaurus*, *Russellosaurus*, *Dallasaurus*, *Yaguarasaurus*, and Halisaurinae) and "later diverging" species (i.e., remaining members of the subfamilies Mosasaurinae, Plioplatecarpinae, and Tylosaurinae), which are morphologically more derived (see [19,23] for more details).

Recently, several paleoneurological studies have used non-invasive imaging techniques to explore mosasaurid internal cranial structures, such as the endosseous labyrinth [24–26] and the neurovascular system [23,27]. However, such paleoneurological investigations remain rare and there are currently no studies delving into the mosasaurid neuroanatomy through computed tomographic scanning. Indeed, brain endocasts in mosasaurids are only known from two latex casts of the endocranial cavity performed by Camp [28] for *Clidastes propython* (UCMP 34535; see Figure 1a) and *Platecarpus* sp. (UCMP 34781; see Figure 1b). The partial brain endocast in *Platecarpus* was reconstructed by joining the dissociated braincase elements and by filling the brain cavity "with liquid rubber backed with sawdust" [28] (p. 40). Due to the open condition of the endocranial cavity, the antero-ventral extension of the *Platecarpus* endocast could not be reconstructed; however, Camp's work did provide valuable information regarding its global morphology [28]. In *Clidastes*, the endocast was reconstructed by using a latex mold on the ventral surfaces of the frontal and parietal only [28] (p. 40). As a result, only the antero-dorsal part of the endocast was reconstructed and Camp provided no additional details about the rest of the structure. Based on these reconstructions, Camp [28] conducted a comparison between the endocasts of *Platecarpus* and *Clidastes* with the brain of a juvenile *Varanus niloticus*. As he observed a significant resemblance among the three species, Camp [28] suggested a close phylogenetic relationship between mosasaurids and varanids.

Several studies have shown that both brain and endocast morphologies in squamates reflect a phylogenetic signal (e.g., [29–33]). However, it is worth noting that Camp's comparison, by mixing brain and endocast, might have introduced some biases. Indeed, it is now known that the brain in squamates fits into the endocranial cavity in ways that vary depending on taxa and ontogeny (e.g., [34–39]). Thus, brains and endocasts in squamates should be considered as distinct structures and not directly compared in order to avoid misinterpretations [40]. In this context, assessing the validity of Camp's hypothesis would require comparing the mosasaurid endocast with that of an adult *Varanus*. In addition, given the lack of consensus regarding the phylogenetic relationships of mosasaurids within Squamata (e.g., [19,41–47]), expanding the scope of comparison to include endocasts of other extant squamates would be more suitable for assessing endocranial morphological affinities between mosasaurids and other squamates.

Using micro-computed tomography, the brain endocast of the early mosasaurid *Tethysaurus nopcsai* [48] from the Turonian of the Goulmima area, southern Morocco, is reconstructed here, described, and compared to the latex reconstitutions made by Camp [28] for the later diverging mosasaurids *Platecarpus* and *Clidastes*. In addition, the endocast of *Tethysaurus* is compared to those of extant squamates of the clade Toxicofera, including anguimorphs, iguanians, and snakes (e.g., [44]). Using landmark-based geometric morphometrics, this study aims to determine which extant squamates the endocast of *Tethysaurus* most closely resembles. The results obtained will allow us to assess the validity of Camp's hypothesis and to discuss the phylogenetic and biological implications.

Institutional Abbreviations—CAS, California Academy of Science, San Francisco, California, USA; FMNH, The Field Museum of Natural History, Chicago, Illinois, USA; KU, University of Kansas, Lawrence, Kansas, USA; LSUMZ, Louisiana State University Museum of Natural Science, Bâton Rouge, Louisiana, USA; MNHN, Muséum National d'Histoire Naturelle, Paris, France; MVZ, Museum of Vertebrate Zoology, University of California, Berkeley, USA; NCSM, North Carolina Museum of Natural Sciences, Raleigh, North

Carolina, USA; TCWC, Texas Cooperative Wildlife Collection, Department of Wildlife Science, Texas A&M University, College Station, Texas, USA; TMM, Texas Memorial Museum, Austin, Texas, USA; TNHC, Texas Natural History Collections, Austin, Texas, USA; SMU, Shuler Museum of Paleontology, Southern Methodist University, Dallas, Texas, USA; UCMP, University of California Museum of Paleontology, Berkeley, CA, USA; UF, The University of Florida, Gainesville, Florida, USA; UMMZ, University of Michigan Museum of Zoology, Ann Arbor, Michigan, USA; UTA, The University of Texas at Arlington, Arlington, Texas, USA; YPM, Yale Peabody Museum, New Haven, Connecticut, USA; ZRC, Zoological Reference Collections, National University of Singapore, Kent Ridge, Singapore.

2. Materials and Methods

2.1. Specimen Sampling and Data Acquisition

The three specimens of the mosasaurid *Tethysaurus nopcsai* analyzed here come from the Turonian in age Unit T2a of the Akrabou Formation, in Goulmima area, Er-Rachidia Province, in southern Morocco [48].

The holotype specimen MNHN GOU 1 was described by Bardet et al. [48] and consists of a nearly complete articulated skull and mandible. Although it is preserved in three dimensions (see Figure 1 in [48]), it is slightly crushed laterally, resulting in the displacement of some bones from their original position. The specimen was scanned at the AST-RX platform of the MNHN (Paris, France) using a v | tome | x L240-180 computed tomograph model from Baker Hughes Waygate Technologies (Huerth, Germany) and reconstructions were performed using DATOX/RES software (phoenix datos | x).

The two other specimens, SMU 76335, which is unpublished, and SMU 75486, of which aspects of the snout and circumorbital series are published [19], consist of nearly complete articulated skulls and mandibles preserved in 3D. They were scanned at the University of Texas High-Resolution X-ray CT Facility (Austin, TX, USA) using an NSI scanner and a 210-kV Feinfocus microfocal source. Voxel size naturally varies depending on specimen size (see Table 1).

Seventy-seven extant species of the clade Toxicofera (See Table 1 and Figures S1–S3) were chosen for the purpose of conducting comparisons with the endocast of *Tethysaurus*. The sample includes 13 iguanians (Figure S1), 29 anguimorphs (Figure S2), and 35 snakes (Figure S3) that were selected to grossly represent the taxonomic diversity of these groups. Computed tomographic scans of these species were obtained from different sources. Scans for eight specimens were sampled from the previous work of Allemand et al. [29] and 69 specimens were acquired from the online database MorphoSource [49] "http://www.MorphoSource.org/" (accessed on 3 August 2024).

Image segmentation and visualization were performed using the software Avizo version 2019.1 (Thermo Fisher Scientific, Waltham, MA, USA). The segmentation tools were used to manually reconstruct the endocast for each species by segmenting the internal surface of the bones or the dura mater when no bones surround the endocranial cavity.

Table 1. Toxicofera list of species analyzed. See Institutional Abbreviations for collection numbers. AH-unnumb, Anthony Herrel (UMR 7179, CNRS/MNHN, Paris, France) personal collection. Ab., Species name abbreviations used in Figure 2. An asterisk (*) indicates specimens coming from the MorphoSource database. Taxonomic classification from [50–52].

Suborder	Family	Species	Ab.	Collection Number	Voxel Size (mm)
Anguimorpha	Anguidae	*Abronia deppii*	Ab.d	CAS:herp:143109 *	0.037
		Abronia graminea	Ab.g	UTA:uta-r:38831 *	0.014
		Abronia taeniata	Ab.t	TCWC:herpetology:4911 *	0.015
		Anguis fragilis	An.f	MVZ:herp:238523 *	0.044
		Barisia imbricata	Ba.i	TNHC:herpetology:76984 *	0.014
		Dopasia harti	Do.h	NCSM:herp:80838 *	0.042
		Elgaria kingii	El.k	UF:herp:74645 *	0.033
		Gerrhonotus infernalis	Ge.i	TNHC: herpetology:92262 *	0.026
		Mesaspis moreletii	Me.m	UF:herp:51455 *	0.022
		Ophisaurus mimicus	Op.m	NCSM:herp:25699 *	0.021
		Pseudopus apodus	Ps.a	KU:kuh:87837 *	0.07

Table 1. *Cont.*

Suborder	Family	Species	Ab.	Collection Number	Voxel Size (mm)
Anguimorpha	Anniellidae	*Anniella grinnelli*	An.g	MVZ:herp:257738 *	0.021
	Diploglossidae	*Celestus costatus*	Ce.c	UF:herp:59382 *	0.027
		Celestus hylaius	Ce.h	UF:herp:75794 *	0.038
		Diploglossus fasciatus	Di.f	UMMZ:herps:115647 *	0.058
		Ophiodes striatus	Op.s	YPM:vz:ypm herr 013348.001 *	0.036
	Xenosauridae	*Xenosaurus grandis*	Xe.g	FMNH:Amphibians and Reptiles:123702 *	X, Y = 0.027/Z = 0.064
		Xenosaurus platyceps	Xe.p	UTA:uta-r:23594 *	X, Y = 0.023/Z = 0.053
	Helodermatidae	*Heloderma horridum*	He.h	UF:herp:42033 *	0.047
	Varanidae	*Varanus acanthurus*	Va.a	UTA:uta-r:13015 *	X, Y = 0.023/Z = 0.051
		Varanus exanthematicus	Va.e	AH_unnumb	0.045
		Varanus gouldii	Va.g	TMM:m:1295 *	X, Y = 0.084/Z = 0.21
		Varanus komodoensis	Va.k	TNHC:herpetology:95803 *	0.163
		Varanus niloticus	Va.n	UF:herp:83764 *	0.041
		Varanus prasinus	Va.p	UF:herp:71411 *	0.037
		Varanus salvator	Va.s	FMNH:Amphibians and Reptiles:35144 *	X, Y = 0.088/Z = 0.201
		Varanus timorensis	Va.t	UF:herp:137865 *	0.058
	Lanthanotidae	*Lanthanotus borneensis*	La.b	FMNH:Amphibians and Reptiles:148589 *	X, Y = 0.022/Z = 0.046
	Shinisauridae	*Shinisaurus crocodilurus*	Sh.c	FMNH:Amphibians and Reptiles:215541 *	X, Y = 0.029/Z = 0.078
Serpentes	Anomalepididae	*Typhlophis squamosus*	Ty.s	MNHN 1997.2042	0.005
	Typhlopidae	*Acutotyphlops kunuaensis*	Ac.k	LSUMZ:herps:93566 *	0.019
		Amerotyphlops brongersmianus	Am.b	FMNH:Amphibians and Reptiles:195928 *	0.033
		Typhlops arenarius	Ty.a	UMMZ:herps:241854 *	0.01
	Aniliidae	*Anilius scytale*	An.s	MNHN 1997.2106	0.01
	Tropidophiidae	*Tropidophis canus*	Tr.c	UMMZ:herps:117024 *	0.017
	Boidae	*Boa constrictor*	Bo.c	FMNH:Amphibians and Reptiles:31182 *	X, Y = 0.078/Z = 0.174
		Candoia carinata	Ca.c	LSUMZ:herps:93576 *	0.035
		Eunectes murinus	Eu.m	UF:herp:84822 *	0.074
		Sanzinia madagascariensis	Sa.m	KU:kuh:183837 *	0.055
	Cylindrophiidae	*Cylindrophis ruffus*	Cy.r	UF:herp:143722 *	0.040
	Uropeltidae	*Rhinophis sanguineus*	Rh.s	UF:herp:78397 *	0.022
	Pythonidae	*Morelia spilota*	Mo.s	UMMZ:herps:227833 *	0.054
		Python bivittatus	Py.b	UF:herp:167549 *	0.086
		Python molurus	Py.m	UF:herp:190353 *	0.052
	Acrochordidae	*Acrochordus javanicus*	Ac.j	KU:kuh:318186 *	0.025
	Viperidae	*Bitis arietans*	Bi.a	UMMZ:herps:61258 *	0.021
		Crotalus molossus	Cr.m	UMMZ:herps:143742 *	0.017
		Vipera aspis	Vi.a	UMMZ:herps:116957 *	0.019
	Homalopsidae	*Cerberus rynchops*	Ce.r	MNHN-RA-1998.8583	0.035
		Gerarda prevostiana	Ge.p	CAS:herp:204972 *	0.015
		Homalopsis buccata	Ho.b	ZRC 2.6411	0.024
	Atractaspididae	*Atractaspis bibronii*	At.b	UMMZ:herps:209986 *	0.012
	Elapidae	*Aipysurus duboisii*	Ai.d	MNHN-RA-1990.4519	0.041
		Bungarus fasciatus	Bu.f	UMMZ:herps:201916 *	0.019
		Emydocephalus annulatus	Em.a	UMMZ:herps:93851 *	0.022
		Hydrophis platurus	Hy.p	AH_MS 64	0.032
		Hydrophis schistosus	Hy.s	ZRC 2.2043	0.021
		Laticauda colubrina	La.c	UMMZ:herps:65950 *	0.017
		Naja nigricollis	Na.n	UMMZ:herps:203814 *	0.025
		Pseudechis porphyriacus	Ps.p	UMMZ:herps:170403 *	0.026
	Colubridae	*Afronatrix anoscopus*	Af.a	CAS:herp:230205 *	0.015
		Drymarchon corais	Dr.c	UMMZ:herps:190326 *	0.018
		Lycodon striatus	Ly.s	UMMZ:herps:123427 *	0.012
		Tropidonophis picturatus	Tr.p	LSUMZ:herps:96093 *	0.028
Iguania	Agamidae	*Agama agama*	Ag.a	UF:herp:180711 *	0.02
		Draco volans	Dr.v	UF:herp:48909 *	0.018
		Physignathus cocincinus	Ph.c	YPM:vz:ypm herr 014378 *	X, Y = 0.023/Z = 0.055
	Chamaeleonidae	*Chamaeleo calyptratus*	Ch.c	UF:herp:191369 *	0.041
	Iguanidae	*Amblyrhynchus cristatus*	Am.c	UF:herp:41558 *	0.052
		Ctenosaura similis	Ct.s	UF:herp:181929 *	0.061

Table 1. *Cont.*

Suborder	Family	Species	Ab.	Collection Number	Voxel Size (mm)
Iguania	Phrynosomatidae	*Sceloporus undulatus*	Sc.u	NCSM:herp:83600 *	0.016
		Uta stansburiana	Ut.s	FMNH:Amphibians and Reptiles:213914 *	X, Y = 0.014/Z = 0.036
	Dactyloidae	*Anolis carolinensis*	An.c	UF:herp:102367 *	0.013
	Corytophanidae	*Basiliscus basiliscus*	Ba.b	FMNH:Amphibians and Reptiles:68188 *	0.068
	Hoplocercidae	*Enyalioides heterolepis*	En.h	UF:herp:68015 *	0.021
	Leiocephalidae	*Leiocephalus carinatus*	Le.c	UF:herp:185239 *	0.029
	Tropiduridae	*Stenocercus roseiventris*	St.r	KU:kuh:214966 *	0.09
Mosasauria	Mosasauridae	*Tethysaurus nopcsai*	Te.n	MNHN GOU 1 SMU 76335 SMU 75486	0.0814 0.0778 0.081

2.2. Landmarks and Statistical Analysis

To compare endocast morphologies, we employed the landmark protocol defined by Allemand et al. [30]. Of the twenty landmarks available in [30], nineteen were selected here. Landmark 8, defined as the 'most ventro-median extent of the endocast at the posterior margin of the optic nerve foramen' [30], could not be located in *Tethysaurus* due to the non-preservation of the orbitosphenoid bone. All landmarks were placed on the virtual endocasts and exported using the software Avizo (version 2019.1). Landmarks were only placed on one *Tethysaurus* specimen (SMU 76335) as the endocast is the most complete and is less deformed compared to the two other specimens (see Supplementary Data S1 for landmark description and position on *Tethysaurus* endocast).

We performed a Generalized Procrustes Analyses (GPA) by using the gpagen function in the R package 'geomorph' [53] to quantify and visualize differences in endocast morphologies captured by the landmarks (see Supplementary Data S2 for the raw coordinates). Phylogenetic structuring of the endocast morphology of extant toxicoferans was assessed by estimating the multivariate K-statistic using the physignal function in the 'geomorph' package [54]. The typology of the phylogenetic tree used to run these analyses is modified from [50,51]. Then, to estimate the occurrences of allometry, the relationships of endocast shape with size were tested based on log-transformed centroid size, using a Procrustes regression with the procD.lm function in 'geomorph' [55]. Given the strong statistical significance of allometry in endocast morphology, the residuals of the regression were used as allometry-corrected shape data in subsequent analyses. A Procrustes analysis of variance (PLM) using the procD.lm function of the 'geomorph' package was performed on allometry-corrected shape data to test patterns of endocast shape variation between iguanians, snakes, and anguimorphs. Pairwise comparisons in the allometry-corrected shape variance between extant toxicoferans were conducted using the pairwise function in the R package 'RRPP' [55]. All tests of statistical significance were based on the distribution of 10,000 iterations.

A Principal Component Analysis (PCA) using the gm.prcomp function in 'geomorph' was performed on the allometry-corrected shape data in order to visualize the pattern of endocranial shape variation in *Tethysaurus*, iguanians, anguimorphs, and snakes. A Linear Discriminant Analysis (LDA) was performed on the PC scores to highlight similarities in endocast morphology between *Tethysaurus* and the three clades of extant toxicoferans. The LDA was carried out using the lda function in the R package 'MASS' [56] on the ten first PCs (accounting for 92% of the variance, see Supplementary Data S3) in order to keep fewer variables than specimens from each group (13 iguanians, 29 anguimorphs, and 35 snakes). The accuracy of the LDA was tested using our dataset for extant species, classifying every species as either iguanian, anguimorph, or snake. Posterior probabilities of the LDA were used to determine to which clade the *Tethysaurus* endocast shows the closest resemblance among snakes, anguimorphs, and iguanians.

3. Results

3.1. Brain Endocast of *Tethysaurus nopcsai*

Among the three specimens of *Tethysaurus* used here, the brain endocast reconstructed in SMU 76335 is the most complete with almost all endocranial regions visible (Figure 1c,d). In the specimens SMU 75486 (Figure 1e,f) and MNHN GOU 1 (Figure 1g,h), only the olfactory bulbs and peduncles as well as the posterior part of the endocast could be reconstructed. Indeed, in both specimens, the state of preservation of the parietal bone prevents the lateral delimitation of the endocast in this area. The endocast in SMU 75486 (Figure 1e,f) is slightly crushed laterally, the antero-dorsal part of the endocast being not aligned with the posterior one. We refer to endocranial regions by their underlying soft-tissue features (i.e., "cerebral hemispheres" rather than "impression of the cerebral hemispheres").

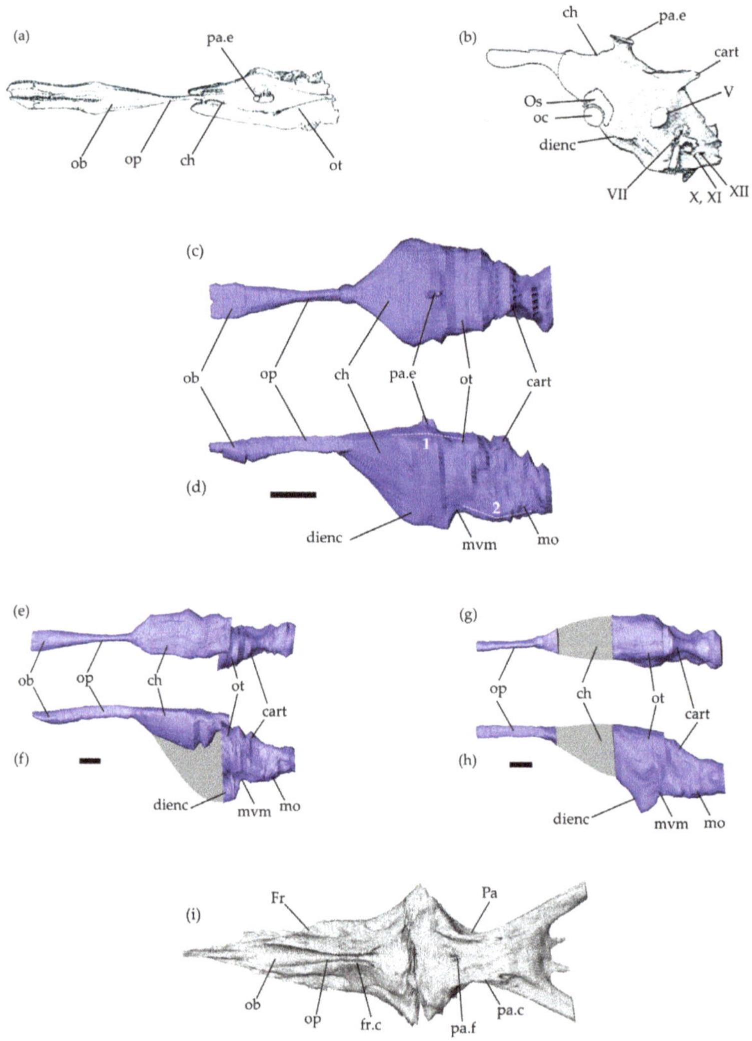

Figure 1. Mosasaurid brain endocasts. Brain endocasts of *Clidastes propython* UCMP 34535 in dorsal view (**a**) and *Platecarpus* sp. UCMP 34781 in left lateral view (**b**), modified from Camp [28] (no scale bar available). (**c**–**h**) Brain endocasts of *Tethysaurus nopcsai* specimen SMU 76335 (**c**,**d**), SMU 75486 (**e**,**f**),

and MNHN GOU 1 (**g**,**h**) in dorsal (**c**,**e**,**g**) and left lateral (**d**,**f**,**h**) views. Scale bars equal 10 mm. (**i**) Virtual reconstructions of the frontal and parietal bones in SMU 76335 in ventral view. Grey areas on the endocasts represent regions that could not be reconstructed. Abbreviations: 1, cephalic flexure; 2, pontine flexure; cart., cartilaginous bridge; ch, cerebral hemispheres; dienc, ventral diencephalon; Fr, frontal; fr.c, frontal cranial crests; mo, medulla oblongata; mvm, mesencephalic ventral margin; ob, olfactory bulbs; oc, optic chiasm; op, olfactory peduncles; Os, orbitosphenoid; ot, optic tectum; Pa, parietal; pa.c, parietal cranial crests; pa.e, parietal eye; pa.f, parietal foramen; V, trigeminal nerve; VII, facial nerve; X,XI, vagus and accessory nerves; XII, hypoglossal nerves.

The endocranial cavity in *Tethysaurus* is surrounded by several bones including the frontal and the parietal from the skull roof (Figure 1i) and the elements forming the braincase: para-basisphenoid, basioccipital, prootic, opisthotic–exoccipital, and supraoccipital. These bones delimit only the dorsal surface as well as the postero-lateral part of the endocranial cavity. The antero-ventral portion of the cavity remains non-ossified, preventing the accurate ventral delimitation of the cerebral hemisphere margins.

The *Tethysaurus* endocast follows the general organizations observed in squamates and other non-avian reptiles (e.g., [40,57]) in that it exhibits a tubular shape that is longer than wide and has a smooth surface (Figure 1b,c–h). In the lateral view, the endocast in *Tethysaurus* appears relatively narrow and flattened (Figure 1d,f,h), showing only weak cephalic and pontine flexures (i.e., angles formed between the telencephalon–mesencephalon and mesencephalon–rhombencephalon, respectively, [34]) that differ from the pronounced flexures figured in *Platecarpus* by Camp [28].

The anterior part of the endocast consists of the olfactory bulbs and the olfactory peduncles (Figure 1c–f). The external morphology of the endocast provides no information that allows the precise delimitation of the olfactory bulbs from the peduncles. The medial limit separating the paired olfactory bulbs and peduncles is not visible on the dorsal surface of the endocast. The olfactory bulbs and peduncles are elongated, representing nearly half of the total endocast length. This is similar to the proportions figured in *Clidastes* by Camp [28] (see Figure 1a), whereas in *Platecarpus* (Figure 1b), the structure seems to be relatively shorter [28,58]. As in *Clidastes* [28], the whole structure (i.e., olfactory bulbs and peduncles) in *Tethysaurus* is wider anteriorly than posteriorly, being mediolaterally compressed in the middle of the antero-posterior length (Figure 1c,e). The olfactory bulbs and peduncles in *Tethysaurus* are roofed by the frontal bone and correspond to the fossa visible on the ventral surface of the bone (Figure 1i). Similar to *Platecarpus* and *Clidastes*, the two structures are not fully enclosed by bones and the frontal cranial crests delimit only their lateral margins (Figure 1i). The ventral margin of the olfactory bulbs and peduncles is delineated here from the dorsa-ventral depth of the frontal cranial crests. The anterior end of the olfactory bulbs is not well delimited and it cannot be distinguished from the course of the olfactory nerves. As in *Clidastes* and *Platecarpus* [58], the imprints left on the ventral surface of the frontal in *Tethysaurus* (Figure 1i) cannot help to delimit the anterior-most extent of the olfactory bulbs.

Posterior to the olfactory peduncles, the cerebral hemispheres expand laterally and ventrally to form the largest part of *Tethysaurus* endocast (Figure 1c,d). The anterior limit of the cerebral hemispheres is difficult to locate and is indicated only by changes in the width of the endocast (Figure 1c). There is no indication allowing us to delimit the cerebral hemispheres posteriorly and the whole structure is poorly differentiated from the neighboring endocast regions. As in *Clidastes* and *Platecarpus* [58], the cerebral hemispheres in *Tethysaurus* are roofed anteriorly by the posterior part of the frontal and posteriorly by the parietal (Figure 1i). The cerebral hemispheres extend between the lateral cranial crests of the parietal (Figure 1i); however, both the lateral and ventral margins of the structure are difficult to delimit due to the lack of osseous elements. Here, the ventral margin of the cerebral hemispheres in SMU 76335 (Figure 1d) was reconstructed by interpolating between the known preserved surfaces of the ventral diencephalon and the preserved olfactory peduncles and then validated against several extant taxa for plausibility.

On the dorsal surface of the *Tethysaurus* endocast (Figure 1c), a small bulge lies dorsomedial to the cerebral hemispheres and is identified as the parietal eye (*sensu* [59]) as it coincides with the median parietal foramen. The parietal eye is relatively small, as in *Clidastes* [28] (Figure 1a), but is different from the large structure reported in *Platecarpus* [28,58] (Figure 1b).

In lateral view, the part of *Tethysaurus* endocast that extends postero-ventral from the cerebral hemispheres, is identified as the ventral diencephalon (Figure 1d). It includes several structures (i.e., hypothalamus, pituitary gland, optic tracts, and optic chiasm [40]) that cannot be observed from the endocast. The whole structure lies dorsal to the rostral process of the parabasisphenoid and the posteroventral part of the ventral diencephalon is situated within the sella turcica and pituitary fossa. The ventral diencephalon in *Tethysaurus* (Figure 1d,f,h) projects more ventrally than the *medulla oblongata*, whereas the opposite condition was figured in *Platecarpus* [28,58] (Figure 1b).

Posterior to the cerebral hemispheres, the optic tectum lies on the ventral surfaces of the posterior parietal. The exact delimitation of the structure is not possible from the endocast and its position is only indicated by a marked change in width, relative to the cerebral hemispheres (Figure 1c,e,g). The optic tectum exhibits a smooth and flattened dorsal surface and lies almost on the same axis as the cerebral hemispheres (Figure 1d). Ventral to the optic tectum, the undifferentiated mesencephalon (i.e., optic tectum, torus semicircularis, and tegmentum) forms a concave margin (Figure 1d,f,h) just posterior to the ventral diencephalon.

Posterior to the mesencephalon, the rhombencephalon forms the posterior-most region of *Tethysaurus* endocast. As in *Clidastes* and *Platecarpus* [28,58], the cerebellum is not discernable from *Tethysaurus* endocast, as a cartilaginous bridge, spanning between the dorsal portions of the otic capsules, covers the structure (Figure 1d,f,h). The medulla oblongata forms a ventrally convex wide arc and exhibits a slight pontine flexure as it extends dorsally to connect with the spinal cord at the level of the foramen magnum (Figure 1d,f,h). The medulla oblongata is ventrally bordered by the posterior part of the para-basisphenoid and the basioccipital, laterally by the prootic and the opisthotic-exoccipital, and dorsally by the supraoccipital.

3.2. Statistical Results and Morphospace Distribution

Tests of the phylogenetic signal revealed a significant effect of evolutionary kinship on patterns of shape variation in extant toxicoferans (K-mult = 0.4012, P = 0.0001, Z_{CR} = 9.4727). However, despite statistical significance, the low K statistic value indicates a weak phylogenetic structuring.

The Procrustes regression analysis showed a significant effect of allometry on endocast shape in extant toxicoferans, with allometry explaining 13.8% of shape variation (Table 2). Procrustes ANOVA (PLM) performed using allometry-corrected shape data revealed significant differences between snakes, anguimorphs, and iguanians, with the clade category explaining 27% of endocast shape variation (Table 2). Pairwise comparisons revealed that the endocast shape in each clade of extant toxicoferans differs from that of the other two clades, with the greatest distance observed between snakes and iguanians (Table 3).

Principal Component Analysis of the endocast morphology of *Tethysaurus*, anguimorphs, snakes, and iguanians resulted in the two first PCs of morphospace accounting for 51.6% of shape variance (PC1 = 35%, PC2 = 16.6%; Figure 2, see Supplementary Data S3 for the variance explained by other PCs). Along the two PCs, the position of *Tethysaurus* is distinct from the different distribution areas of the three clades of extant squamates.

The distribution along PC1 distinguishes the three extant clades—iguanians, anguimorphs, and snakes—from each other (Figure 2), though there are some exceptions. Iguanians show the greatest dispersion across morphospace and represent the extreme values along PC1. *Tethysaurus* falls within the range of values seen in some anguimorphs on the negative side of this axis. PC1 mostly captures morphological variation in the olfactory complex and cerebral hemispheres, with the negative side possessing relatively long and slender olfactory bulbs and peduncles that project antero-ventrally. In these species, the anterior end of the olfactory bulbs is wider than the posterior part of the olfactory peduncles. In the dorsal view, the most lateral point of the cerebral hemispheres in these species is located on the postero-dorsal half of the structure. In contrast, species positioned on the positive side of PC1, mostly serpentes, show relatively short olfactory bulbs and peduncles. In these species, the widest portion of the olfactory complex is situated more posteriorly and maintains a consistent width toward the posterior end of the structure. In the dorsal view, species on the positive side of PC1 display cerebral hemispheres where the most lateral point is located on the anterior half of the structure. Additionally, the ventral diencephalon in the species on the negative side of PC1 projects more ventrally than the medulla oblongata in the lateral view. This differs from species on the positive side of PC1 in which the ventral diencephalon is aligned with the ventral margin of the medulla oblongata.

Table 2. Results of (1) the Procrustes regression for the test of scaling endocast shape data with size based on log-transformed centroid sizes and (2) the Procrustes ANOVA (PLM) obtained from the allometry-corrected shape data for the test of differences in endocast shape between iguanians, snakes, and anguimorphs.

Models	Df	SS	MS	Rsq	F	Z	p
			Procrustes allometric regression				
log(GPA$Csize)	1	0.21119	0.211185	0.13804	12.011	4.972	0.0001
			Procrustes ANOVA				
Clades	2	0.35638	0.178188	0.27024	13.702	6.3235	0.0001

Note: Significance test was based on 10,000 iterations.

Table 3. Pairwise comparisons of allometry-corrected shape disparity between endocasts of extant toxicoferans.

	d	UCL (95%)	Z	Pr > d
Anguimorphs:Iguanians	0.1027768	0.06384204	3.395855	0.0002
Anguimorphs:Snakes	0.1032785	0.04825282	4.346074	0.0001
Iguanians:Snakes	0.1666223	0.06216106	5.140798	0.0001

Note: Significance test was based on 10,000 iterations.

The distribution along PC2 (Figure 2) does not separate iguanians, anguimorphs, and snakes from each other. *Tethysaurus* falls within the most negative range of snakes on that axis. Species along the negative side of PC2 exhibit a relatively narrow endocast in the dorsal view that is flattened in the lateral view. In these species, the anterior end of the olfactory bulbs is distant from the most lateral point of the olfactory complex, the most lateral point of the endocast is on the anterior half of the cerebral hemispheres, and the most dorsal point of the endocast is reached at the level of the rhombencephalon. In contrast, species positioned on the positive side of PC2 exhibit wide endocasts that are dorso-ventrally taller in the lateral view. In these species, the most lateral and dorsal points of the endocast are located on the posterior half of the cerebral hemispheres. In addition, the ventral diencephalon in species on the negative side of PC2 shows no marked ventral projection, being at the same level as the mesencephalic ventral margin. This differs from species on the positive side of PC2 in which the ventral diencephalon is distinct from the mesencephalic ventral margin.

Results obtained from the Linear Discriminant Analysis (LDA) show that the first discriminant function (DF1) explains 86.54% of the total variance, whereas DF2 captures 13.46% (Figure 3). The accuracy of the LDA to classify modern species as either an anguimorph, iguanian, or snake is 93.5%, and only four species, *Lanthanotus borneensis*, *Amerotyphlops brongersmianus*, *Chamaeleo calyptratus*, and *Stenocercus roseiventris* were incorrectly assigned (see Supplementary Data S4 for details). Our results indicate that the *Tethysaurus* endocast shape exhibits more resemblance with the endocast of snakes (0.99 posterior probability) than with anguimorhs' (1.76×10^{-4}) and iguanians' (1.73×10^{-7}).

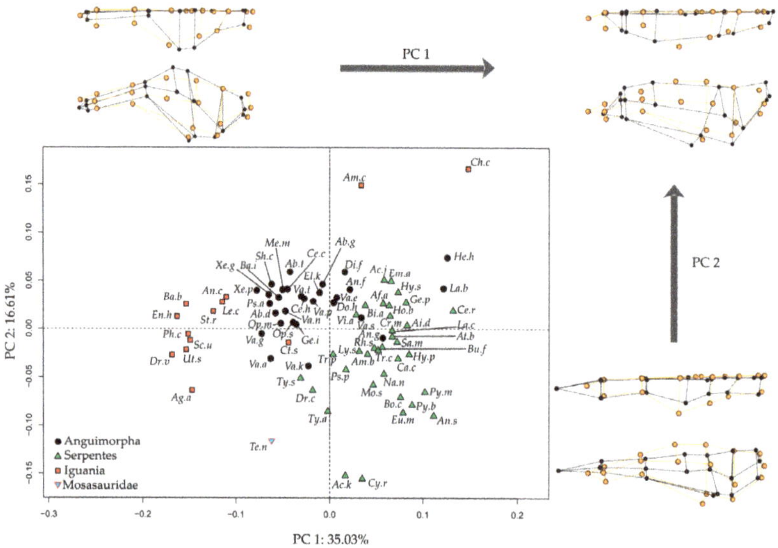

Figure 2. Morphospaces of endocast shape between *Tethysaurus* and extant squamates. The two first principal component axes PC1-PC2 are visualized. The wireframes represent the difference in landmark position between the most extreme endocranial morphologies along PC1 and PC2 (figured in black) as compared to the average (figured in orange), in dorsal (**top**) and lateral (**bottom**) views. Abbreviations of the species names are in Table 1.

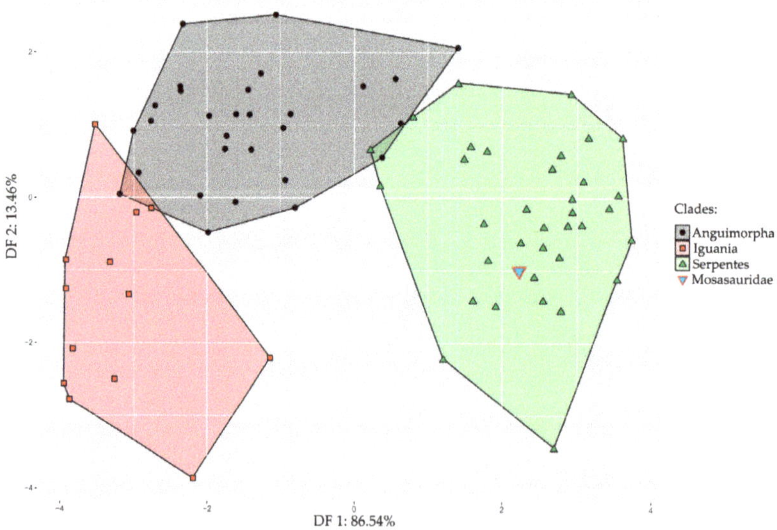

Figure 3. Linear Discriminant Analysis of endocranial shape variation in toxicoferans squamates, constructed from DF1 and DF2. Area of distribution of iguanians, anguimorphs, and snakes are figured in red, black, and green, respectively. The inverted blue triangle indicates the position of the mosasaurid *Tethysaurus nopcsai*.

4. Discussion and Conclusions

Paleoneurological studies dealing with mosasaurid endocasts remain rare. Using micro-computed tomography, for the first time, we reconstruct the brain endocast of a mosasaurid squamate, namely the early *Tethysaurus nopcsai* from the Turonian of Morocco. Our digital reconstructions showed that *Tethysaurus* is characterized by a relatively narrow and flattened endocast with weak cephalic and pontine flexures. The olfactory bulbs and peduncles are relatively long and gracile, with the anterior part of the olfactory complex being wider than the posterior part. The cerebral hemispheres represent the most dorsal and lateral points of the endocast. Posterior to the cerebral hemispheres, the position of the optic tectum is only indicated by gradual changes in the endocast width. The optic tectum lies almost on the same axis as the cerebral hemispheres. The ventral diencephalon projects more ventrally than the ventral margin of the medulla oblongata. Finally, the posterior end of the slightly convex medulla oblongata remains ventral to the antero-posterior axis the olfactory bulbs and peduncles.

Comparisons between *Tethysaurus* and the later *Platecarpus* and *Clidastes* suggest that different endocranial organizations likely occurred among mosasaurids. Both *Tethysaurus* and *Clidastes* exhibit elongated medio-laterally compressed olfactory bulbs and peduncles in the middle of their antero-posterior length, along with a relatively small parietal eye. However, as Camp [28] only figured out the dorsal view of *Clidastes* endocast, comparisons are limited and restricted to these details. In contrast, the endocranial organization in *Tethysaurus* seems different from that reported in *Platecarpus*. The latter, as figured by Camp [28], exhibits short olfactory bulbs and peduncles, a relatively large parietal eye, and a pronounced cephalic flexure differing from *Tethysaurus* and *Clidastes*. It is worth noting that the olfactory bulbs and peduncles in *Platecarpus* were figured in dotted lines by Camp [28] and Russell [58], making the exact length of the whole structure uncertain. Thus, the most notable distinctions between *Tethysaurus* and *Platecarpus* consist of the endocast flexure and the size of the parietal eye.

Pronounced brain flexures generally characterize the early ontogenetic stages of brain development in non-avian reptiles (e.g., [60–62]). When the cranial cavity is relatively small, the brain shows a more pronounced 'S' shape, with the anterior part positioned higher than the posterior one [34]. As the braincase grows more rapidly than the brain itself, there is more space available, allowing the brain to 'unfold' in mature individuals [63–65]. Here, the variable degrees of flexure noticed between *Tethysaurus* and *Platecarpus* might be potentially related to different ontogenetic stages. The endocast flexures observed in *Platecarpus* specimen UCMP 34781 are consistent with the late juvenile stage reported by Sheldon [66]. In contrast, although the ontogenetic stage for *Tethysaurus* remains undetermined [48], the specimen is presumably an adult on the basis of several anatomical characteristics [67] and the nearly straight endocast reported here reinforces such a hypothesis. However, it is worth noting that, as in extant archosaurs [61], possible heterochrony may impact the morphology of the mosasaurid endocast. Thus, the pronounced flexure observed in *Platecarpus* could possibly indicate retention of juvenile endocranial shape (i.e., paedomorphosis) in an adult specimen. In this context, further studies investigating changes in mosasaurid endocast morphology over ontogeny are required. Although ontogenetic series of mosasaurid skulls are rather scarce, future studies could consider the limited material already available (e.g., *Tylosaurus*, [68]) to assess the extent to which endocast morphology in mosasaurids reflects ontogenetic stages.

The variable position and size of the parietal eye have been reported within Mosasauridae (e.g., [69–73]), including differences at the intraspecific level (e.g., *Tylosaurus proriger*, [74]). Such variability in the size of the parietal eye was also reported in various clades of vertebrates (e.g., [75]), suggesting that the common occurrence of a large or small foramen is not a phylogenetic signal but is instead related to differences in the photoreceptive and neuroendocrine functions associated with the structure [76]. The parietal eye in extant squamates detects changes in light levels and this sensory input regulates various seasonal metabolic processes, including behavioral thermoregulation, diurnal rhythms, physical activity, and behavior (e.g., [77,78]). In mosasaurids, the environmental conditions and physiological roles associated with the development of a large parietal eye remain unclear. Connolly [73] found no significant relationships between the size of the parietal eye and the paleolatitude distribution of mosasaurids, nor between the size of the parietal eye and their ability to dive deep. In this context, Connolly [73] suggested that the mosasaurid parietal eye may have functioned primarily for navigation and orientation related to migration. However, such correlations remain unclear and physiological roles associated with the variability in the reptilian parietal eye are needed to better interpret the variability in mosasaurids.

The results obtained here showed that the endocast morphology in extant toxicoferans is linked to phylogeny. However, despite being statistically significant, the low K statistic value indicates weak phylogenetic structuring. Similar to previous studies conducted on snake endocasts [29,33], this indicates that, although significant, the phylogenetic signal remains weak and other factors (e.g., habitat, activity period) could influence the endocast morphology in all extant toxicoferans.

Comparisons with extant toxicoferans highlighted the unique combination of endocranial features in *Tethysaurus*, showing only partial resemblance with anguimorphs, iguanians, or snakes. Indeed, the endocast of *Tethysaurus* does share morphological similarities with each of these three clades, such as (1) the relatively long and gracile olfactory bulbs and peduncles in *Tethysaurus* resemble those observed in iguanians and some anguimorphs, contrasting with the wider and shorter structures seen in snakes; (2) the weak cephalic flexure observed in *Tethysaurus* is similar to that found in anguimorphs and snakes, whereas most iguanians typically exhibit a stronger flexure in their endocast; and (3) posterior to the cerebral hemispheres, the gradual changes in endocast width indicating the position of the optic tectum in *Tethysaurus* resembles the condition observed in anguimorphs but differs from the more abrupt narrowing seen in snakes or the nearly absence of change in iguanians. Overall, the strong endocranial resemblance between varanids and mosasaurids reported by Camp [28] is not observed here. Instead, our results suggested more similarities in endocast shape between *Tethysaurus* and some of the snakes. To assess this result, further comparisons should expand data sampling on mosasaurid endocasts to provide a more comprehensive understanding of the morphological variability within the clade. Thus, the CT data already existing for the mosasaurid *Plotosaurus bennisoni*, available from the online database MorphoSource, could constitute a good starting point for such endocranial studies.

The position of mosasaurids within Toxicofera varies depending on phylogenetic analysis. The clade is positioned either within Anguimorpha and closely related to varanoids (e.g., [45,47]), as the sister group of snakes (e.g., [42,79]), or as the sister group of a clade comprising Anguimorpha and Iguania (e.g., [43,80]). Camp [28] suggested a close phylogenetic relationship between mosasaurids and varanids as he observed a strong resemblance in the morphology of their endocasts. However, this hypothesis is challenged here, as we observe similarities in endocast shape between *Tethysaurus* and each of the three clades included in Toxicofera, with a particular resemblance to some snakes (*Acutotyphlops kunuaensis, Typhlops arenarius, Anilius scytale, Cylindrophis ruffus, Eunectes murinus,* and *Python bivittatus*). These results, based on the digital brain endocast alone, do not enable this study to position mosasaurids within Toxicofera and, furthermore, our findings support no specific phylogenetic hypothesis. In this context, phylogenetic inferences made from endocast morphologies should be treated with caution as other factors may influence the endocast morphology in squamates. In addition, accurate and precise inferences of brain morphology from mosasaurid endocasts require careful consideration. Although certain aspects of brain morphologies in squamates can be extrapolated from endocast morphology, such information varies according to species, clades, and brain regions [40]. Therefore, predicting brain–endocast ratios in mosasaurids, identifying which parts of the endocast accurately reflect brain morphology, and enabling biological inferences require the consideration of a large panel of extant toxicoferans in order to avoid any misinterpretations.

Supplementary Materials: The following supporting information can be downloaded at https://www.mdpi.com/article/10.3390/d16090548/s1. Figure S1: Schematic phylogenetic relationships of iguanians sampled in the study (modified from [50,52]) associated with 3D renderings showing the endocast in dorsal (left) and left lateral (right) views. Figure S2: Schematic phylogenetic relationships of anguimorphs sampled in the study (modified from [50,52]) associated with 3D renderings showing the endocast in dorsal (left) and left lateral (right) views. Figure S3: Schematic phylogenetic relationships of snakes sampled in the study (modified from [50,52]) associated with 3D renderings showing the endocast in dorsal (left) and left lateral (right) views. Supplementary Data S1: Landmark description used in this study (modified from [30]) and position on *Tethysaurus* endocast in dorsal and lateral views. Supplementary Data S2: Raw coordinates of the 19 landmarks placed on each species used in the study. Abbreviations in the main text (Table 1). Supplementary Data S3: Variance explained along the different PCs obtained in the Principal Component Analysis performed on the endocasts of *Tethysaurus* and extant squamates. Supplementary Data S4: Results LDA.

Author Contributions: Conceptualization, R.A. and N.B.; methodology, R.A. and C.L.-A.; software, R.A.; investigation, R.A.; writing—original draft preparation, R.A.; writing—review and editing, M.J.P., A.H., P.V., C.L.-A., and N.B.; visualization, R.A.; supervision, A.H., P.V., and N.B.; funding acquisition, N.B., A.H., and P.V. All authors have read and agreed to the published version of the manuscript.

Funding: This work was supported by a grant from the Agence Nationale de la Recherche under the LabEx ANR-10-LABX-0003-BCDiv, in the program 'Investissements d'avenir' ANR-11-IDEX-0004-02.

Institutional Review Board Statement: Not applicable.

Data Availability Statement: The endocasts (stl files) used in the study will be available on MorphoSource and attached to the corresponding files for each species. The raw scan data of *Tethysaurus* specimen MNHN GOU 1 will be available on the MNHN Digitization Work Portal https://3dtheque.mnhn.fr (accessed on 4 August 2024).

Acknowledgments: We thank the AST-RX, plateau d'Accès Scientifique à la Tomographie à Rayons X du MNHN, UAR 2700 2AD CNRS-MNHN, Paris, for access to the CT scanner and M. Garcia-Sanz (MNHN, UMS 2700 OMSI) for producing the CT scan of specimen MNHN GOU 1. We also thank M. Colbert and the High-Resolution X-ray Computed Tomography Facility at The University of Texas for producing CT scans of the specimens SMU 76335 and SMU 75486. We warmly thank I. Ineich, J. Rosado, K. Lim, and A. Herrel for the loan of CT scans. We would like to acknowledge L.A. Scheinberg, S. Scarpetta, C. Spencer, D. Ledesma, E. Stanley, A. Motta, G. Schneider, N. Rios, S. Grant, J. Maisano, Z. Randall, J. Gray, C. Sheehy, and D. Blackburn for access to CT scans through the Morphosource database. We would also like to thank two anonymous reviewers for their helpful comments and suggestions that improved the paper.

Conflicts of Interest: The authors declare no conflicts of interest.

References

1. Bardet, N.; Falconnet, J.; Fischer, V.; Houssaye, A.; Jouve, S.; Pereda Suberbiola, X.; Pérez-Garcia, A.; Rage, J.-C.; Vincent, P. Mesozoic marine reptile palaeobiogeography in response to drifting plates. *Gondwana Res.* **2014**, *26*, 869–887. [CrossRef]
2. Polcyn, M.J.; Jacobs, L.L.; Araújo, R.; Schulp, A.S.; Mateus, O. Physical drivers of mosasaur evolution. *Palaeogeogr. Palaeoclimatol. Palaeoecol.* **2014**, *400*, 17–27. [CrossRef]
3. Rothschild, B.M.; Martin, L.D. Mosasaur ascending: The phytogeny of bends. *Neth. J. Geosci.* **2005**, *84*, 341–344. [CrossRef]
4. Houssaye, A.; Lindgren, J.; Pellegrini, R.; Lee, A.H.; Germain, D.; Polcyn, M.J. Microanatomical and histological features in the long bones of mosasaurine mosasaurs (Reptilia, Squamata)–implications for aquatic adaptation and growth rates. *PLoS ONE* **2013**, *8*, e76741. [CrossRef] [PubMed]
5. Bell, G.L., Jr. A phylogenetic revision of North American and Adriatic Mosasauroidea. In *Ancient Marine Reptiles*; Callaway, J.M., Nicholls, E.L., Eds.; Academic Press: London, UK; New York, NY, USA; San Francisco, CA, USA, 1997; pp. 293–332.
6. Lindgren, J.; Jagt, J.W.; Caldwell, M.W. A fishy mosasaur: The axial skeleton of Plotosaurus (Reptilia, Squamata) reassessed. *Lethaia* **2007**, *40*, 153–160. [CrossRef]
7. Lindgren, J.; Caldwell, M.W.; Konishi, T.; Chiappe, L.M. Convergent evolution in aquatic tetrapods: Insights from an exceptional fossil mosasaur. *PLoS ONE* **2010**, *5*, e11998. [CrossRef] [PubMed]
8. Schulp, A.S.; Vonhof, H.B.; Van der Lubbe, J.H.J.L.; Janssen, R.; Van Baal, R.R. On diving and diet: Resource partitioning in type-Maastrichtian mosasaurs·. *Neth. J. Geosci.* **2013**, *92*, 165–170. [CrossRef]

9. Bardet, N.; Houssaye, A.; Vincent, P.; Pereda Suberbiola, X.; Amaghzaz, M.; Jourani, E.; Meslouh, S. Mosasaurids (Squamata) from the Maastrichtian phosphates of Morocco: Biodiversity, palaeobiogeography and palaeoecology based on tooth morphoguilds. *Gondwana Res.* **2015**, *27*, 1068–1078. [CrossRef]
10. Holwerda, F.M.; Bestwick, J.; Purnell, M.A.; Jagt, J.W.; Schulp, A.S. Three-dimensional dental microwear in type-Maastrichtian mosasaur teeth (Reptilia, Squamata). *Sci. Rep.* **2023**, *13*, 18720. [CrossRef]
11. Longrich, N.R.; Polcyn, M.J.; Jalil, N.E.; Pereda-Suberbiola, X.; Bardet, N. A bizarre new plioplatecarpine mosasaurid from the Maastrichtian of Morocco. *Cretac. Res.* **2024**, *160*, 105870. [CrossRef]
12. Kear, B.P.; Long, J.A.; Martin, J.E. A review of Australian mosasaur occurrences. *Neth. J. Geosci.* **2005**, *84*, 307–313. [CrossRef]
13. Fernández, M.S.; Talevi, M. An halisaurine (Squamata: Mosasauridae) from the Late Cretaceous of Patagonia, with a preserved tympanic disc: Insights into the mosasaur middle ear. *Comptes Rendus Palevol* **2015**, *14*, 483–493. [CrossRef]
14. Konishi, T.; Ohara, M.; Misaki, A.; Matsuoka, H.; Street, H.P.; Caldwell, M.W. A new derived mosasaurine (Squamata: Mosasaurinae) from south-western Japan reveals unexpected postcranial diversity among hydropedal mosasaurs. *J. Syst. Palaeontol.* **2023**, *21*, 2277921. [CrossRef]
15. Lindgren, J.; Siverson, M. *Tylosaurus ivoensis*: A giant mosasaur from the early Campanian of Sweden. *Trans. Roy. Soc. Edinb Earth Sci.* **2002**, *93*, 73–93. [CrossRef]
16. Martin, J.E. Biostratigraphy of the Mosasauridae (Reptilia) from the Cretaceous of Antarctica. *Geol. Soc.* **2006**, *258*, 101–108. [CrossRef]
17. Fernández, M.S.; Gasparini, Z. Campanian and Maastrichtian mosasaurs from Antarctic Peninsula and Patagonia, Argentina. *Bull Soc Géol Fr.* **2012**, *183*, 93–102. [CrossRef]
18. González Ruiz, P.; Fernández, M.S.; Talevi, M.; Leardi, J.M.; Reguero, M.A. A new Plotosaurini mosasaur skull from the upper Maastrichtian of Antarctica. Plotosaurini paleogeographic occurrences. *Cretac. Res.* **2019**, *103*, 104–166. [CrossRef]
19. Polcyn, M.J.; Augusta, B.G.; Zaher, H. Reassessing the morphological foundations of the pythonomorph hypothesis. In *The Origin and Early Evolutionary History of Snakes*; Gower, D.J., Zaher, H., Eds.; Systematics Association Special Volume Series; Cambridge University Press: Cambridge, UK, 2022; Volume 90, pp. 125–156.
20. Palci, A.; Caldwell, M.W.; Papazzoni, C.A. A new genus and subfamily of mosasaurs from the Upper Cretaceous of northern Italy. *J. Vert. Paleontol.* **2013**, *33*, 599–612. [CrossRef]
21. Madzia, D.; Cau, A. Inferring 'weak spots' in phylogenetic trees: Application to mosasauroid nomenclature. *PeerJ* **2017**, *5*, e3782. [CrossRef]
22. Simões, T.R.; Vernygora, O.; Paparella, I.; Jimenez-Huidobro, P.; Caldwell, M.W. Mosasauroid phylogeny under multiple phylogenetic methods provides new insights on the evolution of aquatic adaptations in the group. *PLoS ONE* **2017**, *12*, e0176773. [CrossRef]
23. Polcyn, M.J.; Bardet, N.; Albright III, L.B.; Titus, A. A new lower Turonian mosasaurid from the Western Interior Seaway and the antiquity of the unique basicranial circulation pattern in Plioplatecarpinae. *Cretac. Res.* **2023**, *151*, 105621. [CrossRef]
24. Cuthbertson, R.S.; Maddin, H.C.; Holmes, R.B.; Anderson, J.S. The braincase and endosseous labyrinth of *Plioplatecarpus peckensis* (Mosasauridae, Plioplatecarpinae), with functional implications for locomotor behavior. *Anat. Rec.* **2015**, *298*, 1597–1611. [CrossRef] [PubMed]
25. Yi, H.; Norell, M.A. The burrowing origin of modern snakes. *Sci. Adv.* **2015**, *1*, e1500743. [CrossRef]
26. Yi, H.; Norell, M. The bony labyrinth of *Platecarpus* (Squamata: Mosasauria) and aquatic adaptations in squamate reptiles. *Palaeoworld* **2019**, *28*, 550–561. [CrossRef]
27. Álvarez–Herrera, G.; Agnolin, F.; Novas, F. A rostral neurovascular system in the mosasaur *Taniwhasaurus antarcticus*. *Sci. Nat.* **2020**, *107*, 19. [CrossRef] [PubMed]
28. Camp, C.L. California Mosasaurs. In *Memoirs of the University of California*; University of California Press: Berkeley, CA, USA; Los Angeles, CA, USA, 1942; Volume 13, p. 67.
29. Allemand, R.; Boistel, R.; Daghfous, G.; Blanchet, Z.; Cornette, R.; Bardet, N.; Vincent, P.; Houssaye, A. Comparative morphology of snake (Squamata) endocasts: Evidence of phylogenetic and ecological signals. *J. Anat.* **2017**, *231*, 849–868. [CrossRef]
30. Allemand, R.; López-Aguirre, C.; Abdul-Sater, J.; Khalid, W.; Lang, M.M.; Macrì, S.; Di-Poï, N.; Daghfous, G.; Silcox, M.T. A landmarking protocol for geometric morphometric analysis of squamate endocasts. *Anat. Rec.* **2023**, *306*, 2425–2442. [CrossRef]
31. De Meester, G.; Huyghe, K.; Van Damme, R. Brain size, ecology and sociality: A reptilian perspective. *Biol. J. Linn. Soc.* **2019**, *126*, 381–391. [CrossRef]
32. Macrì, S.; Savriama, Y.; Khan, I.; Di-Poï, N. Comparative analysis of squamate brains unveils multi-level variation in cerebellar architecture associated with locomotor specialization. *Nat. Commun.* **2019**, *10*, 5560. [CrossRef]
33. Segall, M.; Cornette, R.; Rasmussen, A.R.; Raxworthy, C.J. Inside the head of snakes: Influence of size, phylogeny, and sensory ecology on endocranium morphology. *Brain Struct. Funct.* **2021**, *226*, 2401–2415. [CrossRef]
34. Hopson, J.A. Paleoneurology. In *Biology of the Reptilia*; Gans, C., Northcutt, R.G., Ulinski, P., Eds.; Academic Press: London, UK; New York, NY, USA; San Francisco, CA, USA, 1979; Volume 9, pp. 39–146.
35. Starck, D. Cranio-cerebral relations in recent reptiles. In *Biology of the Reptilia*; Gans, C., Northcutt, R.G., Ulinski, P., Eds.; Academic Press: London, UK; New York, NY, USA; San Francisco, CA, USA, 1979; Volume 9, pp. 1–38.
36. ten Donkelaar, H.J. Reptiles. In *The Central Nervous System of Vertebrates*; Nieuwenhuys, R., ten Donkelaar, H.J., Nicholson, C., Eds.; Springer: Berlin/Heidelberg, Germany, 1998; pp. 1315–1499.

37. Kim, R.; Evans, D. Relationships among brain, endocranial cavity, and body sizes in reptiles. In Proceedings of the Society of Vertebrate Paleontology 74th Annual Meeting, Berlin, Germany, 5–8 November 2014.
38. Triviño, L.N.; Albino, A.M.; Dozo, M.T.; Williams, J.D. First natural endocranial cast of a fossil snake (cretaceous of Patagonia, Argentina). *Anat. Rec.* **2018**, *301*, 9–20. [CrossRef]
39. Perez-Martinez, C.A.; Leal, M. Lizards as models to explore the ecological and neuroanatomical correlates of miniaturization. *Behaviour* **2021**, *158*, 1121–1168. [CrossRef]
40. Allemand, R.; Abdul-Sater, J.; Macrì, S.; Di-Poï, N.; Daghfous, G.; Silcox, M.T. Endocast, brain, and bones: Correspondences and spatial relationships in squamates. *Anat. Rec.* **2023**, *306*, 2443–2465. [CrossRef] [PubMed]
41. Gauthier, J.A.; Kearney, M.; Maisano, J.A.; Rieppel, O.; Behlke, A.D. Assembling the squamate tree of life: Perspectives from the phenotype and the fossil record. *Bull. Peabody Mus. Nat. Hist.* **2012**, *53*, 3–308. [CrossRef]
42. Reeder, T.W.; Townsend, T.M.; Mulcahy, D.G.; Noonan, B.P.; Wood, P.L., Jr.; Sites, J.W., Jr.; Wiens, J.J. Integrated analyses resolve conflicts over squamate reptile phylogeny and reveal unexpected placements for fossil taxa. *PLoS ONE* **2015**, *10*, e0118199. [CrossRef] [PubMed]
43. Simões, T.R.; Caldwell, M.W.; Tałanda, M.; Bernardi, M.; Palci, A.; Vernygora, O.; Bernardini, F.; Mancini, L.; Nydam, R.L. The origin of squamates revealed by a Middle Triassic lizard from the Italian Alps. *Nature* **2018**, *557*, 706–709. [CrossRef]
44. Simões, T.R.; Pyron, R.A. The squamate tree of life. *Bull. Mus. Comp. Zool.* **2021**, *163*, 47–95. [CrossRef]
45. Augusta, B.G.; Zaher, H.; Polcyn, M.J.; Fiorillo, A.R.; Jacobs, L.L. A review of non-mosasaurid (dolichosaur and aigialosaur) mosasaurians and their relationships to snakes. In *The Origin and Early Evolutionary History of Snakes*; Gower, D.J., Zaher, H., Eds.; Systematics Association Special Volume Series; Cambridge University Press: Cambridge, UK, 2022; Volume 90, pp. 157–179.
46. Zaher, H.; Augusta, B.G.; Rabinovich, R.; Polcyn, M.; Tafforeau, P. A review of the skull anatomy and phylogenetic affinities of marine pachyophiid snakes. In *The Origin and Early Evolutionary History of Snakes*; Gower, D.J., Zaher, H., Eds.; Systematics Association Special Volume Series; Cambridge University Press: Cambridge, UK, 2022; Volume 90, pp. 180–206.
47. Zaher, H.; Mohabey, D.M.; Grazziotin, F.G.; Wilson Mantilla, J.A. The skull of *Sanajeh indicus*, a Cretaceous snake with an upper temporal bar, and the origin of ophidian wide-gaped feeding. *Zool. J. Linn. Soc.* **2023**, *197*, 656–697. [CrossRef]
48. Bardet, N.; Pereda Suberbiola, X.; Jalil, N.E. A new mosasauroid (Squamata) from the Late Cretaceous (Turonian) of Morocco. *Comptes Rendus Palevol* **2003**, *2*, 607–616. [CrossRef]
49. Boyer, D.M.; Gunnell, G.F.; Kaufman, S.; McGeary, T.M. Morphosource: Archiving and sharing 3-D digital specimen data. *Pal Soc. Pap.* **2016**, *22*, 157–181. [CrossRef]
50. Pyron, R.A.; Burbrink, F.T.; Wiens, J.J. A phylogeny and revised classification of Squamata, including 4161 species of lizards and snakes. *BMC Evol Biol* **2013**, *13*, 93. [CrossRef]
51. Zheng, Y.; Wiens, J.J. Combining phylogenomic and supermatrix approaches, and a time-calibrated phylogeny for squamate reptiles (lizards and snakes) based on 52 genes and 4162 species. *Mol Phylogenet Evol* **2016**, *94*, 537–547. [CrossRef] [PubMed]
52. Singhal, S.; Colston, T.J.; Grundler, M.R.; Smith, S.A.; Costa, G.C.; Colli, G.R.; Moritz, C.; Pyron, R.A.; Rabosky, D.L. Congruence and conflict in the higherlevel phylogenetics of squamate reptiles: An expanded phylogenomic perspective. *Syst. Biol.* **2021**, *70*, 542–557. [CrossRef] [PubMed]
53. Baken, E.K.; Collyer, M.L.; Kaliontzopoulou, A.; Adams, D.C. Geomorph v4.0 and gmShiny: Enhanced analytics and a new graphical interface for a comprehensive morphometric experience. *Methods Ecol. Evol.* **2021**, *2*, 2355–2363. [CrossRef]
54. Adams, D.C. A generalized K statistic for estimating phylogenetic signal from shape and other high-dimensional multivariate data. *Syst. Biol.* **2014**, *63*, 685–697. [CrossRef]
55. Adams, D.C.; Collyer, M.L. Multivariate comparative methods: Evaluations, comparisons, and recommendations. *Syst. Biol.* **2018**, *67*, 14–31. [CrossRef]
56. Venables, W.N.; Ripley, B.D. *Modern Applied Statistics with S*, 4th ed.; Statistics and Computing Springer: New York, NY, USA, 2002; p. 498.
57. Watanabe, A.; Gignac, P.M.; Balanoff, A.M.; Green, T.L.; Kley, N.J.; Norell, M.A. Are endocasts good proxies for brain size and shape in archosaurs throughout ontogeny? *J. Anat.* **2019**, *234*, 291–305. [CrossRef]
58. Russell, D.A. Systematics and morphology of American mosasaurs (Reptilia, Sauria). *Bull. Peabody Mus. Nat. Hist.* **1967**, *23*, 241.
59. Smith, K.T.; Bhullar, B.A.S.; Köhler, G.; Habersetzer, J. The only known jawed vertebrate with four eyes and the bauplan of the pineal complex. *Curr. Biol.* **2018**, *28*, 1101–1107. [CrossRef]
60. Jirak, D.; Janacek, J. Volume of the crocodilian brain and endocast during ontogeny. *PLoS ONE* **2017**, *12*, e0178491. [CrossRef]
61. Hu, K.; King, J.L.; Romick, C.A.; Dufeau, D.L.; Witmer, L.M.; Stubbs, T.L.; Rayfield, E.J.; Benton, M.J. Ontogenetic endocranial shape change in alligators and ostriches and implications for the development of the non-avian dinosaur endocranium. *Anat. Rec.* **2021**, *304*, 1759–1775. [CrossRef] [PubMed]
62. Roese-Miron, L.; Jones, M.E.H.; Ferreira, J.D.; Hsiou, A.S. Virtual endocasts of *Clevosaurus brasiliensis* and the tuatara: Rhynchocephalian neuroanatomy and the oldest endocranial record for Lepidosauria. *Anat. Rec.* **2023**, *307*, 1366–1389. [CrossRef]
63. Giffin, E.B. Pachycephalosaur paleoneurology (Archosauria: Ornithischia). *J. Vertebr. Paleontol.* **1989**, *9*, 67–77. [CrossRef]
64. Lautenschlager, S.; Hübner, T. Ontogenetic trajectories in the ornithischian endocranium. *J. Evol. Biol.* **2013**, *26*, 2044–2050. [CrossRef] [PubMed]

65. Fabbri, M.; Koch, N.M.; Pritchard, A.C.; Hanson, M.; Hoffman, E.; Bever, G.S.; Balanoff, A.M.; Morris, Z.S.; Field, D.J.; Camacho, J.; et al. The skull roof tracks the brain during the evolution and development of reptiles including birds. *Nat. Ecol. Evol.* **2017**, *1*, 1543–1550. [CrossRef]
66. Sheldon, A. Ontogeny, Ecology and Evolution of North American Mosasaurids (*Clidastes*, *Platecarpus* and *Tylosaurus*): Evidence from Bone Microstructure. Ph.D. Dissertation, University of Rochester, New York, NY, USA, 1995.
67. Houssaye, A.; Bardet, N. A baby mosasauroid (Reptilia, Squamata) from the Turonian of Morocco—*Tethysaurus* 'junior' discovered? *Cretac. Res.* **2013**, *46*, 208–215. [CrossRef]
68. Zietlow, A.R. Craniofacial ontogeny in Tylosaurinae. *PeerJ* **2020**, *8*, e10145. [CrossRef] [PubMed]
69. Bardet, N.; Pereda Suberbiola, X.; Iarochene, M.; Bouya, B.; Amaghzaz, M. A new species of *Halisaurus* from the Late Cretaceous phosphates of Morocco, and the phylogenetical relationships of the Halisaurinae (Squamata: Mosasauridae). *Zool. J. Linn. Soc.* **2005**, *143*, 447–472. [CrossRef]
70. Polcyn, M.J.; Bell, G.L., Jr. *Russellosaurus coheni* n. gen., n. sp., a 92 million-year-old mosasaur from Texas (U.S.A.), and the definition of the parafamily Russellosaurina. *Neth. J. Geosci.* **2005**, *84*, 321–333. [CrossRef]
71. Páramo-Fonseca, M.E. *Eonatator coellensis* nov. sp. (Squamata: Mosasauridae), a new species from the Upper Cretaceous of Colombia. *Revista Acad. Colomb. Ci Exact.* **2013**, *37*, 499–518.
72. Konishi, T.; Caldwell, M.W.; Nishimura, T.; Sakurai, K.; Tanoue, K. A new halisaurine mosasaur (Squamata: Halisaurinae) from Japan: The first record in the western Pacific realm and the first documented insights into binocular vision in mosasaurs. *J. Syst. Palaeontol.* **2015**, *14*, 809–839. [CrossRef]
73. Connolly, A. Exploring the Relationship between Paleobiogeography, Deep-Diving Behavior, and Size Variation of the Parietal Eye in Mosasaurs. Master's Thesis, University of Kansas, Lawrence, KS, USA, 2016.
74. Jiménez-Huidobro, P.; Caldwell, M.W. Reassessment and reassignment of the early Maastrichtian mosasaur *Hainosaurus bernardi* Dollo, 1885, to *Tylosaurus* Marsh, 1872. *J Vertebr Paleontol* **2016**, *36*, e1096275. [CrossRef]
75. Holloway, W.L.; Claeson, K.M.; O'keefe, F.R. A virtual phytosaur endocast and its implications for sensory system evolution in archosaurs. *J. Vert. Paleontol.* **2013**, *33*, 848–857. [CrossRef]
76. Holmes, R.B. Evaluation of the photosensory characteristics of the lateral and pineal eyes of *Plioplatecarpus* (Squamata, Mosasauridae) based on an exceptionally preserved specimen from the Bearpaw Shale (Campanian, Upper Cretaceous) of southern Alberta. *J. Vertebr. Paleontol.* **2023**, *43*, e2335174. [CrossRef]
77. Ralph, C.L.; Firth, B.T.; Turner, J.S. The role of the pineal body in ectotherm thermoregulation. *Am. Zool.* **1979**, *19*, 273–293. [CrossRef]
78. Labra, A.; Voje, K.L.; Seligmann, H.; Hansen, T.F. Evolution of the third eye: A phylogenetic comparative study of parietal-eye size as an ecophysiological adaptation in *Liolaemus* lizards. *Biol. J. Linn. Soc.* **2010**, *101*, 870–883. [CrossRef]
79. Pyron, R.A. Novel approaches for phylogenetic inference from morphological data and total-evidence dating in squamate reptiles (lizards, snakes, and amphisbaenians). *Syst. Biol.* **2017**, *66*, 38–56. [CrossRef]
80. Caldwell, M.W.; Simões, T.R.; Palci, A.; Garberoglio, F.F.; Reisz, R.R.; Lee, M.S.; Nydam, R.L. *Tetrapodophis amplectus* is not a snake: Re-assessment of the osteology, phylogeny and functional morphology of an Early Cretaceous dolichosaurid lizard. *J. Syst. Palaeontol.* **2021**, *19*, 893–952. [CrossRef]

diversity

MDPI

Article

Mosasaurids Bare the Teeth: An Extraordinary Ecological Disparity in the Phosphates of Morocco Just Prior to the K/Pg Crisis †

Nathalie Bardet [1,*], Valentin Fischer [2], Nour-Eddine Jalil [1,3], Fatima Khaldoune [4], Oussama Khadiri Yazami [4], Xabier Pereda-Suberbiola [5] and Nicholas Longrich [6]

1 CR2P Centre de Recherche en Paléontologie de Paris, UMR 7207 CNRS-MNHN-SU, Muséum National d'Histoire Naturelle, 75005 Paris, France; nour-eddine.jalil@mnhn.fr
2 Evolution and Diversity Dynamics Lab, Université de Liège, 4000 Liège, Belgium; v.fischer@uliege.be
3 Muséum d'Histoire Naturelle de Marrakech, Faculté des Sciences Semlalia, Université Cadi Ayyad, Marrakech 40000, Morocco
4 Office Chérifien des Phosphates, Khouribga 25010, Morocco; khaldoune.fatima@ocpgroup.ma (F.K.); o.khadiriyazami@ocpgroup.ma (O.K.Y.)
5 Departamento de Geología, Facultad de Ciencia y Tecnología, Universidad del Pais Vasco/Euskal Herriko Unibertsitatea, 48080 Bilbao, Spain; xabier.pereda@ehu.eus
6 Department of Biology and Biochemistry, University of Bath, Claverton Down, Bath BA2 7AY, UK; nrl22@bath.ac.uk
* Correspondence: nathalie.bardet@mnhn.fr

† In memory of Henri Cappetta (1946–2024), whose selachians swam with mosasaurids in the same Phosphate seas.

Abstract: Mosasaurid teeth are abundant in the fossil record and often diagnostic to low taxonomic levels, allowing to document the taxonomic diversity and ecological disparity through time and with fewer biases than in other marine reptiles. The upper Maastrichtian Phosphates of Morocco, with at least fifteen coeval species representing a wide range of sizes and morphologies, undoubtedly represent the richest outcrop in the world for this clade of iconic Mesozoic squamates and one of the richest known marine tetrapod assemblages. Until now, the methods used to link tooth morphology to diets in marine amniotes were mainly qualitative in nature. Here, using the dental morphology of mosasaurids from Morocco, we combine two complementary approaches—a thorough comparative anatomical description and 2D/3D geometric morphometry—to quantitatively categorize the main functions of these teeth during feeding processes and infer diet preferences and niche-partitioning of these apex predators. Our results from combining these two approaches show the following: (1) Mosasaurids from the upper Maastrichtian Phosphates of Morocco occupy the majority of dental guilds ever colonized by Mesozoic marine reptiles. (2) As seen elsewhere in the Maastrichtian, mosasaurines dominate the regional mosasaurid assemblage, exhibiting the greatest taxonomic diversity (two-thirds of the species) and the largest range of morphologies, body sizes (2 m to more than 10 m) and ecological disparities (participating in nearly all predatory ecological guilds); strikingly, mosasaurines did not developed flesh piercers and, conversely, are the only ones to include durophagous species. (3) Halisaurines, though known by species of very different sizes (small versus large) and cranial morphologies (gracile versus robust), maintain a single tooth shape (piercer). (4) Plioplatecarpines were medium-size cutters and piercers, known by very morphologically diverging species. (5) Tylosaurines currently remain scarce, represented by a very large generalist species; they were largely replaced by mosasaurines as apex predators over the course of the Maastrichtian, as observed elsewhere. Also, when comparing tooth shapes with body sizes, the largest taxa (>8 m long) occupied a restricted area of tooth shapes (generalist, durophagous), whereas small and medium-sized species (<8 m long) range across all of them (generalists, durophagous, cutters, piercers). In other words, and probably related to the specificities and advantages of biomechanical resistance, apex predators are never dedicated piercers, micro-predators are conversely never generalists, and meso-predators show the widest range of dental adaptations. These diversities and disparities strongly suggest that Tethyan mosasaurids evolved strong niche-partitioning in the shallow marine environment of the upper Maastrichtian Phosphates of Morocco. Such a high diversity sensu lato just prior to the K/Pg biological crisis suggests that their extinction was rather sudden, though the exact causes of their extinction remain unknown. Finally, *Gavialimimus* Strong et al., 2020 is systematically reassigned to *Gavialimimus ptychodon* (Arambourg, 1952), and an emended diagnosis (for teeth and dentition) is proposed for this species.

Keywords: mosasaurid squamates; Morocco; phosphates; Late Cretaceous; comparative anatomy; morphometrical analyses; diet preferences; niche-partitioning

1. Introduction

Mosasaurids were a specialized clade of large marine squamates that underwent a spectacular radiation during the Late Cretaceous, before becoming extinct during the K/Pg biological crisis (e.g., [1,2]). They exhibit a high species diversity, as well as a wide range of morphotypes and body sizes (about 1–17 m), suggesting a high ecological disparity. They were also widely distributed from the Santonian up to the end of the Maastrichtian. Among the most diverse and abundant mosasaurid faunas are those from the Santonian-Campanian of the Western Interior Sea of North America, the Campanian–Maastrichtian of New Zealand and the Maastrichtian of the Netherlands (where they were first unearthed at the end of the 18th century), Belgium, Angola and Morocco (e.g., [2,3]). Among them, with sixteen species described up to now, the Maastrichtian Phosphates of Morocco undoubtedly represent the richest outcrop for mosasaurids worldwide.

The exceptional richness of fossil vertebrates in the Maastrichtian–Ypresian Phosphates of Morocco has been known since the pioneering works of the French paleontologist Camille Arambourg [4]. At present, almost 400 species of vertebrates are known from these deposits, of which more than 95% are marine. These include a plethora of selachians, actinopterygians and marine reptiles, as well as scarce continental taxa, including Maastrichtian non-avian dinosaurs and pterosaurs, Paleogene mammals and birds originating from the nearby African Craton (e.g., [5–8]).

After selachians (e.g., [9,10]), marine reptiles are the most common and diverse group in the Phosphates, both in the Maastrichtian and Paleogene [5,11]. Mosasaurid squamates dominate the Maastrichtian ecosystems and, as a mirror, the dyrosaurid crocodyliforms the Paleogene ones [12]. In addition to mosasaurids, the Maastrichtian marine reptile assemblage include elasmosaurid plesiosaurians [13,14], chelonioid turtles [15,16], gavialoid crocodilians [12] and pachyvaranid squamates [17].

Mosasaurids are by far the most abundant marine reptiles in the Maastrichtian Phosphates of Morocco, being known by at least fifteen coeval species (Tables 1 and A1). They are represented by a plethora of specimens of all sizes, morphologies and ecologies. Many of these species were previously unknown [3–5,18–33], and some are typical of the Tethyan Southern Margin [34].

At the interface between predators and their environment, teeth are a major component of food acquisition and processing, and their shape is closely linked to eaten items, making it possible to infer possible diets [35,36]. Despite a widespread belief that reptilian teeth are poorly informative from a systematical and ecological point of view, and except for some overlaps in gross morphology [36], it has been shown that mosasaurids possess series of diagnostic traits in their teeth. These turn out to be useful for systematic identification (at the generic and even specific levels) when only isolated teeth are found (e.g., [3,37]).

In addition, mosasaurid teeth are characterized by a diverse range of external macrostructures (shape, ornamentation, wear) (e.g., [3,38–43]), internal microstructures (enamel/dentin texture and composition) (e.g., [44–47]) and attachment and replacement modes (e.g., [48–50]). These played an important role in feeding process and as such reveal a large spectrum of diets and niche occupation in the water column. The rich fossil record of mosasaurid teeth is therefore capable of documenting both the taxonomical and ecological diversity of mosasaurid faunas and their evolution, with fewer biases than in other marine reptile groups.

Here, focusing on the dental morphology of the mosasaurid assemblages from the upper Maastrichtian Phosphates of Morocco, we combine two complementary approaches, namely, a thorough qualitative comparative anatomy and two quantitative geometric morphometry analyses (2D and 3D), to identify their tooth shape and infer the diet preferences, predation modes and possible niche-partitioning of these predators in the regional marine ecosystem, just prior to the K/Pg biological crisis.

Institutional abbreviations. HUJ, Hebrew University of Jerusalem; IRScNB, Institut Royal des Sciences Naturelles de Belgique, Bruxelles, Belgium; MGUAN, Museo Geológico da Universidade Agostino Neto, Luanda, Angola; MHNM, Muséum d'Histoire Naturelle de Marrakech, Université Cadi Ayyad (UCAM), Marrakech, Morocco; MNHN, Muséum National d'Histoire Naturelle, Paris, France; OCP, Office Chérifien des Phosphates, Khouribga, Morocco; UALVP, University of Alberta Laboratory for Vertebrate Paleontology, Edmonton, Alberta, Canada; ULg, Université de Liège, Belgique; VANPS, Paleontological Museum of the Vancouver Paleontological Society, Richmond, British Columbia, Canada.

2. Geographical and Geological Settings

The Phosphates of Morocco are part of an extensive belt of sedimentary deposits named the 'Mediterranean (Tethyan) Phosphogenic Province' that developed during the Late Cretaceous and early Paleogene epochs. These formed primarily in subtropical paleolatitudes 12–22° in shallow marine paleoenvironments (e.g., [51,52]). Currently, these phosphatic deposits widely crop out in the Middle East and northwest Africa, up to the Pernambuco Province of Brazil, where they are exploited as a valuable economical resource [51,52].

Historically, the Phosphates have been known in Morocco since the beginning of the 20th century [53,54] and have been exploited on a large scale since 1921 by the Office Chérifien des Phosphates [55]. Morocco has the world's largest phosphatic deposits and reserves (more than 70%) and is the leading exporter and second-largest producer of phosphates in the world, after China [56].

The Phosphates currently crop out in four main basins distributed through central Morocco, these being the Oulad Abdoun, Ganntour, Meskala and Souss basins (Figure 1A). Only the Oulad Abdoun and Ganntour basins are economically exploited, especially in the Sidi Daoui, Sidi Chennane (Oulad Abdoun) and Ben Guerir (Ganntour) areas (Figure 1B,C), favoring the local discoveries of vertebrate fossil remains.

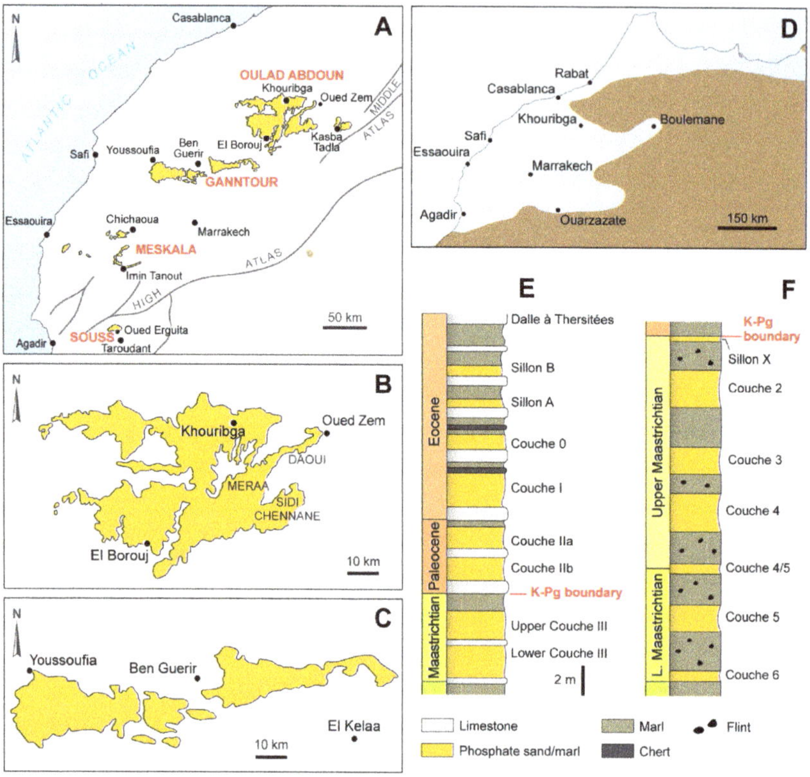

Figure 1. The Phosphates of central Morocco. (**A**) Geographical map showing the main phosphatic basins, from NE to SW: Oulad Abdoun and Ganntour (economically exploited), Meskala and Souss (not exploited). (**B**,**C**) Details of the Oulad Abdoun and Ganntour basins' geography. (**D**) Paleogeographical reconstruction of Morocco during the Late Cretaceous, after [57]. (**E**,**F**) Synthetic stratigraphical column of the phosphatic series in the Oulad Abdoun (Maastrichtian–Paleogene) and Ganntour (Maastrichtian only) basins. All figures modified from [3,5]. Drawings and design © Alexandre Lethiers (CR2P/ISteP, Paris).

Paleogeographically, the phosphatic sediments were deposited in a large shallow marine gulf located on the northwest margin of the African craton (Figure 1D). This large area was characterized by a high productivity upwelling system that developed from the Late Cretaceous up to present times along the western coast of Africa, driven by the trade winds pushing surface waters offshore (e.g., [2,51,52,56]). The upwelling system probably permitted the development of the exceptional local biomass [58], whereas the phosphatogenesis favored a high and selective conservation of vertebrate remains, as opposed to microfossils and invertebrates, which are frequently dissolved and/or recrystallized (e.g., [59]). The two phenomena—upwelling and phosphatogenesis—are probably responsible for the exceptional richness in marine vertebrate remains, both in terms of abundance, preservation and taxonomical diversity, which characterizes the Maastrichtian–Ypresian Phosphates of Morocco.

Stratigraphically, the Phosphates of Morocco range from the base of the Maastrichtian to the top of the Ypresian, spanning about 24 My without major unconformities [60,61]. As such, they represent the most time-expanded phosphatic deposits of the above-mentioned 'Phosphogenic Mediterranean Province' [51,52]. In all basins, the phosphate series include an alternation of soft and hard phosphatic beds (only the soft ones are exploited), marls and limestones, with frequent flint accumulations and chert levels. The series is usually topped by the '*Dalle à Thersitées*', a calcareous reference level (Figure 1E). Note that the different beds are named '*Sillon*' and '*Couche*' ('layer' in French, 'sillon' being a specific mining name for layer) and are numbered in Roman numbers in the Oulad Abdoun Basin and in Arabic ones in the Ganntour Basin, mainly for age differentiation. As an example, *Couche* 2 of the Ganntour Basin is late Maastrichtian in age, whereas *Couche* II of the Oulad Abdoun Basin is Paleocene. Due to subsidence, the thickness of the Maastrichtian series of central Morocco increases from NE to SW, being less than 10 m thick in the NE of the Oulad Abdoun Basin (Figure 1E), about 25 m in the Ganntour Basin (Figure 1F) and reaching 300 m in the Atlasic basins of Meskala and Souss [22,55].

Mosasaurid remains are very frequent in all levels of the Maastrichtian series, as isolated remains (mainly teeth and vertebrae) in levels *Couche* 6 to *Couche* 2 of the Ganntour Basin and as articulated, sometimes complete, specimens in lower and Upper *Couche* III of the Oulad Abdoun Basin.

3. Materials and Methods

The present work is based on the teeth of most of the mosasaurid species known in the Maastrichtian Phosphates of Morocco (Tables 1 and A1).

Each species is briefly introduced, with indication of its global spatiotemporal occurrences and a short discussion of its systematical validity and reassignment, if necessary. The main clades to which they belong are also briefly introduced, using main recent references (see in respective parts).

3.1. Material Acquisition and Selection

The tooth sample has been gathered thanks to several field campaigns of level-by-level systematic collection by some of the authors (NB, XPS) in the *Couche* 6 to *Couche* 2 (lower to uppermost Maastrichtian) of the Ganntour Basin [22] and to more than two decades of fieldwork by the authors in the *Couche* III (upper Maastrichtian) of the Oulad Abdoun Basin. This work was carried out into the framework of the long-term *PhosphaPal* French–Moroccan program of scientific collaboration and of the recent agreement between the University of Bath and the Cadi Ayyad University (see details in Acknowledgments). The specimens personally sampled, studied and analyzed by the authors are kept in the collections of the Office Chérifien des Phosphates (OCP), the Muséum d'Histoire Naturelle de Marrakech (MHNM, UCAM), the Muséum National d'Histoire Naturelle of Paris (MNHN) and the Université de Liège (ULg); some teeth are issued from private collections (controlled by the authors).

From more than a thousand isolated mosasaurid teeth collected, representing hundreds for the most common species, we selected for each taxon a representative 'Sample' of about ten well-preserved median marginal teeth, because they are both the reference ones for systematical identifications (e.g., [62]) and also those mainly used in food procurement (see rationale in [36]). This sample is used for the thorough anatomical descriptions. From this sample, in turn, one tooth is chosen as representing the 'Standard' (Figure 2) of each species and used for the morphometric analyses.

When available, the most complete and best-preserved skulls of each species are also measured to assess their general morphology and proportions, as well as to estimate the overall size of the taxon.

In order to make sensible faunal comparisons, only coeval taxa from the late Maastrichtian were included in the study (Table 1); as such, *Prognathodon giganteus*, from the lower Maastrichtian of the Ganntour Basin [22], was excluded. The few species not sampled nor described personally were included in the anatomical descriptions (using the original bibliography as a support) but excluded from the morphometric analyses see below Section 3.3.

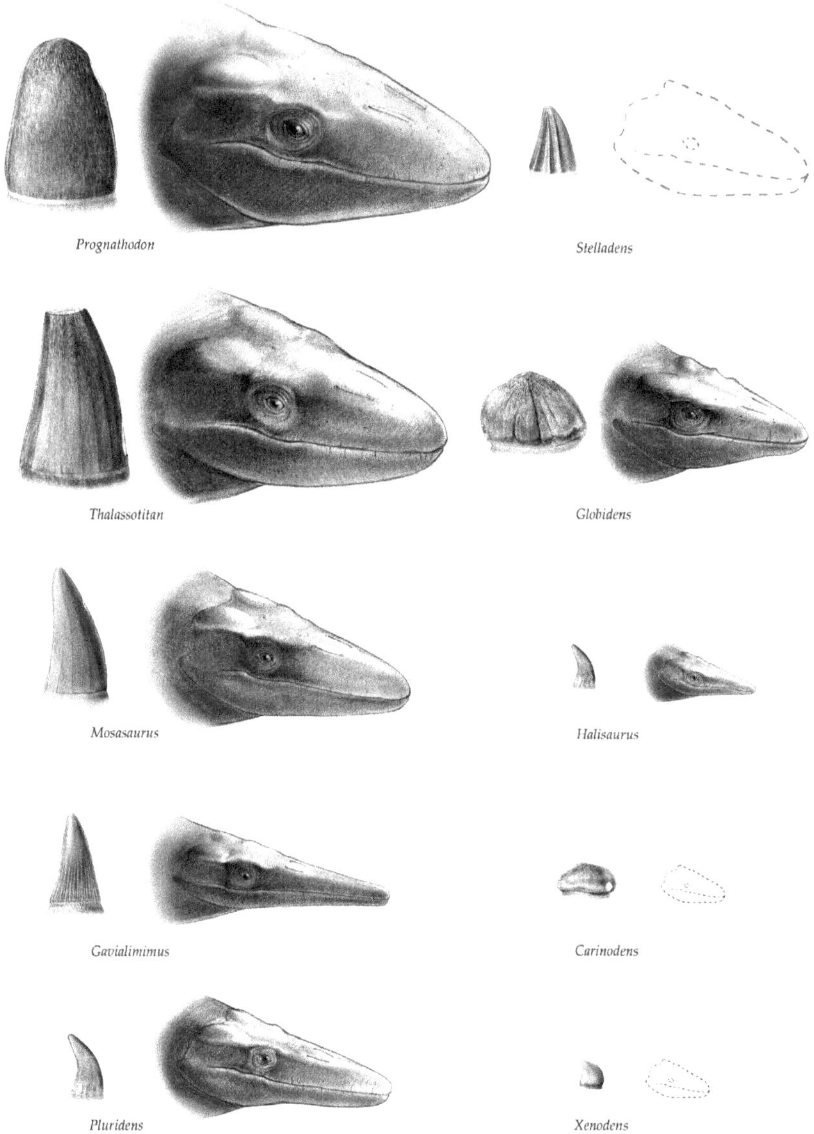

Figure 2. Mosasauridae from the Maastrichtian Phosphates of Morocco: head reconstructions and 'Standard' tooth drawings. The head reconstructions are on scale (with *Mosasaurus* = 1 m) and based on a selection of main representative species and most-complete specimens (see Tables A1 and A2). Living monitors, especially *Varanus niloticus*, a highly aquatic species, were chosen as a model. Skulls are deliberately reconstructed 'snout wide shut' in order to better appreciate the proportions of the main regions (jaw, orbit, temporal zones) and, above all, to highlight two constant features of extant squamates (assumed to be identical in mosasaurids, too often reconstructed archosaur-like): a high fleshy area above the gums, giving a thicker aspect to the jaws, and no teeth protruding from them. Tooth drawings on scale (with *Halisaurus* = 1 cm). Paleoartistic reconstructions, drawings and design © Charlène Letenneur (CR2P, MNHN, Paris).

Table 1. Mosasaurids from the Maastrichtian Phosphates of Morocco. Species known from the Upper *Couche* III (upper Maastrichtian) of the Oulad Abdoun Basin. See details in Table A1.

Mosasaurinae
Mosasaurus beaugei Arambourg, 1952 [4]
Carinodens belgicus (Woodward, 1891) [63]
Carinodens minalmamar Schulp et al., 2009 [33]
Xenodens calminechari Longrich et al., 2021a [25]
Eremiasaurus heterodontus LeBlanc et al., 2012 [23]
Thalassotitan atrox Longrich et al., 2022 [27]
Prognathodon currii Christiansen and Bonde, 2002 [64]
Globidens phosphaticus Bardet and Pereda-Suberbiola, 2005b [20]
Globidens simplex LeBlanc et al., 2019 [24]
Stelladens mysteriosus Longrich et al., 2023 [28]
Halisaurinae
Halisaurus arambourgi Bardet and Pereda-Suberbiola, 2005a [19]
Pluridens serpentis Longrich et al., 2021b [26]
Plioplatecarpinae
Gavialimimus ptychodon (Arambourg, 1952) [4,32]—**new combination**
Khinjaria acuta Longrich et al., 2024b [29]
Tylosaurinae
Hainosaurus boubker Rempert et al., 2022 [31]

3.2. *Comparative Anatomy and Morphofunctional Interpretation*

Teeth of the above-defined 'Sample' are thoroughly described for each species, using a suite of comparable and reproductible characters such as general size, shape of the crown, basal cross-section, curvature, ornamentation of the enamel, carinae occurrence and crenulations, wear facets, etc. This aims, firstly, to define the often unique tooth morphology of each species and, secondly, to qualitatively approach to which general dental guilds sensu Massare [38] it belongs.

Two ratios using the height (H) and width (W) compared to the length (L) of the crown are used (see Table A2): H/L ratio estimates whether the crown is high (H/L > 1), as high as long (H/L = 1) or low (H/L < 1); W/L ratio estimates whether the crown is strongly labiolingually compressed (W/L < 0.5), moderately compressed (W/L > 0.5) or not compressed (round section, W/L =1). Measurements were taken on the 'Sample' teeth, and the ratios correspond to the average value (or to a range of values if notable differences are observed).

Global characteristics of the entire dental series (including pterygoid teeth)—when preserved—are given, in order to estimate both the degree of homodonty/heterodonty and the respective size between marginal and pterygoid teeth. They have importance in seizing and manipulating prey.

The general morphology of the skull (or skull portions) is briefly described and is focused on characteristics allowing the inference of size, robustness and proportions of some key regions. These also have importance in predation and feeding processes. As for teeth, if several skulls are measured, the ratios correspond to the average value (or to a range of value if notable differences are observed). A paleoartistic fleshy reconstruction of the heads of main species based on close comparisons with extant varanids is proposed (Figure 2).

The following ratios were calculated, using skull length (measured from the premaxilla tip to a perpendicular line passing by the skull posterolateral corners (formed by squamosal–supratemporal)), as the reference measurement (see Table A2):

- W/L ratio (skull width/skull length): approximates robustness. Skull width is measured between postorbitofrontal/frontal lateral margins and/or the posterolateral corners of the skull. The higher the ratio value, the more robust the skull (and vice versa). It should be noted, however, that the width of a mosasaurid skull is difficult to access and often biased because of dorso-ventral crushing during taphonomical processes of burial and fossilization that artificially increases the real width. This ratio should thus be taken with caution. The external margins of the frontal are however considered as a relatively good proxy, as, this bone being flat, its natural morphology is de facto poorly affected by dorso-ventral crushing.
- PreO/L ratio (preorbital length/skull length): approximates snout elongation compared to skull length. Preorbital length is measured from the premaxilla tip to the anterior orbital margin. The higher the ratio value, the more elongated the snout in relation to skull length (and vice versa).
- PostO/L ratio (postorbital length/skull length): approximates adductor muscle volume. Postorbital length is measured from the posterior orbital margin (or, if not preserved, from a line passing by the frontal–parietal–postorbitofrontal suture (dorsally) or the postorbitofrontal–jugal suture (laterally)) to the perpendicular line passing by the skull posterolateral corners. The higher the ratio value, the more elongated the posterior part of the skull in relation to skull length (and vice versa).
- Orb/L ratio (orbit length/skull length): approximates the capability for vision in low-light or turbid environments. Measured in the median part of the orbit, from the posterior margin of the prefrontal to the anterior margin of the postorbitofrontal–jugal bar. The higher the ratio value, the larger the orbit (and vice versa).

The overall size (or size range) of each species is estimated, using as proxy the length of the skull or of isolated bones permitting the estimation of it (e.g., maxilla, dentary, posterior mandibular unit) (see Table A2).

A morphofunctional interpretation of the feeding process and inferred prey preferences, using Massare's dental guilds [38], is given for each species (Figure 3), to be complemented by quantitative analyses (see below).

3.3. *Morphometric Analyses*

We quantitatively analyzed the disparity of crown shapes in mosasaurids from the Phosphates of Morocco using two methods. The first one is two-dimensional and uses a modification of the Fourier transform to analyze shapes [65]. The second one is three-dimensional and uses surface semi-landmarks, in a geometric morphometrics framework.

Some mosasaurids species were excluded from the morphometric analyses because we did not have access to their teeth, either described by other authors (*H. boubker*, *G. simplex*) or very scarce and/or poorly preserved (*C. minalmamar*). In total, the teeth of eleven species on fifteen were analyzed (see Table A3). We also assembled a dataset of skull lengths (see Table A2).

3.3.1. 2D Shape Analyses Using Fourier Transform

We gathered photographs (or snapshots of 3D models in orthographic view) of teeth from the median part of the jaw. We used these images to draw the basal and the labiolingual outlines of the teeth, creating full, closed black masks over a white background. Each of these images was then imported and transformed as an outline in the R statistical environment v4.3.1 using the package Momocs v1.4.1 [65]. Each closed outline was centered and scaled. The teeth of *Globidens* and *Xenodens* are labiolingually longer than apicobasally high, resulting in erroneous alignments (and thus incorporation of a rotation factor in PC1) if we normalized Fourier coefficients. Therefore, all images were aligned prior to importation in R, and 'norm' was set to FALSE in the efourier function.

Then, a principal component analysis was run on the Fourier coefficients, yielding one morphospace of labiolingual shape and another of basal crown section shape. Because both analyses of shapes concentrated most of the variance (≈80%) in their respective first axes, we extracted these coordinates to produce a composite morphospace, where the x-axis is the PC1 of the labiolingual PCA, and the y-axis is the PC1 of the basal PCA. We also extracted theoretical shapes along each of these axes for visualization purposes.

Finally, we computed the density of morphospace occupation using a Kernel density estimator. We provide the 2D masks as .jpg files and the R script as Supplementary Files to fully replicate the procedure.

3.3.2. 3D Geometric Morphometric Analyses

For each tooth, we used the semi-automated high-density 3D geometric morphometric protocol established by Fischer and collaborators [36]. Essentially, this method pseudo-landmarks a 3D 'dome' shape (2000 surface semi-landmarks) and uses it as an atlas to patch these surface semi-landmarks onto each crown 3D model, using five fixed landmarks as anchor points.

First, we obtained 3D models from a previous study [36], which sampled particularly well-preserved crowns from the median part of the jaws. We then complemented this sample by digitizing the crowns of additional species, still focusing our efforts on the median part of the jaws, using a handheld laser scanner (Creaform Handyscan 300, 0.2 mm scan resolution, Creaform, Lewis, Canada). Then, we placed five fixed landmarks (one at the apex of the tooth and four at the base of the crown: labially, lingually, distally and mesially) on the dome and on each crown model, using Stratovan Checkpoint v20.10.13.0859.

The rest of the procedure was fully automatized by an R script, using the packages geomorph v4.0.6, Morpho v2.12 and MASS v7.3-60.0.1 for data treatment and the packages ggplot2 v3.4.4, plotly v4.10.3, ggrepel v0.9.4 for visualisations, all in R v4.3.1. Basically, the meshes and their coordinates were imported, and the atlas was created and patched onto each crown model. The resulting 3D coordinates of the crowns were subjected to a generalized Procrustes superimposition, then in turn subjected to a principal component analysis. We computed the density of morphospace occupation using a Kernel density estimator (see [36]). We provide the atlas, the 3D models, the fixed landmark coordinates and R script as Supplementary Files to fully replicate the procedure.

4. Results

4.1. *Comparative Anatomy and Morphofunctional Interpretation (Figures 2 and 3, Tables A1 and A2)*

4.1.1. Mosasaurinae

With around 11 genera and 35–40 species known worldwide, mosasaurines are the most diverse clade of mosasaurids. They range from the Turonian up to the end of the Maastrichtian and reach a worldwide distribution by Campanian–Maastrichtian time (e.g., [2,3,27,66,67]). Mosasaurinae include two clades: Mosasaurini as a sister group of an unamed clade (previously referred to as Globidensini (sensu [68,69]), including the tribes Prognathodontini and Globidensini [27]. As did previous analyses (e.g., [66,67]), this last work recognized *Mosasaurus* and *Prognathodon* as paraphyletic taxa. With a wide range of sizes (about 2–17 m), morphologies and tooth shapes, mosasaurines exhibit a much higher ecological plasticity than other mosasaurid subfamilies [3].

Eight genera and 10 species of Mosasaurinae are so far recognized in the upper Maastrichtian Phosphates of Morocco.

- Mosasaurini

Mosasaurus beaugei Arambourg, 1952 is based on an isolated large antero-median tooth crown (MNHN PMC 7) from the Maastrichtian of the Oulad Abdoun Basin ([4], p. 282, pl. 39, fig. 13). In Arambourg's time, only isolated teeth were known, coming from all the Maastrichtian phosphatic basins of Morocco [3,4,18,70]. In the last two decades, however, more complete specimens, including skulls and mandibles, have been unearthed in the Upper *Couche* III (upper Maastrichtian) of the Daoui area (Oulad Abdoun Basin) and described [18]. Additional teeth have also been collected in the *Couches* 3 and 2 (upper Maastrichtian) of the Ben Guerir area (Ganntour Basin) [22]. These new specimens permit to significantly improve our knowledge of this species, both anatomically (diagnosis emended) and stratigraphically (range precised); it is considered a valid taxon among Mosasaurini [18,71]. Compared to other taxa and despite its large size, *M. beaugei* remains relatively scarce in the Phosphates of Morocco, a point already mentioned by Arambourg [4]. This species has been found only in the upper Maastrichtian Phosphates of Morocco and contemporaneous formations of the Southern Margin of the Mediterranean Tethys/Southern Atlantic (Brazil, ?Egypt) [34].

Description—Median teeth are large (crowns can reach 4 cm in height), robust and faceted, with a sharp, pointed apex. They are almost twice as high as long (H/L ratio 1.8–1.9), slightly laterally compressed (W/L ratio 0.6–0.7) and weakly posteriorly recurved. The anterior surface is slightly convex and the posterior one slightly concave. Both bear over their entire height a sharp carina bearing minute serrations. The labial surface is almost flat and smaller than the strongly convex lingual one, resulting in a U-shape cross-section, which is characteristic of the genus *Mosasaurus*, sensu [62]. The labial surface bears 3–5 large facets (average number, 4) separated by sharp edges; the lingual surface bears 6–9 less-distinct ones (average number, 8). In both surfaces, most facets do not reach the apex; they are less numerous and discernable on the largest and posteriormost teeth. The enamel is thick but smooth, with a shiny appearance. Wear facets are rare and, when present, mostly occur along the carinae as very narrow elongated zones and at the apex as a small, rounded area.

The dentition is subhomodont along the jaws, exhibiting only the minor variations usual in mosasaurids: the median teeth are the largest; anterior ones are slender; posterior ones are lower and wider; maxillary teeth are slightly larger than dentary ones [18]. Pterygoid teeth are hook-like and much smaller than the marginal teeth (about one-third the height of median teeth); they are strongly posteriorly recurved, with a rounded cross-section and a finely ridged enamel [18].

The skull is 80 cm to 110 cm long. It is robustly built but long and relatively narrow, with its width almost one-fourth of the length, giving it a rather gracile appearance. The jaws are noteworthy in being long, more than one-half the skull length, straight and narrow. The postorbital part of the skull remains short, being only about one-fourth of the skull length; the orbit is relatively small. The body size of *M. beaugei* is estimated to be about 8–10 m long [3].

Paleoecology—The sharp general aspect of the teeth of *M. beaugei*, with a pointed apex, several marked cutting structures (carinae, large prisms separated by cutting edges) and a smooth enamel, coupled to subhomodont dentition and small pterygoid teeth, indicate that *M. beaugei* mainly performed 'cutting' sensu Massare [38]. However, from the near-lack of wear facets and breakages, as well as the elongate, narrow morphology of the skull and the short postorbital region, leverage was probably reduced, indicating low biting force. Food manipulation and processing were probably limited to a preliminary phase of seizing, followed by simple cutting and ingestion of small prey or large flesh pieces. Its diet probably consisted of relatively soft fleshy items, possibly cephalopods, large fish like *Enchodus* Agassiz, 1833–1845 [72] and small marine reptiles. Given its large size, estimated at 8–10 m, this species was one of the local apex predators but probably occupied a more distinct niche than *Thalassotitan atrox*, *Prognathodon currii* or *Hainosaurus boubker*.

Carinodens belgicus (Woodward, 1891) is a very rare mosasaurid, based on an incomplete toothed dentary (IRScNB R 43) from the upper Maastrichtian of the Maastricht area, Limburg, The Netherlands [73]. Since that time, scarce additional remains, mostly consisting of isolated teeth and a couple of dentaries, have been found in the upper Maastrichtian of Europe (including Russia), South America, the Middle East and Africa [21,33,69,74]. Despite its scarcity, *C. belgicus*'s paleobiogeographical distribution is extensive, including both Northern/Southern margins of the Mediterranean Tethys and Southern Atlantic; however, its stratigraphic range is limited to the upper Maastrichtian [34]. In the upper Maastrichtian Phosphates of Morocco, this species has been unearthed in the Upper *Couche* III of the Oulad Abdoun Basin as isolated teeth (Daoui) and dentaries (Sidi Chennane) and in the *Couche* 3 of the Ganntour Basin as isolated teeth (Ben Guerir) [21,22,33]. *Carinodens* Thurmond, 1969 [75] has been classically considered a member of Globidensini [33,69], but recent work suggests closer affinities with Mosasaurini [27].

Description—The crowns are subrectangular, half as high as long (H/L ratio about 0.5) and strongly laterally compressed (W/L ratio about 0.55). The base is swollen, with a strong constriction between the crown and the slender root, giving the teeth a mushroom aspect. Both labial and lingual surfaces are convex and subequal in size, so that the cross section is oval. The anterior surface of the crown is much larger than the posterior one, rising up almost vertically, then curving at an almost right angle to form a long horizontal 'plateau' up to the apex, that is, de facto posteriorly displaced from the middle of the crown. The posterior surface is slightly concave posterior to the apex, then strongly convex. There are two faint, curved, unserrated carinae developed only on the upper half of the crown. The enamel is thick and ornamented by coarse, anastomosed ridges that become larger and thicker on the upper third of the crown, especially near the apex. The apex usually bears a large and rounded wear facet. Worn areas are also frequent on the anteroventral corner of the labial surface and on the posteroventral corner of the lingual one.

The dental series of *C. belgicus* is strongly heterodont, both in morphology and size, with small, pointed, recurved anterior teeth; large, blunt median teeth; smaller and lower posterior teeth. Also noteworthy, with 17–18 versus 13 teeth (on holotype IRScNB R 43), the dentaries from Morocco show that this bone was longer than previously thought, which has implications for its possible bite force and inferred diet [33]. Pterygoid teeth remain unknown on *C. belgicus*, but *Carinodens palistinicus* Kaddumi, 2009 from Jordan exhibits pterygoid teeth [76], which are hook-like and smaller than the marginal teeth, as in most mosasaurids (e.g., [62]).

The dentaries are long, straight and slender; the skull was probably long and gracile, like that of *C. palistinicus* [76]. Based on the dentary size, *C. belgicus* skull length was probably less than 40 cm long [3], suggesting a total body length of about 2.5–3 m.

Paleoecology—Both the small size of *C. belgicus* and its highly heterodont and specialized dentition suggest that food processing was basically suitable for 'crushing' sensu Massare [38] but probably also included a preliminary gripping phase and a final one of finer crushing [33]. If pterygoid teeth are present in *C. belgicus*, like in *C. palistinicus*, where they are small hooks, they could have been used in helping the engulfment process. The inferred diet probably consisted of small mollusks and arthropods [77]; however, the re-estimation of the greater length of the dentary suggests that the biting force was probably less than expected, suggesting also that prey were probably softer (with thinner shells?) than previously thought [33].

Carinodens minalmamar Schulp et al., 2009 is based on two dentaries from the same individual (OCP DEK/GE 453)—one with two in situ teeth—coming from the upper Maastrichtian *Couche* III of probably Sidi Chennane, Oulad Abdoun Basin [33]. It should be noted that this species was first described in the Phosphates of Morocco by Arambourg ([4], pl. 40, fig. 3) on the basis of an isolated tooth, for a long time considered lost but recently found (MNHN PMC 29), from Oued Meskoura, north of El Borouj (Oulad Abdoun Basin). The tooth was first referred by [4] to *Globidens aegyptiacus* Zdansky, 1935 [78] (a species originally described in Egypt; see [20] for details), later reassigned to *C. belgicus* [3,20,21] and then to *C. minalmamar* [33]. Up to now, *C. minalmamar* is known only from the upper Maastrichtian of Morocco and Russia [33,74].

Description—The crowns are generally comparable to those of *C. belgicus* in being rectangular in lateral view, twice as long as high, basally swollen and with a finely anastomosed enamel and faint carinae. However, they differ in being lower (H/L ratio about 0.4) and much more compressed (W/L ratio less than 0.4, reaching 0.3 on MNHN PMC 29), and in bearing two pronounced vertical sulci on the median part of the labial surface, reminiscent of those found in *G. phosphaticus* Bardet and Pereda-Suberbiola, 2005b [20,33], as well as faint, straight carinae. On MNHN PMC 29, because of these sulci, the labial surface is slightly concave, whereas the lingual surface is gently convex. As in *C. belgicus*, rounded wear facets are located on the apex, as well as on the anteroventral corner of the labial surface and on the anterior extremity of the horizontal 'plateau' of the anterior surface.

As for *C. belgicus*, the teeth preserved on the holotype OCP DEK/GE 453 show that the dentition was also probably strongly heterodont.

The dentary is long, straight and narrow but smaller and slender than that of *C. belgicus*, indicating a more gracile taxon; also, the alveoli indicate that teeth were more imbricated along the dentary than in *C. belgicus* [33]. The skull length was possibly around 35 cm and the estimated total length around 2.5 m.

Paleoecology—Compared to *C. belgicus*, the smaller and slender dentary of *C. minalmamar*, as well as its more packed and laterally compressed teeth, indicate that its diet was slightly different from that of *C. belgicus*, possibly including smaller and/or softer prey [33].

Xenodens calminechari Longrich et al., 2021a is a very rare and small-sized mosasaurid, so far known only by a complete maxilla with four teeth (MHNM.KHB.331) [25]. The holotype was unearthed in the Upper *Couche* III (upper Maastrichtian) of Sidi Chennane, Oulad Abdoun Basin [25]. *Xenodens* is thus far known only known from the upper Maastrichtian Phosphates of Morocco. The unique tooth structure and implantation suggest affinities with *Carinodens* among Mosasaurini [25].

Description—The crowns are very distinct from those of any other mosasaurid. They are quadrangular and low, about as high as long (H/L ratio about 1), with a sharp hooked apex, and are very laterally compressed (W/L ratio about 0.4). The anterior surface is strongly convex and twice the size of the slightly-concave-to-straight posterior one. Both surfaces are almost flat, bearing only two subtle ridges and sharp unserrated carinae. The enamel is totally smooth and shiny; wear facets are absent. The teeth preserved are closely packed and laid obliquely to the jaw margin, whereas their roots are anteroposteriorly expanded and fused, forming a continuous ridge of bone. As a whole, the teeth and their root arrangement are remarkable in forming a single sawblade-shaped cutting edge, not only unique among mosasaurids but also among tetrapods [25].

The maxilla size and morphology suggest that *Xenodens* was a small, gracile species, with a skull length not exceeding 30 cm and a total size estimated around 2 m [25]. *Xenodens* is so far the smallest known mosasaurid from the Maastrichtian Phosphates of Morocco but also worldwide, being even smaller than some basal mosasaurids like *Tethysaurus* Bardet et al., 2003 [79].

Paleoecology—The flattened, bladelike and closely packed teeth of *Xenodens* have never been observed in any other mosasaurid nor any known reptile, but comparable dental apparati are known in dogfish sharks and in several bony fishes [25]. This highly specialized 'cutting' dentition—but of very small size and lacking wear facet, contrary to Massare's [38] characteristics of cutting dentition—indicates previously unknown feeding strategies, diet preferences and niche occupation. *Xenodens* probably used its sawblade battery to remove pieces of flesh from large prey during scavenging or predation, but its diet could also have included smaller prey like fish, crustaceans and cephalopods [25].

- Prognathodontini

Eremiasaurus heterodontus LeBlanc et al., 2012 is based on the syntypes OCP DEK/GE 112, a partial skeleton with precisely known spatiotemporal occurrences (Upper *Couche* III of Sidi Daoui, upper Maastrichtian, Oulad Abdoun Basin) and UALVP 51744, a subcomplete skeleton whose occurrence remains uncertain (probably *Couche* III of the Oulad Abdoun Basin) [23]. Among the isolated teeth that Arambourg [4] described indistinctly as *Mosasaurus* (*Leiodon*) cf. *anceps* (Owen, 1841) [80]—a taxon now considered a nomen dubium by [81]—the smaller and slenderer ones (MNHN PMC41, 42, 45, 48, 50, respectively pl. 38, figs. 1, 25, 8, 10 of [4]) match remarkably well with those of *Eremiasaurus* [23], while the others match with *Thalassotitan* [27] (see below). *Eremiasaurus* has been recovered in the whole Maastrichtian series (*Couche* 6 to *Couche* 2) of Ben Guerir, in the Ganntour Basin [22]. This species is rather common in the Maastrichtian Phosphates of Morocco, known by countless isolated teeth and some more complete remains. It is a typical taxon from the Southern Margin of the Mediterranean Tethys, having also been found in the Negev Desert and in Brazil [34]. *Eremiasaurus* has initially been considered a sister group of *Mosasaurus* Conybeare, 1822 [82] and *Plotosaurus* Camp, 1942 [83] among Mosasaurini [23], but it has also been recovered as a sister group of *Prognathodon kianda* Schulp et al., 2008 [81] within Prognathodontini [27].

Description—The crowns are high and robust pointed cones (H/L ratio about 1.9), only slightly laterally compressed (W/L around 0.8) and posteriorly recurved. The anterior surface is gently convex and the posterior one gently concave, both having aligned sharp carinae running along their height and bearing minute serrations. The apex is sharp and pointed. The labial and lingual surfaces are subequal and convex, resulting in an oval cross-section. The enamel is totally smooth, thin and shiny; faint traces of facets expanded along two-thirds of the crown height are sometimes present. Wear facets are rare and, when present, are reduced to a small rounded apical area and slender zones at the base of the anterior carina, mostly as in *M. beaugei*. As a whole, median teeth are blade-like.

The dentition exhibits pronounced heterodonty (hence its specific name) with straight and conical anterior teeth, large, blade-like; more compressed median teeth; strongly posteriorly recurved posterior teeth [23]. Also, the upper and lower teeth strongly interdigitate anteriorly, fitting into interdental pits on the jaws, but astonishingly become superimposed medially and posteriorly, the maxillary teeth masking the dentary ones [23]. Pterygoid teeth, though smaller than the median marginal ones, are large, approaching the size of the posterior marginal teeth; they are typically hook-like, rounded in cross-section and smooth [23].

The skull as a whole is robust but relatively narrow (width about one-fourth of the length), with almost straight and slender jaws giving it a relatively gracile appearance, reminiscent of *M. beaugei* and *P. kianda* [23]. The jaws are surprisingly gracile compared to the large teeth they housed: in UALVP 51744, the anterior teeth are nearly as high as the dentary, but this becomes less marked posteriorly as dentaries become deeper [23]. The preorbital portion is long, being half the total skull length, the postorbital region is rather short [3] and the orbits rather small. The syntype skulls are around 65–70 cm long, and the total body length of *Eremiasaurus* is estimated to be around 4.5–5 m [3,23].

Paleoecology—The bladelike and closely packed teeth of *Eremiasaurus* indicate that their main function was probably 'cutting' sensu Massare [38], which is coherent with its long and narrow, but robust skull. However, teeth are almost devoid of wear facets and breakages, indicating probably relatively soft prey. The estimated body length (around 5 m) of *Eremiasaurus* suggests it was a meso-predator in the trophic network of the Maastrichtian Phosphates of Morocco. It occupied a distinct niche from ecologically distinct predators of the same size (*Globidens*, *Pluridens*) and was rather comparable in shape to the apex predator *M. beaugei*. However, its smaller size suggests that it probably preyed on different items, such as smaller fish, cephalopods or small marine reptiles. Noteworthy, the increase in the number of pygal vertebrae suggest that this was probably a high-speed pursuit predator, comparable to *Plotosaurus* [23,84].

Thalassotitan atrox Longrich et al., 2022 is based on the complementary syntypes OCP DEK/GE 417, a partial skeleton (crushed skull and mandible, cervical and dorsal vertebrae, girdle and limb elements) and MHNM.KHB.231, a well-preserved complete and articulated mandible (plus some dorsal vertebrae), both unearthed in the upper Maastrichtian Upper *Couche* III of Sidi Daoui, Oulad Abdoun Basin [27]. It should be noted that: (1) this taxon has for a long time been called *Prognathodon* nov. sp. (e.g., [3,5,34]); (2) as previously mentioned for *Eremiasaurus*, among the isolated teeth that Arambourg ([4], pl. 38) described indistinctly as *Mosasaurus* (*Leiodon*) cf. *anceps*, the largest and stoutest ones, namely MNHN PMC 43, 44, 46, 47, 49, 51 (respectively figs. 3, 4, 6, 7, 9 and 11 of [4]) most probably belong to *Thalassotitan* [27]. *Thalassotitan* is a very common taxon in the Phosphates of Morocco, known both by countless isolated teeth and several more complete specimens [27]. Articulated skeletons come from the Upper *Couche* III of most areas of the Oulad Abdoun Basin (Sidi Daoui, Meraa Lharach, Sidi Chennane) [27], whereas isolated teeth come from the same areas as well as from the *Couches* 3 and 2 of the Ben Guerir area, in the Ganntour Basin [22]; its stratigraphical range appears thus limited to the upper Maastrichtian. *Thalassotitan* is a characteristic taxon from the Southern Margin of the Mediterranean Tethys, having been found in Jordan, Negev, Egypt, Angola and Brazil, but possibly also in Poland, underlining a broader paleobiogeographical distribution [34]. Among Prognathodontini mosasaurines, *Thalassotitan* is considered a close relative of *Prognathodon currii* Christiansen and Bonde, 2002 [64] and *Prognathodon saturator* Dortangs et al., 2002 [27,85].

Description—Teeth are very large and robust (crowns can reach 5–6 cm high and 3–4 cm long). Median crowns are conical and massive, of medium height (H/L ratio 1.5 to 1.8). The basal cross-section is ovoid, almost rounded (W/L about 0.8–0.9) and slightly swollen, as in *P. currii*. The apex is moderately pointed and very often broken. The crowns are weakly posteriorly recurved, with a slightly convex anterior surface becoming more convex in its upper half, and a nearly straight posterior surface. Both bear a marked carina with fine denticulations, which are notably 'pinched' from the shaft of the crown, creating a strong cutting edge. This 'pinching' is more pronounced in the upper half of the carinae, making them protrude more than their lower part, giving the crown this particular upper inflated shape in lateral view. The labial and lingual surfaces are equal in size and regularly convex. They are smooth, lacking facets or prisms, except from some few indistinct vertical ridges present variably on the lingual surface in largest specimens. The enamel is thick and smooth, with a silky aspect, except apically, where it is coarsely anastomosed. In complete teeth, the wrinkles are reduced to the upper third of the crown but are not as pronounced as in *P. currii*. One of the main characteristics of both the marginal and pterygoid teeth of *Thalassotitan* is the presence of heavy spalling, breakage and wear facets, to a degree (both in frequency and extent) not found in any other mosasaurid [27]. Apical breakage is large and irregular, often removing up to one-fifth of the crown height and seem to result from violent impacts due to strong bite forces; sometimes, the broken apex is also worn by use, indicating that despite being damaged, the tooth was still functional. Large wear facets also occur along the carinae and on the lingual surface as large zones exposing dentine.

The dentition as a whole exhibits a developed heterodonty, with anterior teeth narrow, high and basally circular; median teeth conical, being the largest and most robust; posterior teeth smaller, as broad as tall, laterally compressed and posteriorly recurved. The dentary teeth are slightly more laterally compressed than the maxillary teeth. The pterygoid teeth are astonishingly very large and robust, about the same size and shape as the marginal ones, like in other prognathodontins and *Plesiotylosaurus* Camp, 1942 [27,83].

The skull is overall massively built and wide (W/L ratio about 0.38, the largest with that of *P. currii*), remarkably akinetic, with short and very robust jaws, a short postorbital region but a relatively large orbit. The dentary is short, deep and bowed, with a reduced tooth count compared to other mosasaurids [27]. The mandible mirrors the skull in being short, high and very massively built but with a still-functional intramandibular joint [27]. The skull of this species is one of the largest of the Morocco phosphatic fauna, being 1.20–1.30 m long, for a total body length estimated at 9–10 m [27].

Paleoecology—The large, short, conical teeth of *Thalassotitan*, ornamented with thick enamel and exhibiting numerous and large breakage/wear zones, indicate they were specialized to resist large forces when handling, biting and shredding large bony prey [27]. They range into the 'crush–cut' guild of [3]. The akinetic, short, massive and robust skull combined with the short, massive but still kinetic mandible increased biting force and the general withstanding of the skull, permitting the ingestion of large prey items. The reduction in tooth number observed in *Thalassotitan* also occurs in carnivorous apex predators like extant orcas, the extinct whale *Livyatan* Lambert et al., 2010 [86] and the theropod dinosaur *Tyrannosaurus* Osborn, 1905 [27,87]. The ecology of *Thalassotitan* was probably comparable to that of the extant white shark and killer whale [27]. All these characteristics suggest that *Thalassotitan* was probably the largest apex predator of the Maastrichtian Phosphates of Morocco, highly adapted for carnivory and probably preying on any of the abundant regional marine vertebrate faunas, including large sharks, large bony fish like *Enchodus* and other marine reptiles [27].

Prognathodon currii Christiansen and Bonde, 2002 is based on a complete skull and some vertebrae (HUJ.OR 100) found in the Main Phosphorite bed of the Mishash Formation, Oron phosphatic mine, Negev Desert [64]. Long dated as late Campanian-early Maastrichtian, the age of the Main Phosphorite bed has been reevaluated to early Maastrichtian on the basis of selachian faunas [88]. In Morocco, *P. currii* is very rare and known up to now only by a dozen isolated teeth found in the *Couche* 6 to *Couche* 2—i.e., along the Maastrichtian stage—of Ben Guerir, Ganntour Basin [20,22]. This species is however typical of the Southern Mediterranean Tethys Margin, and, in addition to Morocco and Negev, it possibly also occurs in the Maastrichtian of Angola [34]. Among Prognathodontini, *P. currii* is considered a sister group of *P. saturator* and *Thalassotitan* [27].

Description—The teeth are unique among mosasaurids in being straight and robust cones. They are large (crowns about 4–5 cm high), relatively high (H/L ratio about 1.3) and weakly laterally compressed (W/L ratio 0.77) with a generally blunt apex. The anterior and posterior surfaces are equal in size, straight and parallel along their height except when they converge to the apical nubbin. Anterior and posterior strong serrated carinae extend over the whole height of the crown and are more pronounced in its apical half because they are 'pinched' from the shaft of the crown, as occurs in *Thalassotitan*. The labial and lingual surfaces are also equal, being regularly convex, giving the crown an ovoid to almost rounded cross section. The enamel is very thick and anastomosed, this texture being more marked on the upper half of the crown. The apex is rounded. Wear facets are frequent, located mainly at the apex and along the carinae as large rounded or elongated ovoid zones, exposing the underlying dentine.

On the holotype, the preserved marginal teeth are homogenous in size and shape, indicating that the dentition was probably subhomodont, with only the small variations expected for the mosasaurid marginal series. The dentition is reduced (12 dentary teeth). Pterygoid teeth are very large, heavy and markedly recurved teeth, subequal in size to the marginal teeth [64].

The holotype skull of *P. currii* is probably the largest mosasaurid skull ever found (L = 1.40 m; W/L about 0.4). It is very robustly built and looks mostly akinetic. The jaws are massive, and the dentary is deep and bowed, relatively short but not as much as in *Thalassotitan*. The postorbital region is also relatively short, and the orbit very small. The size of the animal is estimated to be 10 m or more.

Paleoecology—The unique teeth of *P. currii* range into the 'crunch' guild of Massare [38]. These high, robust, anastomosed and straight cones are reminiscent of those of the large teleosauroid thalattosuchian *Machimosaurus* Meyer, 1837 [89,90]. The very robustly constructed skull and jaws of *P. currii*, coupled with these resistant, massive, conical, blunt teeth, indicate a powerful, crushing bite. Its large size makes it one of the apex predators of the Maastrichtian Phosphates of Morocco. It was probably adapted for predominately hunting large, bony prey, including large sharks, fish and marine reptiles, occupying as such an ecological niche close to that of *Thalassotitan*, but, like *Machimosaurus*, probably with a component of armored prey in the diet, such as large ammonites and turtles [91].

- Globidensini

Globidens phosphaticus Bardet and Pereda-Suberbiola, 2005b is based on thirteen isolated teeth representing a hypothetical dental series, including holotype (OCP DEK/GE 361, OCP DEK-GE 338-343) and referred specimens (OCP DEK/GE 344-348, MNHN PMC 17) [20]. Most of the teeth come from the *Couche* 3 (and some from the *Couches* 6, 4 and 2) of Ben Guerir (Ganntour Basin), ranging thus from the lower to upper Maastrichtian [20,22]. This species was firstly described in the Phosphates of Morocco as *Globidens aegyptiacus* by Arambourg, on the basis of isolated teeth (MNHN PMC 17, 18, 19) from the *Couche* III (upper Maastrichtian) of several localities of the Oulad Abdoun Basin ([4], pl. 40, figs. 1, 2). More recently, cranial and axial elements from the Maastrichtian of Angola have significantly improved our knowledge of this species and confirm its validity [92]. *G. phosphaticus* was a typical tropical to subtropical (10° N–25° S paleolatitudes) species from the Southern Margin of the Mediterranean Tethys (North Africa, Middle East) and South Atlantic (Brazil, Angola) [20,34,92].

Description—Median teeth are large, inflated and low compared to other mosasaurids except *Carinodens*, giving them a bulbous appearance. They are roughly as high as long (H/L ratio 1 to 0.7) and labiolingually expanded in an irregular fashion (W/L ratio 0.8 anteriorly to 0.65 posteriorly). The apex is large, rounded and usually worn. It lies near the middle of the crown or is slightly posteriorly displaced, and is slightly posteriorly and—very unusually—labially (and not lingually, as usual in mosasaurids) oriented. In apical view, the crown has a very irregularly oval cross section, being divided into a large inflated anterior surface and a smaller posterior one [20]. In lateral view, the anterior surface is regularly convex, the posterior one being slightly concave just posterior to the apex (reminiscent of what is observed in *Carinodens*), then strongly convex. Both surfaces are totally devoid of carinae (very discrete ones are still present on anteriormost teeth). The crown is typically swollen at the base, as in other Globidensini, giving it a mushroom-like shape. Both labial and lingual surfaces bear unique deep vertical sulci (average number 2–3) in their median part, similar to those observed in *C. minalmamar*. These sulci deeply notch the labial surface, resulting in its irregular shape in apical view (inflated—concave—inflated). The lingual sulci are less marked, so that this surface is roughly convex. The enamel is thick and ornamented by crude anastomosed ridges that become larger (and the enamel thicker) from the upper half of the crown up to the apex. Wear facets are frequent and usually round to oval large zones, located mainly at the apex but also irregularly on labial and lingual surfaces.

The dentition of *G. phosphaticus* is strongly heterodont, with subconical anterior teeth being higher than long (H/L ratio about 1.5), posteriorly recurved and prognathous; median teeth large, bulbous and lower (1.5 < H/L ratio < 1); posterior teeth large, bulbous and very low (H/L ratio around 0.5). The largest teeth are probably those occupying the 6th to 8th position of the dentary [20,92]. There is so far no evidence of pterygoid teeth in *Globidens* Gilmore, 1912 [93], a unique characteristic in mosasaurids [62]. Interestingly, as noted by [92], most of the teeth of the specimen from Angola were found complete (crowns and attached roots) but displaced in, or out of their respective sockets.

The cranial elements of *G. phosphaticus* (from Angola) show that the skull was short, robust and probably wide (according to frontal width), with robust bowed dentary about one-half the total length, being unique among mosasaurids in being shorter than the posterior mandibular unit [24]. The orbit was probably relatively small (estimated from coronoid dorsal curvature) and the postorbital portion was of medium size. The skull length is estimated to be about 75–80 cm, and the total body size to be 5–6 m.

Paleoecology—The highly specialized dentition of *G. phosphaticus*, composed of low and bulbous teeth with thick anastomosed and often worn enamel, indicates that its main function was 'crushing' sensu Massare [38]. The conical anterior teeth were probably used in grasping prey. The lack of pterygoid teeth suggests that the prey were ground into small pieces directly by the powerful crushing battery, as such not requiring prey to be held before being engulfed. Most of the teeth (seen on the Angola specimen) have been found complete and preserved out of the jaws, which suggests that the periodontal ligaments did not mineralize in life [94]. This is interpreted as a possible adaptation to increased loading due to durophagous habits [92] (see also below for *G. simplex*). The robustness and shortness of the skull (known from the Angola specimen) imply that bite force was probably high.

All these characteristics suggest a diet based on rather small but hard prey such as small-shelled cephalopods, echinoderms, large bivalves, etc., a view confirmed by stomach contents preserved in other *Globidens* species [24]. However, compared to them, especially the coeval *G. simplex*, the highly heterodont dentition and irregular shape of the median teeth of *G. phosphaticus* indicate a specific, slightly different durophagous diet.

Globidens simplex LeBlanc et al., 2019 is based on a partial disarticulated but three-dimensionally preserved skull and associated cervical vertebrae (formerly UALVP 51746, now MHNM.KHB.221) [24]. Though its exact provenance remains unknown, the rock matrix suggests it was probably unearthed from the Upper *Couche* III of the Oulad Abdoun Basin [24].

Description—Median teeth are typically low and bulbous (H/L ratio 0.65 to 0.77) and slightly labiolingually compressed (W/L ratio 0.73 to 0.82), giving them an ovoid cross-section. This compression is more marked in the posterior teeth. The crown is broadly triangular in lateral view, with symmetrical, slightly convex to almost straight anterior and posterior surfaces converging to the apex. The apex is rounded, not recurved and often worn. The base is strongly swollen, giving the teeth a mushroom aspect. The enamel is thick and strongly anastomosed. There are no carinae or sulci [24].

The dental series of *G. simplex* is strongly heterodont, with anterior teeth conical, rounded, higher than long and posteriorly recurved; median teeth about as high than long, bulbous and slightly compressed labiolingually; and posterior teeth bulbous, low (height about half the length) and more compressed. The general shape of the teeth is more regular than in *G. phosphaticus*, being globally ovoid and without carinae or sulci. As in *G. phosphaticus*, most of the teeth of *G. simplex* are preserved complete (crowns and roots) and separated from the jaws [24].

The skull is overall short and robust, completely akinetic and probably wide. Unlike *G. phosphaticus* (and other *Globidens* species), the dentary is very straight and relatively slender, but, as in *G. phosphaticus*, it is shorter than the posterior mandibular unit [24]. As in *G. phosphaticus*, the orbit is relatively small and the postorbital portion of medium size. The skull length is estimated to be about 75–80 cm, for a total body size of about 5–6 m.

Paleoecology—The dentition and skull morphology of *G. simplex* clearly indicate a durophagous species ranging into the 'crush' guild of Massare [38], which probably fed on shelled invertebrates (mollusks, echinoderms). Unlike most mosasaurids (including the Campanian species of *Globidens*) and squamates in general, whose teeth are firmly ankylosed by their roots to the jaws, most of the teeth of *G. simplex*, as in *G. phosphaticus*, are preserved complete and out of the jaws, indicating a dental gomphosis (ligamentous tooth attachment, as in archosaurs and mammals) [24]. This unique dental attachment shared only by the two Maastrichtian species of *Globidens* is interpreted as an adaptative response for resisting and absorbing shocks when biting [24,92,94]. A dentary shorter than the posterior mandibular unit is also interpreted as an adaptation to increase the mechanical advantage for shell-crushing [24]. The akinetic skull as well as the straight and slender dentary of *G. simplex* are however unusual for a durophagous species, as all extant shell-crushing squamates possess kinetic skulls and a bowed dentary [24]. *G. simplex* exhibits the same size range as *G. phosphaticus* but differs in having a different dentition and a slenderer/straighter dentary; this suggests that both species had a slightly different diet and niche.

- Mosasaurinae incertae sedis

Stelladens mysteriosus Longrich et al., 2023 is a very scarce taxon based on a dentary fragment and two associated teeth (MHNM.KHB.1436) found in the *Couche* III (upper Maastrichtian) of Sidi Chennane, Oulad Abdoun Basin [28]. So far, *Stelladens* is only known from the Maastrichtian Phosphates of Morocco. The global morphology of the teeth and dentary suggest mosasaurine affinities [28].

Description—The crowns are triangular and rather low (L/H ratio about 1.3–1.4), weakly laterally compressed (W/L ratio about 0.7–0.8) and posteriorly recurved, with a convex anterior surface and a slightly concave posterior one. The apex is pointed. There are two prominent sharp and serrated carinae, the posterior one being stronger and more 'pinched' than the anterior one. The enamel is ornamented with minute anastomosing ridges, giving it a silky aspect. Wear facets are present at the apex of the teeth and on the posterior carina as large but narrow zones. The labial surface of the teeth is almost flat, and the lingual surface is strongly convex, giving the crown a U-shaped section typical of many mosasaurines. Whereas the labial surface bears five to eight poorly differentiated ridges, the lingual surface bears two to four large protruding cutting ridges, separated by deep, V-shaped grooves. These large ridges are unique among mosasaurids (and not pathological, as they are present on all teeth), giving the lingual surface a zig-zag aspect, and the crown in dorsal view the shape of the tip of a Phillips screwdriver. Another remarkable and unique among mosasaurids fact is that these ridges are serrated, like the carinae. The serrations extend full length on the anterior ridges, but only apically on the posterior ones.

The relative size and robustness of the dentary and associated teeth suggest *S. mysteriosus* was a medium-sized species, with a skull length estimated at about 80 cm and a total body length at about 5 m [28].

Paleoecology—The presence of prominent sharp serrated ridges is not only entirely unique among mosasaurids but also has no extant or extinct analogues, indicating a highly specialized and previously unknown diet [28]. This battery of large, sharp and serrated carinae (main carinae and accessory ridges acting as carinae) coupled with relatively low crowns bearing a subtle anastomosed enamel and large wear facets suggests that the teeth were probably used for both cutting and crushing, but also holding the prey, being as such globally generalists. *Stelladens'* diet probably included semi-hard prey, either bony (medium size fish) or with a thin carapax/shell (crustaceans, cephalopods) [28].

4.1.2. Halisaurinae

Halisaurines are basal mosasaurids known by only four genera and about ten species, with a stratigraphic range from the Coniacian to the end of the Maastrichtian and a worldwide distribution (e.g., [2,3,19,26,66,67,95,96]). During the Maastrichtian, the clade exhibits its greatest diversity and distribution, being common in subtropical assemblages of Africa and Middle East but remaining scarcer in higher latitudes (North America, Europe, Japan) [26,95,96]. Halisaurines include two tribes with very different morphology and sizes and therefore different habits and niches: the gracile and small Halisaurini (around 3 m in body length) and the large and robust Pluridensini (up to 8 m) [26].

Both Halisaurini and Pluridensini are present in the Phosphates of Morocco and are represented by one species each. As previously pointed out [11], despite the great abundance of halisaurine teeth in all Maastrichtian phosphatic deposits of Morocco, Arambourg [4], who collected in these outcrops hundreds of mosasaurid teeth (including very small ones), surprisingly did not mention or even collect a single tooth belonging to this taxon, nor mistake them with fish ones (NB. pers. obs. on the Arambourg collection; MNHN, Paris); the reason for this collecting bias remains unknown.

- Halisaurini

Halisaurus arambourgi Bardet and Pereda-Suberbiola, 2005a is based on a partial skeleton (MNHN PMC 14), including a skull, mandible and part of a vertebral column, found in the Upper *Couche* III (upper Maastrichtian) of Sidi Daoui, Oulad Abdoun Basin [19]. It is among the most common species in the Phosphates of Morocco, being known by countless isolated teeth in all phosphatic basins and by articulated skeletons in the Oulad Abdoun Basin [19]. The species ranges from *Couche* 6 to *Couche* 2 of the Ben Guerir area (Ganntour Basin), thus spanning the Maastrichtian stage [22]. It is a typical species of the Southern Margin of the Mediterranean Tethys, being known in the Maastrichtian of Syria, Jordan, Negev and Angola [34]. Among Halisaurini, *H. arambourgi* is recovered as a sister taxon of *Halisaurus platypsondylus* Marsh, 1869 [97] from the Maastrichtian of North America [26,96].

Description—The crowns are small (height 0.5 to 1 cm maximum), delicate, hook-like, and have a sharply pointed apex. They are high (H/L ratio 2, the highest in proportion of all the Phosphate mosasaurids) and weakly labiolingually compressed, being ovoid to rounded in cross section (W/L around 0.8). The anterior surface is strongly convex and the posterior one strongly concave; at about mid-height, both surfaces are abruptly recurved posteriorly, giving the crown its typical hooked shape. Both anterior and posterior surfaces bear a small unserrated carina, present only on the upper two-thirds of the crown and more marked above the median-height abrupt curvature. The anterior carina is more marked than the posterior one, which is slightly displaced labially. Both labial and lingual surfaces are regularly convex, but the labial one is slightly flatter. The enamel is ornamented by minute striae, giving it a silky aspect. These striae are present on two-thirds of the crown and are more marked and slightly anastomosed around mid-height. The upper third of the enamel crown is smooth and shiny. In the lower third of the crown, some underlying 'folds' or flutings, variable in number, position and extension, are present on some teeth. Wear facets are lacking and, when present, reduced to a tiny apical rounded area.

The dentition of *H. arambourgi* is homodont, comprised of numerous teeth (16 Mx, 19 De, 12 Pt) and poorly differentiated both in shape and size along the jaws [19]. Anteriormost and posteriormost teeth are slightly smaller than median teeth, but their shape is similar; the position of isolated teeth on the jaws is thus difficult to assess. The pterygoid teeth differ from the marginals only in being about half as high and even more hooked.

The skull of *H. arambourgi* is long and slender, gracile, with long, straight jaws occupying one-half of the total skull length, a short postorbital region and large orbits (about one-fifth of the total length). It is also highly kinetic, skulls being often found completely disarticulated. Its average size is around 35–40 cm long for a total observed body length (on OCP specimens, NB, pers. obs.) of 3–4 m [19].

Paleoecology—*H. arambourgi*, with numerous small, hook-like teeth devoid of wear facets, is clearly placed into the 'piercing I' guild of Massare [38]. The teeth resemble those of *Plioplatecarpus* (Dollo, 1882) [98], although they are much slender, and also those of some plesiosauroids (e.g., [99]). This suggest a diet based on soft and small prey (cephalopods, fish), probably harpooned and captured in the trap made by these numerous, needle-sharp teeth, before being swallowed whole. Like other Halisaurini, the large orbits (and supposed eyes) of *H. arambourgi* and a degree of binocularity suggest a visual hability for detecting prey in low-light conditions, either at night or at depth [26,95]. However, *Halisaurus* rarely exhibits the avascular necrosis of bone tissue common to deep divers [100], suggesting open sea nocturnal habits such as looking for phosphorescent cephalopods, as in *Phosphorosaurus* Dollo, 1889b [95], rather than deep diving habits [101].

- Pluridensini

Pluridens serpentis Longrich et al., 2021b is based on the complementary syntypes OCP DEK/GE 548 (complete skull) and MHNM.KHB.262 (complete skull with articulated mandible and some cervical vertebrae), both from the upper Maastrichtian Upper *Couche* III of Sidi Daoui, Oulad Abdoun Basin [26]. Compared to other mosasaurids, this species remains uncommon, and articulated material has primarily been found so far in the Daoui area (Oulad Abdoun Basin), though isolated teeth are known in all basins. The teeth of *P. serpentis* being indistinguishable from those of *H. arambourgi* except being twice their size (NB, pers. obs.), large teeth interpretated previously as belonging to *H. arambourgi* [19,22] may therefore belong to *Pluridens*. As such, both its stratigraphical range (lower to uppermost Maastrichtian) and paleobiogeographical distribution (Southern Margin of the Mediterranean Tethys) could be the same that of *H. arambourgi* [34]. Among Pluridensini,

P. serpentis is considered a sister group of the two *Pluridens* Lingham-Soliar, 1998 [102] species from the Maastrichtian Iullemmeden Basin of Niger and Nigeria [26,102].

Description—The teeth of *P. serpentis* are identical to those of *H. arambourgi*; see above for detailed description. The only notable difference is that, because they are twice the size (1.5 cm to 2 cm high), features such as carinae, striations, basal flutings, etc. appear more marked/visible than on *H. arambourgi*.

As in *H. arambourgi*, the dentition of *P. serpentis* is characterized by a high tooth count (18 Mx, 26–28 De) and is remarkably homodont. However, it differs in that tooth roots are obliquely (and not vertically) oriented with respect to the long jaw axis, and in that replacement pits are positioned medianly (and not posteriorly) on the roots; both are autapomorphies of *Pluridens* [26].

Despite similar tooth shape, increase of tooth count and homodont dentition, the *P. serpentis* skull Bauplan differs drastically from the small, delicate one of *H. arambourgi*. It is overall elongated but massively built, especially in its anterior two thirds. The jaws are almost half the skull length, very robust and deep, contrasting with the small and delicate teeth they bear, and have numerous neurovascular foramina. The orbits are small (Orb/L ratio 0.12, the smallest of all Phosphate mosasaurids), with a large dorsal prefrontal–postorbitofrontal contact, forming a kind of reinforced protective flange. The postorbital part of the skull is, on the contrary, rather long and slender, about one-third of the skull length. The skull is large, 70 to 90 cm long, and the estimated body size is 5–6 m [26].

Paleoecology—*Pluridens* is unusual among mosasaurids in combining a large and robust skull—rivalling other coeval meso-predators like *Eremiasaurus* or *Gavialimimus*—with a dentition indicating as a whole 'piercing I' habits sensu Massare [38] and a diet probably composed of small prey like fish and cephalopods [26]. However, these teeth being twice the size of those of *H. arambourgi*, the prey were probably larger (though still soft).

Pluridens is also characterizsed by small orbits, reinforced dorsally by a rim of bone, and by numerous neurovascular foramina on the jaws. These characteristics suggest both a possible adaptation to diving and a feeding strategy probably relying on chemo- and mechanoreception rather than on visual cues [26]. If *Halisaurus* (large orbits) and *Pluridens* (small orbits) were probably adapted to prey on dark waters and to feed in small soft items, their hunting strategies were probably quite different, revealing very different niche occupations: *Halisaurus* was probably a visual hunter, preying at night in open sea (see above), whereas *Pluridens* was probably a chemo-tactile predator, looking for hidden prey at depth, in burrows and crevices [26,96].

4.1.3. Plioplatecarpinae

With about 14 genera known from the Turonian to the end of the Maastrichtian, plioplatecarpines were diversified and widespread mosasaurids (e.g., [2,3,29,66,67,103]). They reach a maximum diversity during the Campanian and achieved a worldwide distribution during the Maastrichtian, where they are known in Europe, Africa, the Middle East, North and South America and Antarctica [29,104]. Plioplatecarpinae include several basal taxa and two main tribes: the small to medium-sized (3–6 m) Plioplatecarpini of 'piercing' type, distributed worldwide in medium-high paleolatitudes; and the larger (around 8 m), more robust Selmasaurini of 'cutting' type, found in low paleolatitudes of both the Western Interior Seaway (North America) and the Arabo-African platform (northwestern Africa and the Middle East) [29].

In the Phosphates of Morocco, plioplatecarpines are so far represented only by Selmasaurini, which are known by two genera.

- Selmasaurini

Gavialimimus ptychodon (Arambourg, 1952) **new combination** (see below) was originally described as *Platecarpus* (?) *ptychodon* by [4]. The species was based on an isolated median tooth (MNHN PMC 30) from the Maastrichtian *Couche* III of Sidi Daoui, Oulad Abdoun Basin ([4], pl. 39, fig. 2). Several isolated teeth and some caudal vertebrae were also attributed to this species ([4], pl. 39, figs. 1, 3–7, 12). Both the teeth bearing ridges and striae and the vertebrae with unfused chevrons are russellosaurine synapomorphies [68,105]. The discovery in the last decade of more complete specimens in both Morocco and Angola confirmed it was a new plioplatecarpine genus [106], as glimpsed by Arambourg [4]. Then, the new genus and species *Gavialimimus almaghribensis* Strong et al., 2020 was described, on the basis of a complete skull and mandible with teeth (MHNM.KHG.1231) from the Maastrichtian Phosphates of Morocco [32]. If the creation of the genus *Gavialimimus* is not in doubt, the proposal of the new species name *almaghribensis* to replace Arambourg's original one *ptychodon* is here challenged (see below). Teeth of this species are very abundant in the Maastrichtian Phosphates of Morocco, having been unearthed both in the *Couche* III of the Oulad Abdoun Basin and in the whole Maastrichtian series (*Couche* 6 to *Couche* 2) of the Ganntour Basin [22]. *G. ptychodon* is a characteristic taxon of the Southern Margin of the Mediterranean Tethys and of the Southern Atlantic, having been found in the Maastrichtian of the Middle East (Syria, Jordan, Negev) and northwestern Africa (Morocco, Angola) [34,106]. It represents one of the few plioplatecarpines known from the Arabo-African platforms. Among plioplatecarpines, *Gavialimimus* has been recovered as a Selmasaurini, a sister taxon of either *Selmasaurus* Wright and Shannon, 1988 [32,107] or *Goronyosaurus* Azzaroli et al., 1972 [108] and *Khinjaria* [29].

Description and emended diagnosis (for teeth and dentition)—Median teeth are of medium size (1 to 2.5 cm high), triangular in shape, with a relatively long base and medium height (H/L ratio around 1.5), and bear a sharp and pointed apex. They are distinctly labiolingually compressed (W/L ratio around 0.7), with a convex labial surface and a flatter lingual one, resulting in an irregular ovoid basal cross section. They are slightly posteriorly recurved, with slightly recurved anterior and posterior surfaces, both bearing a sharp unserrated carina. Lingual and labial surfaces are ornamented by both strong protruding ridges (about 12–15 labially and twice lingually) extending about one-half to two-thirds of the crown height (slightly lower lingually), and shorter and thinner striae, variably inserted between the strong ridges and less extended upwards. As none reach the apex, the upper third to half of the crown height is smooth and shiny. All these ridges and striae are separated by shallow grooves, giving them a fluted aspect that may continue on the upper part of the root. The enamel is thin, and wear facets are very rare, limited when present to a small apical point.

The dentition is almost homodont (both in shape and size) along the jaws, so that isolated teeth are difficult to position in the jaws. Teeth are few (13 Mx, 16 De), despite the very elongated gavial-like jaws, and are spaced well apart from each other. The pterygoid teeth are typically hook-like, half the size of the marginal ones, similarly ornamented but slightly labiolingually compressed.

The skull as a whole is large (about 90 cm) and narrow (W/L = 0.25), with remarkably long, straight and slender jaws about two-thirds of the skull length. The jaws bear widely spaced, interfingering teeth, with accommodation pits on the opposite jaw. The appearance of the skull and jaws is gharial-like (hence the generic name) [32]. The orbital region is short, with a very small isosceles-triangle-shaped frontal bone. The postorbital region is long and slender. *Gavialimimus* is also unique among mosasaurids for its extremely retracted nares [32,106]. The body size is estimated at 6 m.

Paleoecology—The high, sharp, ridged teeth without wear facets range into the 'piercing II' guild of Massare [38], indicating a diet composed of possibly small fish and soft cephalopods. However, they also approach the 'cutting' guild by the presence of sharp carinae, lateral compression and the characteristic protruding vertical ridges present on two-thirds of both lingual and labial surfaces, which may have helped to both cut and penetrate the flesh [3]. Alternatively, these features were interpreted by [32] as rather indicative of 'pierce II' close to 'pierce I' functions, and the tooth flutings an adaptation to piscivory (this last interpretation being not contradictory with [3] arguments). *Gavialimimus* also shows a suite of characters indicating advanced adaptations to aquatic life, the most spectacular being the very retracted nares and the shortening of the skull posterior to the snout, reminiscent of the 'telescopation' observed in cetacean evolution (e.g., [109]). *Gavialimimus* was probably an open-sea meso-predator that fed on small fish and soft cephalopods [3,32], both representing a particularly abundant biomass in this area of intense upwelling [58]. The elongate jaws suggest the prey was seized using rapid strikes of the jaws, similar to gharials and river dolphins, rather than high-speed pursuit as in porpoises and dolphins.

Systematic reassignment—The main arguments of Strong and collaborators [32] to justify the creation of the new species *almaghribensis* to the detriment of the already-existing *ptychodon* were these: (1) the teeth of *P.* (?) *ptychodon* are not diagnostic, looking similar by convergence to many other mosasaurids; (2) Arambourg's original diagnosis was unprecise and applied to many taxa, such as the mosasaurines *Mosasaurus lemonnieri* Dollo, 1889b [101] and *Prognathodon solvayi* Dollo, 1889a [110], the plioplatecarpine *Platecarpus somenensis* Thévenin, 1896 [111], as well as the tylosaurines *Tylosasaurus ivoensis* (Persson, 1963) [112] and *Taniwhasaurus* Hector, 1874 [113]. As a result, [32] considered the species *P.* (?) *ptychodon* as a nomen dubium, because diagnosis and holotype were insufficient to characterize it.

Here, we challenge this view by thoroughly examining and comparing the *Gavialimimus* teeth to those of not only *P.* (?) *ptychodon* and the species considered by [32] as 'similar' by convergence but also to those of mosasaurid subfamilies as a whole.

First of all, the teeth of the holotype skull (MHNM.KHG.1231) of *G. almaghribensis* are indistinguishable from the holotype (MNHN PMC-30) and referred teeth (MNHN PMC-31-34) of *P.* (?) *ptychodon*, not because they exhibit vague convergent traits but, on the contrary, a suite of diagnostical ones, as described above.

Second, all the characteristics mentioned by Arambourg [4] in his original diagnosis of *P.* (?) *ptychodon*—that are: '*teeth relatively low and wide at the base of the crown, slightly compressed, with obtuse anterior and posterior carinae without crenulations; lingual and labial surfaces ornamented by numerous irregular vertical ridges extending from the base of the crown only two-thirds of the way up*' (translated from French)—are found in our description. It thus appears more parsimonious to emend this original diagnosis rather than to consider this species (cited many times since Arambourg) as a nomen dubium.

Third, [32] did not provide any precise descriptions of the teeth of the species they consider as 'similar' to *P.* (?) *ptychodon*. The morphological overlap they mention results only from very superficial observations.

As mentioned above, the presence of both ridges and striae on labial and lingual surfaces of the crowns of *Gavialimimus* is a russellosaurine (Plioplatecarpinae + Tylosaurinae) synapomorphy (e.g., [68,105]), de facto never present in Mosasaurinae (including *Mosasaurus lemonnieri* and *Prognathodon solvayi* used by [32]). Conversely, all *Mosasaurus* species exhibit a characteristic U-shape cross section [62], and both *M. lemonnieri* and *P. solvayi* have large smooth facets and serrated carinae; none of these characteristics are observed in *Gavialimimus* or any other plioplatecarpine, even on replacement and worn teeth.

The teeth of basal mosasaurids like *Tethysaurus* [79], *Dallasaurus* Bell and Polcyn, 2005 [114] and *Russellosaurus* Polcyn and Bell, 2005 [105] and of Halisaurinae [19,26] are small and slender posteriorly recurved sharp cones, finely striated or smooth, diverging greatly from *Gavialimimus*.

Tylosaurinae teeth differ from *Gavialimimus* in being large, robust, high cones, either asymmetrical (*Tylosaurus* Marsh, 1872 [115]) or flattened and symmetrical (*Hainosaurus* Dollo, 1885 [116]) [37,117]. Though their teeth bear ridges and striae, they are present only basally and lingually, often reach the apex, but not continues basally as flutings. *Taniwhasaurus* teeth are smaller and slender, bearing marked striations that disappear near the apex, resembling as such *Gavialimimus*; however, *Taniwhasaurus* differs in lacking posterior carinae on anterior teeth [118] and, as is noteworthy, in having higher and more-slender crowns bearing large wear facets (NB, pers. obs.).

Among plioplatecarpines, Plioplatecarpini [29] bear small teeth with rounded to oval cross sections that are strongly posteriorly recurved at the mid-height of the crown [68,105], differing greatly from *Gavialimimus*. As mentioned by [32], *Platecarpus somenensis* teeth are similar at first glance to *Gavialimimus* in being high and slender, of comparable size (1–2.5 cm high), poorly posteriorly recurved, laterally compressed and exhibiting two unserrated carinae, as well as labial and lingual surfaces ornamented by ridges and striae that do not reach the apex. However, in the details, *P. somenensis* holotype (MNHN 1895-7) shows obvious differences: the median crowns are straighter and more compressed than in *Gavialimimus* (W/L = 0.5 versus 0.6–0.8) and, as is noteworthy, the labial and lingual surfaces bears few 'facets' (seven labially, about seven lingually) reminiscent of those of mosasasaurines, rather than numerous ridges (12–15 labially, twice lingually) as in *Gavialimimus*. Between these 'facets', minute striae extending over one-fourth to one-third of the crown height are interspersed, forming a basal striated 'ring'. These minute striae are also present on pterygoid teeth. Finally, wear facets are frequent and relatively large. Among Selmasaurini, *Gavialimimus* teeth strongly differs from those of *Goronyosaurus* and *Khinjaria* that are large, straight, smooth and dagger-like [29]. Although the teeth of *Selmasaurus* show some resemblances with those of *Gavialimimus* (poorly posteriorly recurved, subequal labial and lingual surfaces, two unserrated carinae, very rare wear facets and basal flutings), they however differ in being higher, slender (H/L ratio about 2) and ornamented by strong but few ridges (five to six labially, six to seven lingually) [119] contrary also to the numerous ones of *Gavialimimus*.

To sum up, the median standard teeth of *Gavialimimus almaghribensis* are similar to those of *P.* (?) *ptychodon* but differ from those of any other mosasaurids. They bear a suite of diagnostical characteristics (described above)—some of them already mentioned by Arambourg in his original diagnosis of *P.* (?) *ptychodon* [4]—that make them unique and recognizable at first sight among hundreds of mosasaurid teeth. As a result, we propose to accomplish the following:

(1) Rehabilitate the name *ptychodon* as a valid species, *almagribensis* being considered its junior synonym;
(2) Emend Arambourg's original diagnosis (for teeth and dentition only), using the description detailed above;
(3) Consider for nomenclatural stability MNHN PMC 30 (holotype of *P.* (?) *ptychodon*, with well-known geographical and stratigraphical occurrences) and MHNM.KHG.1231 (holotype of *G. almaghribensis*, with geographical origin uncertain and stratigraphical occurrence obtained second hand) as complementary syntypes of the new combination *G. ptychodon* (Arambourg, 1952).

Khinjaria acuta Longrich et al., 2024b is known by a unique incomplete skull (MHNM.KHG.521) found in the Lower *Couche* III of Sidi Chennane, Oulad Abdoun Basin [29]. Lower *Couche* III being considered as middle-late Maastrichtian in age [61],

Khinjaria is probably slightly older than the other mosasaurids from the Upper *Couche* III assemblage; as such, it has not been included in the morphometric analyses of this work. *Khinjaria* is up to now only known in the Phosphates of Morocco. Among Selmasauri, it shows close affininities with *Goronyosaurus* from the Maastrichtian Iullemmeden Basin of Niger and Nigeria [120–122], with which it forms a clade, sister group of *Gavialimimus*, all knownfrom Northwest Africa and the Middle East. These genera represent a distinct clade of specialized mosasaurids, so far unknown outside of the Arabo-African platform [29].

Description—The crowns are very high, slender (H/L ratio estimated to 1.8) and dagger-like in shape. Though they cannot be measured (because the holotype is still partly imbedded in matrix), they are clearly labiolingually compressed, with a lenticular cross section. Strikingly, they are not posteriorly recurved, with the anterior and posterior surfaces being almost straight along most of their height, except apically, where they converge to the rounded tip. This character is particularly marked in the large anterior and median teeth. The smaller posterior teeth are slightly posteriorly recurved. The anterior and posterior surfaces of the crowns bear a prominent, unserrated carina, 'pinched' from its shaft, which reinforces its sharp aspect, particularly in the anterior and median teeth. The enamel is completely smooth and shiny, without any ornamentation except some variably subtle ridges labially and lingually. There is no trace of wear facets on any of the functional teeth.

The dentition is strongly heterodont in size rather than in shape (most teeth are dagger-like), with anterior teeth forming large fangs (a characteristic shared with *Goronyosaurus*), about twice the length of the median ones. This condition is unique in mosasaurids, where median teeth are always the largest [29,62]. The posteriormost teeth are small, being one-quarter the height of the anterior teeth, another unusual character. The tooth count is low (10–11 Mx, 12 De), and the teeth are well-spaced from each other. They are strongly interlocking, and large interdental pits for the housing of the opposite teeth during occlusion are present on both the maxilla and dentary, as in *Gavialimimus*. Pterygoid teeth remain unknown.

The skull of *Khinjaria* is highly akinetic and is characterized by a short and robust rostrum, a long postorbital region with large temporal fenestrae and probably reduced orbits [29]. A noteworthy feature is that the dentary deepens in its symphyseal region (probably mirrored on the premaxilla) to accommodate the large anterior teeth. This is also unique among mosasaurids, where the jaws always gently taper anteriorly [29,62]. These reinforcement of the anterior part of the rostrum, coupled with the large anterior fangs they hold (share with *Goronyosaurus*), are reminiscent of the 'spatulated' symphysis of pliosaurs, which also housed the largest teeth of the dental series [29]. *Khinjaria* was a medium to large taxon, with a skull estimated to be 90 cm long and a total body size of 8 m [29].

Paleoecology—The large dagger-like teeth and strong interlocking dentition of *Khinjaria* indicate a clear adaptation for 'cutting', though they lack wear facets. Its short and robust jaws were suitable to resist high bite forces, favored by large mandibular adductor muscles housed in expanded temporal fenestrae [29]. As a whole, the large size of *Khinjaria* and its noteworthy short robust skull armed with large sharp teeth indicate it was an apex predator, with a diet probably composed of large prey, but that adapted a different strategy to avoid competition with the larger coeval *Thalassotitan*, *P. currii*, *Mosasaurus* and *Hainosaurus*. Indeed, the lack of wear facets on the teeth indicates that prey must not have been very hard - like the fish *Enchodus* or other marine vertebrates of comparable size and consistency - and that their manipulation was probably simple and restricted to a rough cut into large pieces swallowed whole. Also, the apparently small eyes and lack of neurovascular foramina on jaws (characteristics shared with *Goronyosaurus*) suggest non-visual hunting, probably based on olfaction [29], as previously suggested for *Goronyosaurus* [123].

Non-visual hunting is probably linked to ambush predators with feeding strategies carried out by night, at great depth or in shallow waters with low visibility (reefs, crevices) [29].

4.1.4. Tylosaurinae

Tylosaurinae are a clade of poorly diversified (two to three genera, according to the authors) mosasaurids, known from the Turonian to the early Maastrichtian [123,124]. They were mainly widespread during Santonian–Campanian times, but during the Maastrichtian were largely replaced as apex predators by mosasaurines [2,125]. Except for the Turonian taxa found in subtropical paleolatitudes, younger tylosaurines all occurred in higher paleolatitudes (35–70°) of both hemispheres, mainly in North America and Europe, but also in Japan, New Zealand and Antarctica [123–127]. However, these spatiotemporal distributions have been challenged recently by the discovery of tylosaurine remains in the late Maastrichtian Phosphates of Morocco [31], representing not only the unique post-Turonian subtropical occurrence of the clade but also its youngest record. Tylosaurines include some of the largest mosasaurid taxa (10 to 12 m), rivaling in size the largest mosasaurines.

Tylosaurines remain extremely scarce in the Maastrichtian Phosphates of Morocco, being up to now known from one species only.

Hainosaurus boubker Rempert et al., 2022 is based on upper and lower jaw fragments and isolated teeth (syntypes VANPS 13.0120-13.0121) originating from the upper Maastrichtian Upper *Couche* III of Sidi Chennane, southern part of the Oulad Abdoun Basin [31]. It should be noted that the specific name *boubker*, dedicated to Mr. Boubker Chaibi (discoverer and donator of the specimens) should preferably have been *boubkeri* (*Hainosaurus* 'from Boubker', Latin genitive) following the recommendations of the ICZN. Though the referral of some isolated teeth to tylosaurines is questionable (NB, pers. obs.), most specimens bear two of the main tylosaurine synapomorphies, that are premaxilla and dentary bearing a large edentulous rostrum, and large, robust teeth ornamented by both facets and striations. Pending the discovery of more complete specimens, this species is considered here as valid.

Description—The crowns are large (average height 5 cm), robust high cones (H/L about 1.6), moderately posteriorly recurved and strongly laterally compressed (W/L ratio about 0.7, see [31]). The apex is sharp. The anterior surface is regularly convex and the posterior one concave, both bearing a marked cutting carina with tiny serrations, more marked on the anterior carina. The labial and lingual surfaces are subequal, convex and strongly compressed, resulting in an elliptical cross-section. The labial surface bears five to six large but poorly marked facets. The lingual surface bears also five to seven large facets, even more difficult to distinguish. In addition to these facets, fine striae are superimposed around the basal part (about one-fourth of the total height) of the crown, a typical tylosaurine character [37,117]. As in *M. beaugei*, wear facets occur both at the apex of the crown and on the carinae, especially on the anterior one (tooth cutting leading edge).

Contrary to [31], we consider the heterodonty to be only moderate, falling within the range of variation along the jaws shown in most mosasaurids, with slender recurved anterior teeth, large blade-like median teeth, and lower and more robust posterior teeth.

On the basis of the jaw fragments, the prenarial part of the premaxilla indicates a skull possibly about 1,20 m and a total body length estimated at 8–12 m [31].

Paleoecology—With *T. atrox*, *P. currii* and *M. beaugei*, *H. boubker* is part of the apex predators of the Maastrichtian Phosphates of Morocco. However, unlike these three taxa, whose remains are commonly and jointly found, *H. boubker* remains very scarce, a fact that cannot be explained by collection bias. *H. boubker* possibly occupied a quite different ecological niche than the previously mentioned taxa, both in term of adaptation and of habitat (more offshore species?). The teeth of *H. boubker* indicate that their function was 'cutting' sensu Massare [38]. They are roughly comparable to those of *M. beaugei* in their sharp appearance but are larger and more robust. The diet of *H. boubker* could therefore have included comparable but larger items than that of *M. beaugei*, such as large fish and small marine reptiles.

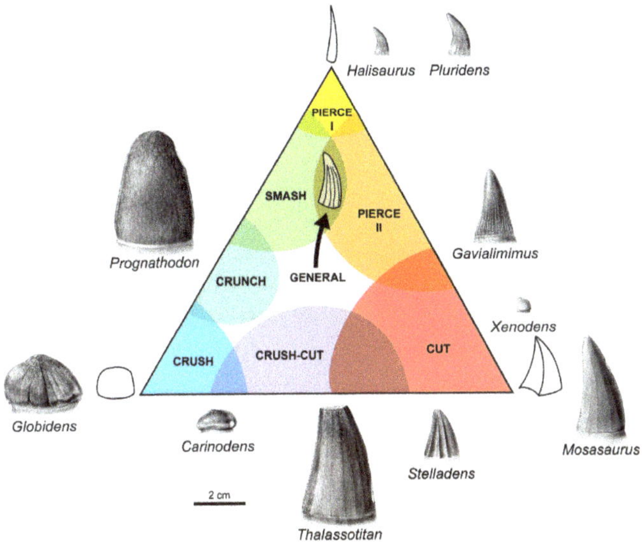

Figure 3. Mosasauridae from the Maastrichtian Phosphates of Morocco: niche partitioning using Massare's dental guilds [38]. Based on the same taxon selection as in Figure 2. Teeth on scale to appreciate size and proportion differences (see measurements and ratios in Table A2). Modified from [3]. Teeth drawings © Charlène Letenneur (CR2P, MNHN, Paris); canvas and design © Alexandre Lethiers (CR2P, ISTeP, Paris).

4.2. Morphometric Analyses (Figures 4 and 5)

4.2.1. High-Density 3D Geometric Morphometrics

Our semi-automated high-density 3D geometric morphometric (HD3DGM) procedure recovers a signal similar to that carried by a global set of marine amniote tooth crowns [36]. Most of the shape signal is indeed dominated by the aspect ratio of the crown and its distal concavity (Figure 4A). In this mosasaurid-only dataset, the presence of cutting edges is also captured by PC1 (which accounts for 86.2% of the total variance): crowns in positive values are pointed, recurved and possess two cutting edges; crowns in negative values are bulbous, at least in labial or lingual view. PC2 (accounting for 9% of the total variance) captures the labiolingual flattening of the crown, as well as the direction of curvature: distally in negative values and distolingually in positive values. Most teeth of the sample are conical and slightly recurved; indeed, this region concentrates most clades and a range of crown volumes/sizes (Figure 4A), with large prognathodontins (*T. atrox*, *E. heterodontus*), one mosasaurin (*M. beaugei*), the halisaurines (*H. arambourgi*, *P. serpentis*) and the only plioplatecarpine of our sample (*G. ptychodon*), the last two subfamilies having fairly small crowns. Three species are isolated from the dense region of phenotypes, disseminated in the negative quadrant of PC1 and dictate most of the signal captured by PC2: the giant 'crush–cut' prognathodontin *P. currii*, as well as the durophagous mosasaurin *C. belgicus* and the globidensin *G. phosphaticus*.

4.2.2. Fourier Transforms

Principal component analyses of Fourier transform coefficients yielded a dominant first axis for both the labiolingual and the basal outline analyses (PC1 = 84.8% of the total variance in the labiolingual outline analysis and PC1 = 79% of the total variance of the basal outline analysis; Figures 4B, A1 and A2). The labiolingual signal is dominated by the aspect ratio, as evidenced by the shape variation along PC1 (Figures 4B, A1 and A2). The first principal component also carries a signal of distal concavity, giving bulbous outlines in negative values and pointed, recurved cones in positive values. The signal of basal outline analysis is dominated by labiolingual compression, giving flattened outlines with straight, parallel labial and lingual surfaces in negative values and rounded outlines in positive values.

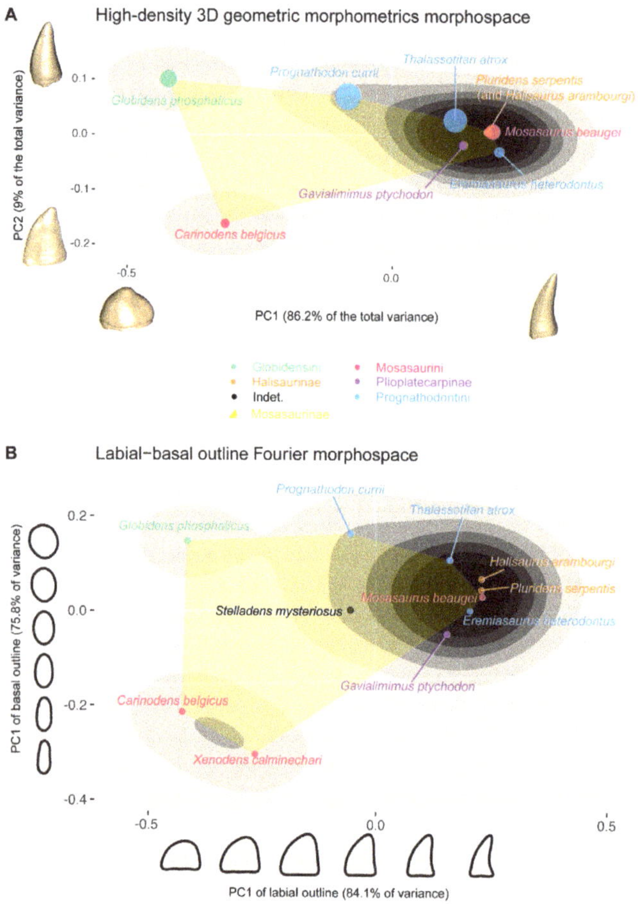

Figure 4. Mosasauridae from the Maastrichtian Phosphates of Morocco: crown shape morphospaces. (**A**) Morphospace (PC1 and PC2) resulting from the principal component analysis of high-density 3D geometric morphometrics, using Fischer and collaborators method [36]. The diameter of each dot is directly proportional to centroid size. We grouped *P. serpentis* and *H. arambourgi* because their teeth are morphologically uncannily similar. To visualize the morphological variation captured by each axis, we generated 3D meshes at the extremes of each axis (20% further than the sampled extremes) using thin-plate splines. (**B**) Composite morphospace using the PC1 of the Fourier analyses of the labiolingual and basal outlines. Kernel density of occupation in both morphospaces is visualized by shades of grey (darker = higher density). Teeth data in Tables A2 and A3.

Combining the PC1s of both outline analyses yields two axes representing most of the signal, which can therefore be used as axes of a two-dimensional composite morphospace. Kernel density estimation indicates a main region of recurring phenotypes, centred on slightly recurved conical crowns with an oval basal cross section. This region concentrates most clades, with representatives of mosasaurin (*M. beaugei*) and prognathodontin (*T. atrox*, *E. heterodontus*) among mosasaurines, halisaurines (*H. arambourgi*, *P. serpentis*) and a plioplatecarpine (*G. ptychodon*). Two highly peculiar crown shapes are located on the border of this region, the gigantic prognathodontin *P. currii* and the mosasaurine *S. mysteriosus*. Three taxa are located clearly outside of the 'common' shape region; all have semioval to square labial outlines, with varying degrees of labiolingual compression: *G. phosphaticus* (rounded labial outline and basal cross section), *C. belgicus* (labiolingually flattened cross section) and *X. calminechari* (squared labial outline, extreme labiolingual compression).

4.2.3. Comparison of Geometric Morphometric Methods and Caveats

Though the very high ecological disparity morphologically exhibited by the mosasaurid assemblages from the upper Maastrichtian Phosphates of Morocco, our HD3DGM and Fourier transform protocols, despite using markedly different techniques and slightly dissimilar datasets (the two highly peculiar species *Stelladens* and *Xenodens* are absent from the HD3DGM analysis), recovered uncannily similar signals, morphospaces and phenotype densities (Figure 4). This means that the main signals present in the dataset are clear and easy to capture. Indeed, most of the shape signal can be summarized as a combination of crown aspect ratio, crown distal curvature and basal cross section. Because most marine amniote crowns seem to vary along the same traits [36], the similarity between the results of both techniques suggests that our protocol using Fourier transform offers a fast, cheap and easy-to-implement alternative to high-density 3D geometric morphometrics of simple conical objects such as marine tetrapod tooth crowns.

Namely, a dense region of 'common' phenotypes (teeth conical, pointed, recurved and bearing two carinae) concentrates most of the species in positive values, independently of their volume and size, varying from gigantic mosasaurines (*Thalassotitan*, *Mosasaurus*) to medium and small-size mosasaurines (*Eremiasaurus*), plioplatecarpines (*Gavialimimus*) and halisaurines (*Halisaurus*, *Pluridens*). Noteworthy are *Halisaurus* and *Pluridens*, which occupy the same position despite their drastically different skull morphologies. Their teeth are morphologically indistinguishable except for a difference in size (1:2 ratio), indicating a fully isometric tooth-growth difference between both species. Three taxa corresponding to bulbous or squared and variably labiolingually compressed teeth (*Globidens*, *Carinodens*, *Xenodens*) are always found isolated from the 'common' shape region and disseminated in negative values. The gigantic *Prognathodon curri*, combining features of both 'crushing' and 'cutting' teeth sensu [3], occupies a median position, as does the mysterious *Stelladens*.

Yet, this also means that both techniques, because they rely on principal component analyses and, for Fourier transform, simple outlines, cannot discern the fine details of crown shape and texture in the first (main) axes of variation. Indeed, when the sample includes teeth as different as the crushing teeth of *Globidens* and the elongate, slightly recurved teeth of *Eremiasaurus*, most of the variation in X, Y, Z coordinates or outline shapes will evidently be the aspect ratio and the labiolingual flattening. Therefore, features like the precise position of cutting edges (which are still captured in PC1 in the HD3DGM method), the rugose enamel texture in *Globidens* or the apicobasal ridges in *Stelladens* will result in only slight variations of X, Y, Z coordinates in HD3DGM and, depending on how they are oriented in the crown, no change in outline shape in the Fourier transform method. Therefore, the shape signal carried by these traits will either be very small compared to other features (and thus captured by PC axes accounting for a very small amount of the total variance in HD3DGM) or not captured at all (Fourier transform). Despite being minor modifications of the global shape, these features can be of importance when trying to assess diet and the finer functional capabilities of teeth [36,38,128]. Therefore, even though it goes without saying, quantitative analyses in isolation are not sufficient to fully discuss and interpret the complexity of functional anatomy.

4.2.4. Skull Size/Crown Shape Relationships

As observed in a broader dataset of aquatic amniotes [36], there is no obvious skull size–tooth shape correlation, in the sense that both small and medium-sized mosasaurids (i.e., with skull lengths less than 1 m) have a wide range of tooth shapes (bulbous, conical, recurved) (Figure 5A,B,D,E), whereas larger taxa seem to occupy a restricted area of tooth shapes, being close to origin of main axis of variance (here, *Thalassotitan* and *Prognathodon currii*). These taxa indeed have 'intermediate' morphologies, with robust crowns still possessing protruding carinae; this prompted [3] to create the crush–cut guild for animals like *Thalassotitan*. Similarly, *P. currii* possesses large, blunt teeth with blunt carinae forming two opposite apicobasal ridges.

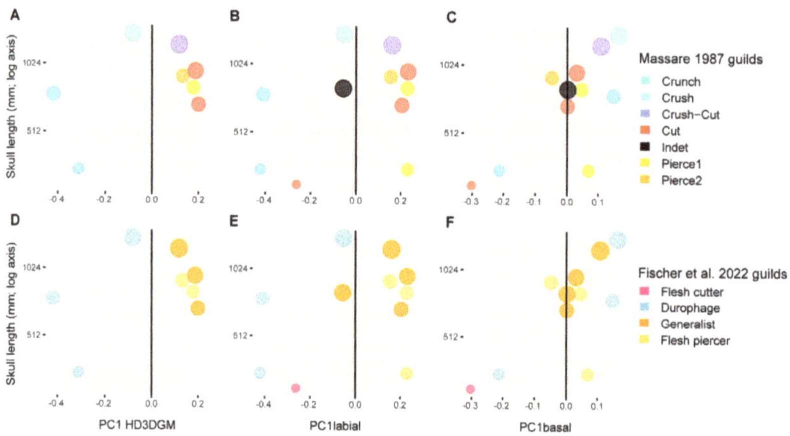

Figure 5. Mosasauridae from the Maastrichtian Phosphates of Morocco: skull size–crown shape relationships. (**A**,**D**) Skull size versus PC1 of the high-density 3D geometric morphometrics. (**B**,**E**) skull size versus PC1 of the Fourier analysis of the labiolingual outline. (**C**,**F**) Skull size versus PC1 of the Fourier analysis of the basal outline, with a mapping of the guilds of Massare [38] (**A**–**C**) and the guilds of Fischer and collaborators [36] (**D**–**F**). Data about skull size in Table A2.

This zone of the morphospace, close to the center of the axis, is weakly populated in smaller forms. The only medium-sized taxon with such a 'median' tooth shape is the highly peculiar *Stelladens* (Figure 5B,E), which actually possesses a series of highly unusual ornamentations that are not captured in the main axis of variation of dental shape. Another zone appears unpopulated: labiolingually compressed teeth in medium and large taxa (Figure 5C,F). Indeed, highly compressed teeth are restricted to the smallest taxa (*Carinodens*, *Xenodens*) (Figure 5C,F); the drivers and effects of these size–shape relationships are discussed below.

5. Discussion

5.1. Paleobiodiversity (Figure 6)

Despite the description over the last decade of many new mosasaurid taxa in the upper Maastrichtian Phosphates of Morocco, the respective diversity of each sub-family expressed as a percentage of the total species number remains surprisingly constant compared to previous work (compare ([3], fig. 6) to this work Figure 6). This stability, despite the increased sampling, suggests it is a genuine paleobiodiversity pattern, rather than collection or description biases.

Notably, Mosasaurinae (eight genera, ten species) accounts for about 60% of the genera and almost 70% of the species described so far in these Moroccan Phosphates. Such a proportion also corresponds to patterns observed in other coeval localities worldwide [2,3,129]. Also, the Moroccan genera represent about half to three-quarters (depending on the taxa considered valid) of the mosasaurines known worldwide, whereas most of the species are solely known in the Southern margin of the Mediterranean Tethys [34].

Halisaurinae and Plioplatecarpinae are less diverse (two genera and two species each), representing about 13% of the species and 15% of the genera currently known in the upper Maastrichtian Phosphates of Morocco. However, this represents about half of the total halisaurine genera versus one-seventh of the plioplatecarpine genera known worldwide, emphasizing the greater global diversity of this last clade during Maastrichtian times and a poor representation in Morocco.

Finally, Tylosaurinae remain extremely scarce (one genus, one species), representing about 6% of the species and 7% of the genera of the Phosphates of Morocco mosasaurid faunas and one-third of the worldwide tylosaurine genera.

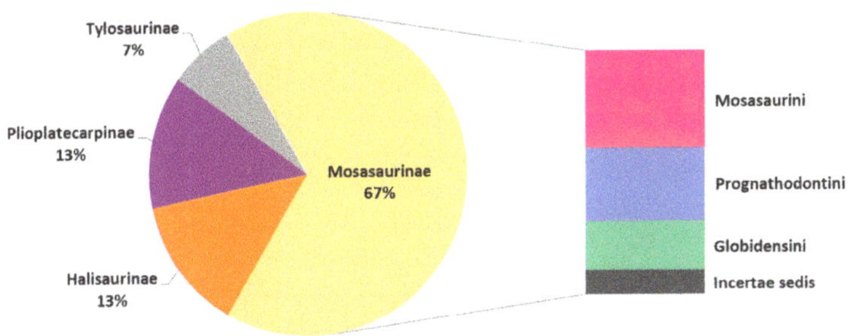

Figure 6. Mosasauridae from the Maastrichtian Phosphates of Morocco: taxonomic paleobiodiversity, expressed for each subfamily as a percentage of the total species number (15). The same for mosasaurine tribes (on the right). Colors are the same as in Figure 4 for ease of comparison.

5.2. Paleoecology and Niche-Partitioning (Figure 7)

In addition to their high taxonomic diversity, the mosasaurid faunas from the upper Maastrichtian Phosphates of Morocco exhibit a wide range of phenotypes (body sizes, skull morphologies, dental shapes) (Figures 2 and 3) (e.g., [this work, [3,130]]), strongly suggesting diverse diet preferences and predation modes, as well as niche-partitioning in the regional environment. In the present work, when using the framework of Massare [38] and subsequent modifications and additions by Chatterjee and Small [130] (trap guild) and Bardet and collaborators [3] (crush–cut guild), Moroccan mosasaurids (10/15 taxa used) occupied a wide range of dental guilds: crunch (*Prognathodon*), crush (*Globidens*, *Carinodens*), crush–cut (*Thalassotitan*), cut (*Mosasaurus*, *Eremiasaurus*, *Xenodens*, ?*Khinjaria*, ?*Hainosaurus*), pierce I (*Halisaurus*, *Pluridens*) and pierce II (*Gavialimimus*) (Figures 3, 5A–C and 7B). The peculiar *Stelladens* occupied a unique, so-far unknown guild, possibly located between 'cut' and 'crush' in that canvas. Without taking *Stelladens* into account, mosasaurids of this ecosystem therefore occupied 75% (6/8) of Massare's guilds ever colonized by marine reptiles.

Only two guilds—'smash' and 'trap'—were not occupied by mosasaurids; this is true not only for the upper Maastrichtian Phosphates of Morocco but also worldwide. The 'trap' guild was defined by [130] to characterize the long, slim and labially directed teeth of long-necked plesiosauroids, notably aristonectine elasmosaurids during the Maastrichtian [131]. None of the marine reptiles known in the upper Maastrichtian Phosphates of Morocco appears to precisely occupy it. Though the anteriorly protruding and usually unworn teeth of the local elasmosaurid plesiosaurian *Zarafasaura* [13] could probably act as a trap, their robust shape and large size proportionally to the skull rather match with the 'pierce' guild of Massare [38]. The 'smash' guild, occupied during the Mesozoic by several ichthyosaurians, is characterized by teeth resembling small, straight cones, with acute but round apexes and without carinae [38]. In the upper Maastrichtian Phosphates of Morocco, such a dental shape is found in crocodylomorphs, more especially in the small gavialoid *Ocepesuchus* [12]. It should be pointed out that sea-going crocodylomorphs remain very scarce and small in the regional Maastrichtian, becoming abundant and larger only after the extinction of the mosasaurids during the K/Pg biological crisis [12]. Their rarity suggests that they may have inhabited freshwater or marine ecosystems, rather than being fully marine.

When using the guilds of Fischer and collaborators [36], Maastrichtian Moroccan mosasaurids occupied 100% (4/4) of the main raptorial feeding guilds (generalist, durophage, cutter, piercer), covering a wide range of crown sizes and shapes: three durophages (*Carinodens*, *Globidens*, *Prognathodon*), one flesh cutter (*Xenodens*), three flesh piercers (*Halisaurus*, *Pluridens*, *Gavialimimus*) and four generalists (*Thalassotitan*, *Mosasaurus*, *Eremiasaurus* and *Stelladens*), two of them being very large (*Prognathodon*, *Thalassotitan*) (Figures 5D–F and 7B). The main difference with Massare's canvas is the presence of more generalist species (resulting from the inclusion in that guild of most of the large 'cutting' species), whereas species specializing in crushing and piercing remain roughly the same (Figure 7B). Specifically, we interpret *Thalassotitan atrox* as a generalist and *Prognathodon currii* as a durophagous predator of large, shelled animals like turtles, even though its size and robustness, comparable to *Thalassotitan*, probably allowed it to kill and consume a wide range of prey items (see [36] for a discussion of diet–size relationships). These ecological differences are also reflected by a different positioning ('crunch' versus 'crush–cut') in Massare's qualitative canvas (Figure 3). *Hainosaurus* and *Khinjaria* were not quantitatively analyzed here, but they are expected to occupy, respectively, the generalist and the flesh cutter guilds.

To sum up, by combining both canvases as well as our Fourier-transform analyses, we observe for each mosasaurid subfamily grosso modo the same trends in ecological disparity as those observed for taxonomic diversity (compare Figures 3, 4 and 7):

(1) Mosasaurinae exhibit the largest disparity of tooth guilds (generalists, durophagous, flesh cutters), combined with a widest range of body sizes (2 m in *Xenodens* to more than 10 m in *P. currii*). However, the clade did not evolve flesh piercers, either here or worldwide. Mosasaurins (*Mosasaurus*, *Eremiasaurus*, *Carinodens*) usually have longer and more gracile skulls armed with labio-lingually compressed teeth either to cut or to crush, whereas prognathodontins (*Thalassotitan*, *Prognathodon*) and globidensins (*Globidens*) are characterized by shorter and more-robust brevirostrine skulls and teeth, indicating stronger bite force and more durophagous habits [130] (Figure 2). Also noteworthy is that Mosasaurinae are the only mosasaurid clade to have developed durophagous species distributed over its three tribes, which exhibit a wide range of both body sizes and crushing tooth shapes (compressed in the 3 m long mosasaurin *Carinodens*, bulbous in the 6 m long globidensin *Globidens*, conical in the up-to-10 m long prognathodontin *P. currii*) and this repeatedly (at least two coeval species of *Carinodens* and two of *Globidens* in the Phosphates of Morocco), once again indicating a larger plasticity of this clade.

(2) Halisaurinae retain the mosasaurid primitive tooth shape and, despite their very different skull morphologies and sizes (3 m gracile *Halisaurus* versus up to 8 m robust *Pluridens*) (Figure 2), have similar teeth, indicating they were flesh piercers in the 'grasping' group of [129].

(3) Plioplatecarpinae are represented in Morocco only by two medium-sized selmasaurins (Figure 2), exhibiting astonishing and drastically opposed skull and tooth morphologies: the longirostrine flesh-piercer *Gavialimimus* and the brevirostrine flesh-cutter *Khinjaria*.

(4) Finally, Tylosaurinae, though poorly known so far in Morocco, are represented by a generalist taxon, whose skull and tooth morphologies are reminiscent of those of Mosasaurini like *Mosasaurus* and *Eremiasaurus*, although much larger.

Also, when comparing tooth shapes with global sizes (Figure 7A), the same trends as those observed with skull sizes (Figure 5) are found, with the largest taxa (more than 8 m long) occupying a restricted area of tooth shapes (generalist, durophagous), as previously shown by [36], whereas small- and medium-sized species (less than 8 m long) occupy all tooth morphologies (generalists, durophagous, cutters, piercers). In the details, flesh piercers appear to have a restricted size-range between 4 and 6 m, whereas generalists are usually larger than 5 m long, and durophages range along all sizes (Figure 7A). In other words, and probably related to the specificities and advantages of biomechanical resistance, apex predators are never dedicated piercers, micro-predators are conversely never generalists, and meso-predators occupy the widest range of dental adaptations (compare Figures 5 and 7A). This signal in the upper Maastrichtian Phosphates of Morocco, where some of the size-shape possibilities are unexplored, also possibly results from the reduction of generalist predators among Northern Hemisphere mosasaurids during the Maastrichtian [130]. Finally, strictly cutting species remain rare among these mosasaurid faunas; these ecological niches were probably occupied by other predators, most likely selachians [22].

A

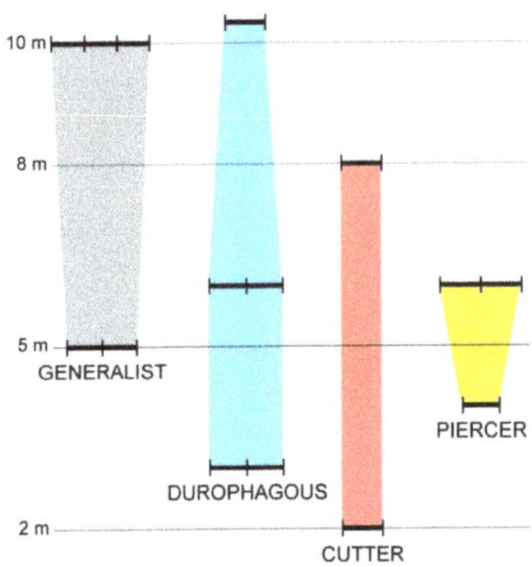

B

TAXON	SIZE	Massare 1987	Fischer et al. 2022
Prognathodon currii	> 10 m	Crunch	Durophagous
Hainosaurus boubker	10 m	Cut	(Generalist ?)
Thalassotitan atrox	10 m	Crush-Cut	Generalist
Mosasaurus beaugei	10 m	Cut	Generalist
Khinjaria acuta	8 m	Cut	(Cutter ?)
Globidens phosphaticus	6 m	Crush	Durophagous
Globidens simplex	6 m	Crush	(Durophagous)
Gavialimimus ptychodon	6 m	Pierce II	Piercer
Pluridens serpentis	6 m	Pierce I	Piercer
Stelladens mysteriosus	5 m	Cut ?	Generalist
Eremiasaurus heterodontus	5 m	Cut	Generalist
Halisaurus arambourgi	4 m	Pierce I	Piercer
Carinodens belgicus	3 m	Crush	Durophagous
Carinodens minalmamar	3 m	Crush	(Durophagous)
Xenodens calminechari	2 m	Cut	Cutter

Figure 7. Mosasauridae from the Maastrichtian Phosphates of Morocco: niche-partitioning. Expressed for each species by combining body sizes and dental guilds (**A**) of Massare [38] and Fischer and collaborators [36]. List of taxa and inferred guilds (**B**), with diets suggested by anatomy, but not quantitatively analyzed, indicated in parentheses. Colors are the same as in Figure 3 for ease of comparison.

Despite these constraints, mosasaurids from the upper Maastrichtian Phosphates of Morocco likely were widespread in a wide range of niches in this shallow marine ecosystem, both vertically (water column) and horizontally (coast to open sea). Apex predators (*P. currii, T. atrox, M. beaugei*) probably fed on any other large bony marine vertebrates, with a harder component in *Thalassotitan* and even harder in *P. currii*, which possibly was a turtle consumer. Meso-predators were the most ecologically diverse and probably the most widely distributed in the water column, avoiding as such some niche overlaps. They include piscivores (*G. ptychodon, P. serpentis*), generalists (*E. heterodontus, S. mysteriosus*), flesh cutters (*K. acuta*) and consumers of hard invertebrates (*G. phosphaticus, G. simplex*). Some taxa like *Pluridens* were probably ambush hunters in dark waters, either at night or in the depths of crevices. *Globidens* is represented by two species with different dentition, indicating the hard-shelled invertebrates they fed on were probably different. Finally, the smallest mosasaurids of the assemblage were either piscivorous (or consumers of soft invertebrates) (*H. arambourgi*), scavengers (*X. calminechari*) or durophagous (*C. minalmamar, C. belgicus*). Here also, the niche-partitioning was pronounced, with *Halisaurus* being possibly a rapid night hunter of soft phosphorescent cephalopods, whereas *Carinodens* (just like *Globidens*) developed two species with different dentition, indicating slightly different semi-hard invertebrate diets.

Compared to the other marine vertebrate predators with which they interacted and shared trophic resources, mosasaurids were much more systematically and ecologically diverse than were coeval plesiosaurians and chelonioid turtles (see list in [5]) but much less so than selachians (e.g., [10]). The coexistence of this totally astonishing plethora of marine vertebrate predators in the same rather restricted environment of the Phosphate Sea (Figure 1D) implies that niche partitioning was particularly high, not only among mosasaurids but also among marine vertebrates as a whole, probably as an evolutionary answer to resist competition pressure.

The intense upwelling system that developed along the western coast of Africa from the Cretaceous [2] is probably responsible for this extraordinary paleobiodiversity and, more specifically, for that abundance of large predators coexisting locally [58]; this mirrors what is observed nowadays in upwelling zones that attract, by their high productivity, a huge range of marine predators (selachians, fishes, marine mammals) [27,58]. The use of Calcium isotopes as a diet marker has however revealed that, despite their huge ecological disparity, most of the marine vertebrate predators of the upper Maastrichtian Phosphates of Morocco (selachians, marine reptiles including mosasaurids) were feeding on a single decipherable calcium source [58]. This therefore suggests that, as also observed in extant upwelling zones, even if all these predators were able to hunt on a wide range of prey, as illustrated by very different morphologies and tooth shapes, the upwelling system facilitated their feeding by providing an important single biomass, likely schools of small teleost fish (like those of today, sardines and anchovies), whose remains are innumerable throughout the series of the Phosphates of Morocco.

6. Conclusions

The anatomical descriptions and quantitative methods used in this work have highlighted their complementarity to tackle the vast array of tooth morphologies in upper Maastrichtian mosasaurids from Morocco and infer the diets of their bearers. They also show their limits when used alone: while comparative anatomy fails to precisely locate teeth into a dental guild, morphometric analyses fail to capture the complex details of tooth ornamentations, which in reptiles probably play a role as important in feeding process as the multiple cusps of mammals. The two methods combined permit to approach the complexity of tooth shape, structure and ornamentation in its entirety.

With at least 15 species known so far, exhibiting a wide range of sizes, morphologies and tooth shapes covering the majority of dental guilds defined for marine amniotes, the mosasaurid faunas from the upper Maastrichtian Phosphates of Morocco reveal an extraordinarily high taxonomical diversity and ecological disparity, unknown elsewhere. Small spike hook-like teeth (*Halisaurus arambourgi, Pluridens serpentis, Gavialimimus ptychodon*) to spear soft fish and invertebrates, large robust pointed teeth to catch large marine vertebrates (*Thalassotitan atrox, Mosasaurus beaugei, Hainosaurus boubker, Eremiasaurus heterodontus*), blade-like teeth to cut large vertebrates (*Khinjaria acuta*), robust cones to smash large bony prey like turtles (*Prognathodon currii*), low and blunt teeth to crush variable hard-shelled prey such as mollusks, crustaceans and echinoids (*Carinodens belgicus* and *Carinodens minalmamar, Globidens phosphaticus* and *Globidens simplex*) and, finally, totally unique morphologies (*Xenodens calminechari, Stelladens mysteriosus*) whose ecological function remain mysterious: this plethora of tooth shapes exhibited by the faunas of the upper Maastrichtian Phosphates of Morocco highlight the great plasticity of mosasaurid teeth in general.

This extreme diversity and disparity indicate that mosasaurids were still markedly diversified in the late Maastrichtian of Morocco, just prior to the K/Pg biological crisis. The intense upwelling system that developed in this Phosphate Sea at this time is probably responsible for this high biodiversity and, more specifically, for the unusual abundance of large coeval predators in the region. However, upwelling ecosystems (at least the current ones) are structured on the wasp-waist food web model, making them particularly fragile to environmental changes. It is thus possible that this upwelling system became a double-edged sword when disruptive abiotic events (whatever they are) occurred at the end of the Cretaceous, leading to the mass extinction of the K/Pg boundary.

Supplementary Materials: The following supporting information can be downloaded at: https://www.mdpi.com/article/10.3390/d17020114/s1, SuppInfo 2DFourier File: tooth crown outlines of the 11 mosasaurid species studied, in basal and lingual views for Fourier analyses; SuppInfo HD3DGM File: 3D models of tooth crowns of the 8 mosasaurid species studied, the fixed landmark coordinates for each specimen, the 3D template for the automated patching procedure, R scripts, specimen metadata.

Author Contributions: Conceptualization, N.B.; methodology, N.B. and V.F.; investigation, N.B., V.F., N.-E.J., N.L. and X.P.-S.; writing, review and editing, all authors; specimen curation, N.B. (MNHN specimens), F.K., O.K.Y. (OCP specimens) and N.-E.J. (MHNM specimens). All authors have read and agreed to the published version of the manuscript.

Funding: Research of N.B. and V.F. was funded by the French–Belgian Program Hubert Curien—Tournesol (project n° 495778). Research of X.P.-S. is supported by the Spanish Ministry of Science, Innovation and Universities, the European Regional Development Fund (FEDER) (research project PID2021-122612OB-I00) and the Basque Country Government (research group IT1485-22).

Data Availability Statement: All data are available in the main body of the paper and on the appendices and Supplementary Files.

Acknowledgments: We warmly thank Charlène Letenneur (CR2P, MNHN, Paris) for producing the anatomical and artistical reconstructions of mosasaurid teeth and fleshy heads, as well as Alexandre Lethiers (CR2P/ISTeP, Paris) for the drawings and designs of the other illustrations. N.B. thanks Anthony Herrel (MECADEV, MNHN, Paris) for interesting discussions and useful data concerning details of the fleshy anatomy of modern monitor lizards, as well as Marta Fernández (Museo de la Plata, La Plata, Argentina) for providing photos of *Taniwhasaurus* for comparison. We warmly thank the referees for their constructive comments that permitted to improve our work. This work was carried out in the following framework: (1) the long-term *PhosphaPal* French–Moroccan program of scientific collaboration between the Muséum National d'Histoire Naturelle (MNHN, France), the Centre

National de la Recherche Scientifique (CNRS, France), the Office Chérifien des Phosphates (OCP, Morocco), the Ministère de la Transition Energétique et du Développement Durable (MTEDD—formerly Ministère de l'Energie, des Mines, de l'Eau et de l'Environnement (MEMEE—Morocco), the Université Cadi Ayyad (UCAM, Morocco) and the Université Chouab Doukkali (UCD, Morocco); (2) the recent agreement between the University of Bath (UB, UK) and the Université Cadi Ayyad (UCAM, Morocco). We are grateful to MTEDD for administrative authorizations and to OCP officials for logistical support during fieldwork. We also thank Blaïd Bougadir (President of the UCAM) and Mohamed Ghamizi (Director of the Natural History Museum of Marrakech (MHNM)), from whom we found the necessary help for the management and conservation of the important paleontological collection acquired within the framework of the UB—UCAM agreement.

Conflicts of Interest: The authors declare no conflicts of interest.

Appendix A

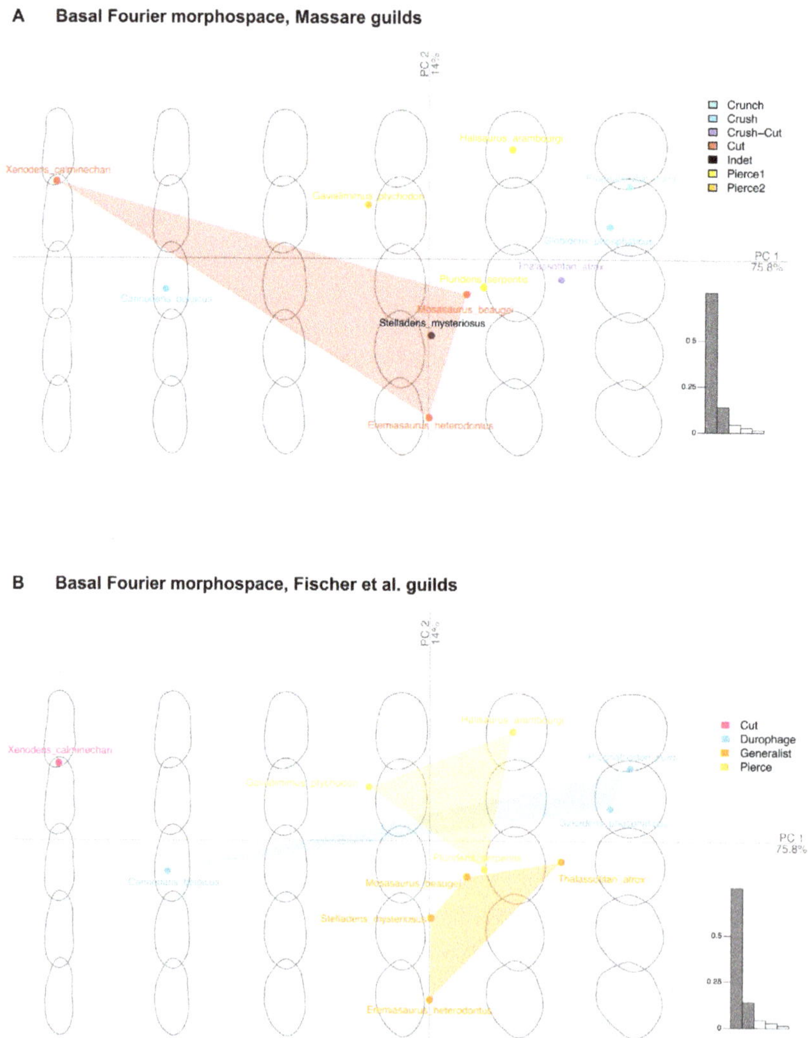

Figure A1. Mosasauridae from the upper Maastrichtian Phosphates of Morocco: PCA basal. Using Massare guilds [38] and Fischer and collaborators guilds [36] for comparisons.

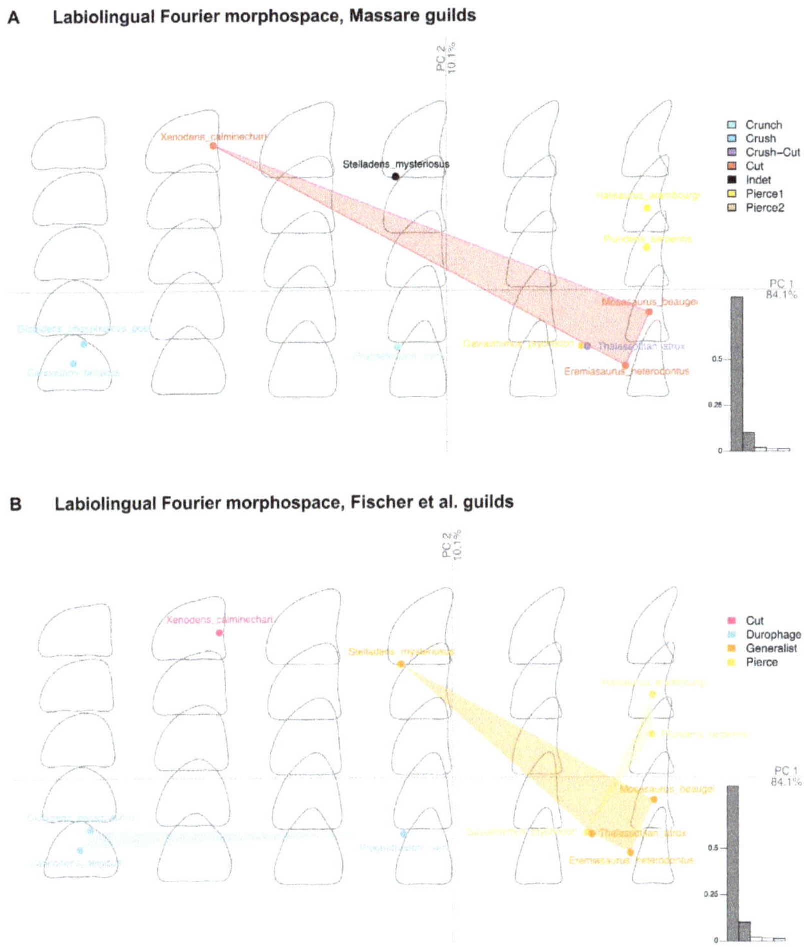

Figure A2. Mosasauridae from the upper Maastrichtian Phosphates of Morocco: PCA labial. Using Massare guilds [38] and Fischer and collaborators guilds [36] for comparisons.

Table A1. Mosasauridae from the upper Maastrichtian Phosphates of Morocco: list of species and main information concerned. Systematic abbreviations: Mosa, Mosasaurinae; Hali, Halisaurinae; Plio, Plioplatecarpinae; Tylo, Tylosaurinae. Stratigraphical abbreviations: C, *Couche* (layer in French). Collection number abbreviations: (h), holotype; (p), paratype; (r), referred specimen, (s), syntype. Institutional abbreviations: MHNM, Muséum d'Histoire Naturelle de Marrakech, Université Cadi Ayyad, Marrakech, Morocco; MNHN, Muséum National d'Histoire Naturelle, Paris, France; OCP, Office Chérifien des Phosphates, Khouribga, Morocco; UALVP, University of Alberta Laboratory for Vertebrate Paleontology, Edmonton, Alberta, Canada; ULg, Université de Liège, Belgium; VANPS, Paleontological Museum of the Vancouver Paleontological Society, Richmond, British Columbia, Canada.

CLADE		TAXON	GEOGRAPHY	STRATIGRAPHY	SPECIMENS	REFERENCES
MOSA	Mosasaurini	*Mosasaurus beaugei* Arambourg, 1952	Sidi Daoui, Oulad Abdoun; Ben Guerir, Ganntour	Upper CIII, C2, upper Maastrichtian	MNHN PMC 7 (h), 8–13, 55, 60 (r); OCP DEK/GE 83, 303, 551, 660 (r)	Arambourg, 1952 [4]; Bardet et al., 2004 [18]; Cappetta et al., 2014 [22]
		Carinodens belgicus (Woodward, 1891)	Sidi Daoui, Sidi Chennane, Oulad Abdoun; Ben Guerir, Ganntour	CIII, C3, upper Maastrichtian	OCP DEK/GE 445–447, 454–455 (r); MNHN casts 6314, 6338, 6340–6342	Bardet et al., 2008 [21]; Schulp et al., 2009 [33]; Cappetta et al., 2014 [22]
		Carinodens minalmamar Schulp, Bardet & Bouya, 2009	?Sidi Chennane, Oued Meskoura, Oulad Aboun	CIII, upper Maastrichtian	OCP DEK/GE 453 (h); MNHN PMC 29 (r)	Arambourg, 1952 [4]; Schulp et al., 2009 [33]
		Xenodens calminechari Longrich et al., 2021a	Sidi Chennane, Oulad Abdoun	Upper CIII, upper Maastrichtian	MHNM.KH.333 (h)	Longrich et al., 2021a [25]
	Prognathodontini	*Eremiasaurus heterodontus* LeBlanc, Caldwell & Bardet, 2012	Sidi Daoui, Oulad Abdoun; Ben Guerir, Ganntour	Upper CIII, C6-C2, lower to upper Maastrichtian	UALVP 51744 + OCP DEK/GE 112 (s); OCP DEK/GE 663, inedites (r); MNHN PMC41, 42, 45, 48, 50 (r)	Arambourg, 1952 [4]; LeBlanc et al., 2012 [23]; Cappetta et al., 2014 [22]
		Thalassotitan atrox Longrich et al., 2022	Sidi Daoui, Meraa Lahrach, Sidi Chennane, Oulad Abdoun; Ben Guerir, Ganntour; Meskala	Upper CIII, C4-C2, upper Maastrichtian	MNHM.KHG.231 + OCP DEK/GE 417 (s); OCP DEK/GE 10, 90, 98, 109, 497, 665 (r); MHNM.KHB.324-326, 330, 396, 1047, 1051, 1253 (r); MNHN PMC 43, 44, 46, 47, 49, 51 (r)	Arambourg, 1952 [4]; Cappetta et al., 2014 [22]; Longrich et al., 2022 [27]
		Prognathodon currii Christiansen & Bonde, 2002	Ben Guerir, Ganntour	C6-C2, lower to upper Maastrichtian	OCP.DEK/GE 349, 350, 5 inedite teeth (r)	Bardet et al., 2005b [20]; Cappetta et al., 2014 [22]
	Globidensini	*Globidens phosphaticus* Bardet & Pereda-Suberbiola, 2005b	Ben Guerir, Ganntour	C3 (h, p), C6-C2, lower to upper Maastrichtian	OCP.DEK/GE 361 (h), 338–343 (p), 346–348 (r), inedites; MNHN PMC 17–19 (r)	Arambourg, 1952 [4]; Bardet et al., 2005b [20]; Cappetta et al., 2014 [22]
		Globidens simplex LeBlanc, Mohn & Caldwell, 2019	Unknown locality, ?Oulad Abdoun	? Upper CIII, upper Maastrichtian	MHNM.KHG.221 (ex UALVPPPP 51746) (h)	LeBlanc et al., 2019 [24]
	Incertae sedis	*Stelladens mysteriosus* Longrich et al., 2023	Sidi Chennane, Oulad Abdoun	Lower CIII, upper Maastrichtian	MHNM.KHG.1436 (h)	Longrich et al., 2023 [28]

Table A1. *Cont.*

CLADE		TAXON	GEOGRAPHY	STRATIGRAPHY	SPECIMENS	REFERENCES
HALI	Halisaurini	*Halisaurus arambourgi* Bardet & Pereda-Suberbiola, 2005a	Sidi Daoui, Oulad Abdoun; Ben Guerir, Ganntour	Upper CIII, C6-C2, lower to upper Maastrichtian	MNHN PMC 14 (h), PMC 15–16, OCP DEK/GE 100–103, etc. (r)	Bardet et al., 2005a [19]; Cappetta et al., 2014 [22]
	Pluridensini	*Pluridens serpentis* Longrich et al., 2021b	Sidi Daoui, Oulad Abdoun; Ben Guerir, Ganntour	Upper CIII, C6-C2, lower to upper Maastrichtian	OCP DEK/GE 548 + MHNM.KH.262 (s); OCP DEK/GE 662, MHNM.KH.386-395 (r)	Bardet et al. 2005a [19]; Cappetta et al., 2014 [22]; Longrich et al., 2021b [26]
PLIO	Selmasaurini	*Gavialimimus ptychodon* (Arambourg, 1952)	Sidi Daoui, Oulad Abdoun; Ben Guerir, Youssoufia, Ganntour	Upper CIII, C6-C2, lower to upper Maastrichtian	MNHN PMC 30 (h) + MHNM.KHG.1231 (h?) + MNHN PMC 31–34, 53 (r); OCP DEK/GE 304, 356, 476, 560, 661, inedites (r)	Arambourg, 1952 [4]; Cappetta et al., 2014 [22]; Bardet et al., 2015, 2017 [3,5]; Strong et al., 2020 [32]
		Khinjaria acuta Longrich et al., 2024	Sidi Chennane, Oulad Abdoun	Lower CIII, upper Maastrichtian	MHNM.KHG.521 (h)	Longrich et al., 2024 [29]
TYLO		*Hainosaurus boubker* Rempert, Martens & Melchers, 2022	Sidi Chennane, Oulad Abdoun	Upper CIII, upper Maastrichtian	VANPS 13.0120 + 13.0121 (s), 13.0122-165 (r)	Rempert et al., 2022 [31]

Table A2. Mosasauridae from the Maastrichtian Phosphates of Morocco: comparative anatomy data. Measurement and ratio definitions provided in text. Abbreviations: L, skull length; Orb, orbit length; PreO, preorbital length; PostO, postorbital length; W, skull width.

CLADE		TAXON	TEETH			SKULL							SIZE
			H/L	W/L	SPECIMENS	L	W	W/L	PreO/L	PostO/L	Orb/L	SPECIMENS	
MOSA	Mosasaurini	*Mosasaurus beaugei*	1.8–1.9	0.6–0.7	MNHN PMC 8, 55, 60	80–110 cm	25–28 cm	0.28	0.55	0.25	0.17	OCP DEK/GE 83, 551	8–10 m
		Carinodens belgicus	0.5	0.55	OCP DEK/GE 447, 454	<40 cm	?	?	?	?	?	OCP DEK/GE 454	2.5–3 m
		Carinodens minalmamar	0.4	0.4	MNHN PMC 29	?	35 cm	?	?	?	?	OCP DEK/GE 453	2.5 m
		Xenodens calminechari	1	0.4	MHNM.KHG.333	30 cm	?	?	?	?	?	MHNM.KHG.333	2 m
	Prognathodontini	*Eremiasaurus heterodontus*	1.8–1.93	0.8–0.85	OCP DEK/GE 663, inedites	65–70 cm	17 cm	0.24	0.49	0.3	0.19	OCP DEK/GE 112; UALVP 51744	4.5–6 m
		Thalassotitan atrox	1.5–1.83	0.82–0.9	OCP DEK/G 665, inedites	120–130 cm	45–55 cm	0.38	0.5	0.32	0.18	MHNM.KHB.231; OCP DEK/GE 497, 109, 417, 10 (pt)	9–10 m
		Prognathodon currii	1.3	0.77	OCP DEK/GE 349	140 cm	49 cm (est)	0.4	0.48	0.32	0.14	HUJ.OR 100 (Negev)	>=10 m

Table A2. *Cont.*

CLADE		TAXON	TEETH			SKULL							SIZE
			H/L	W/L	SPECIMENS	L	W	W/L	PreO/L	PostO/L	Orb/L	SPECIMENS	
	Globidensini	*Globidens phosphaticus*	1 (ante)-0.69 (post)	0.83 (ante)-0.65 (middle)	OCP DEK/GE 361, 343, 346, 492, inedites	75–80 cm	?	0.55 (est)	0.31 (est)	0.17 (est)	?	OCP DEK/GE 492, inedite; PA 24 (Angola)	5–6 m
		Globidens simplex	0.65–0.77	0.73–0.82	MHNM.KHB.221 (figs. E-F and G-H)	75–80 cm	?	0.46 (est)	0.36 (est)	0.16 (est)	?	MHNM.KHB.221	5–6 m
	Incertae sedis	*Stelladens mysteriosus*	1.33–1.44	0.73–0.8	MHNM.KHB.1436	?	80 cm	?	?	?	?	MHNM.KHB.1436	5 m
HALI	Halisaurini	*Halisaurus arambourgi*	2	0.77–0.85	MNHN PMC 14, 15	35 cm	10 cm	0.29	0.53	0.28	0.18	MNHN PMC 14, 15	3–4 m
	Pluridensini	*Pluridens serpentis*	1.87–2.1	0.7–0.9	OCP DEK/GE 662; MHNM.KHB.389, 394	70–90 cm	20–25 cm	0.3	0.48	0.35	0.12	OCP DEK/GE 548, MHNM.KHB.262	5–6 m
PLIO	Selmasaurini	*Gavialimimus ptychodon*	1.44–1.8	0.62–0.85	MNHN PMC 30, 31, OCP DEK/GE 661	90 cm	22 cm	0.25	0.62	0.27	0.13	MHNM.KHG.1231, OCP DEK/GE 560	6 m
		Khinjaria acuta	?	1.8	MHNM.KHG.521	90 cm	?	0.39 (est)	0.5 (est)	?	?	MHNM.KHG.521	8 m
TYLO		*Hainosaurus boubker*	1.62	0.71–0.72	VANPS 13.0133-13.0147, OCP DEK/GE inedite	?	120 cm	?	?	?	?	VANPS 13.0120-13.022	8–12 m

Table A3. Mosasauridae from the Maastrichtian Phosphates of Morocco: morphometric analysis data. Species and specimens used and their digitalization procedures. Abbreviations: NA, non-available.

CLADE		TAXON	FILE_ID	SPECIMENS	SOURCE	REGION
MOSA	Mosasaurini	*Mosasaurus beaugei*	Mosasaurus_beaugei	MNHN - Collection Bardet	This paper, laser, 0.2 mm precision	NA
		Carinodens belgicus	Carinodens_belgicus	MNHN - Cast 6341	Fischer et al., 2022 [36]	NA, Mirrored
		Xenodens calminechari	Xenodens_calminechari	MHNM.KHB.333	Longrich et al., 2021a, Fig. 5 [25]	Maxilla
	Prognathodontini	*Eremiasaurus heterodontus*	Eremiasaurus_heterodontus	Sparla collection, #2	Photogrammetry	NA
		Thalassotitan atrox	Thalassotitan_atrox	OCP DEK-GE 665	This paper, laser, 0.2 mm precision	NA, Mirrored
		Prognathodon currii	Prognathodon_currii	ULg PA20220209-1	Fischer et al., 2022 [36]	NA, sediment removed
	Globidensini	*Globidens phosphaticus*	Globidens_phosphaticus_post	MNHN - Collection Bardet	This paper, laser, 0.2 mm precision	NA, posterior tooth
	Incertae sedis	*Stelladens mysteriosus*	Stelladens_mysteriosus	MHNM.KHG.1436	Longrich et al., 2023, Fig. 4 [28]	NA
HALI	Halisaurini	*Halisaurusarambourgi*	Halisaurus_arambourgi	OCP DEK/GE inedite	This paper, caliper measurement	NA
	Pluridensini	*Pluridens serpentisi*	Pluridens_serpentis	OCP DEK-GE 662	This paper, laser, 0.2 mm precision	NA
PLIO	Selmasaurini	*Gavialimimus ptychodon*	Gavialimimus_ptychodon	OCP DEK-GE 661	This paper, laser, 0.2 mm precision	NA, Mirrored

References

1. Bardet, N.; Falconnet, J.; Fischer, V.; Houssaye, A.; Jouve, S.; Pereda-Suberbiola, X.; Perez-García, A.; Rage, J.-C.; Vincent, P. Mesozoic marine reptile palaeobiogeography in response to drifting plates. *Gondwana Res.* **2014**, *26*, 869–887. [CrossRef]
2. Polcyn, M.J.; Jacobs, L.L.; Araújo, R.; Schulp, A.S.; Mateus, O. Physical drivers of mosasaur evolution. *Palaeogeogr. Palaeoclimatol. Palaeoecol.* **2013**, *400*, 17–27. [CrossRef]
3. Bardet, N.; Houssaye, A.; Vincent, P.; Pereda-Suberbiola, X.; Amaghzaz, M.; Jourani, E.; Meslouh, S. Mosasaurids (Squamata) from the Maastrichtian phosphates of Morocco: Biodiversity, palaeobiogeography and palaeoecology based on tooth morphoguilds. *Gondwana Res.* **2015**, *27*, 1068–1078. [CrossRef]
4. Arambourg, C. Les vertébrés fossiles des gisements de phosphates (Maroc-Algérie-Tunisie). *Notes Mém. Serv. Géolo. Maroc.* **1952**, *92*, 1–372.
5. Bardet, N.; Gheerbrant, E.; Noubhani, A.; Cappetta, H.; Jouve, S.; Bourdon, E.; Pereda-Suberbiola, X.; Jalil, N.-E.; Vincent, P.; Houssaye, A.; et al. Les Vertébrés des phosphates crétacés-paléogènes (72-47,8 Ma) du Maroc. In *Paléontologie des Vertébrés du Maroc: État des Connaissances*; Zouhri, S., Ed.; Mémoires de la Société géologique de France: Paris, France, 2017; Volume 180, pp. 351–452.
6. Labita, C.; Martill, D.M. An articulated pterosaur wing from the Upper Cretaceous (Maastrichtian) phosphates of Morocco. *Cretac. Res.* **2021**, *119*, 104679. [CrossRef]
7. Longrich, N.R.; Pereda-Suberbiola, X.; Bardet, N.; Jalil, N.-E. A new small duckbilled dinosaur (Hadrosauridae: Lambeosaurinae) from Morocco and dinosaur diversity in the late Maastrichtian of North Africa. *Sci. Rep.* **2024**, *14*, 3665. [CrossRef]
8. Gheerbrant, E. Ancestral radiation of paenungulate mammals (Paenungulatomorpha)-New evidence from the Paleocene of Morocco. *J. Vertebr. Paleontol.* **2023**, *42*, e2197971. [CrossRef]
9. Cappetta, H. Chondrichthyes (Mesozoic and Cenozoic Elasmobranchii: Teeth). In *Handbook of Paleoichthyology, Chondrichthyes*; Schultze, H.-P., Ed.; Verlag F. Pfeil: München, Germany, 2012; Volume 3E, pp. 1–512.
10. Noubhani, A.; Cappetta, H. Les Orectolobiformes, Carcharhiniformes et Myliobatiformes (Elasmobranchii, Neoselachii) des bassins à phosphate du Maroc (Maastrichtien-Lutétien basal). Systématique, biostratigraphie, évolution et dynamique des faunes. *Palaeo Ichthyol.* **1997**, *8*, 1–327.
11. Bardet, N.; Pereda-Suberbiola, X.; Jouve, S.; Bourdon, E.; Vincent, P.; Houssaye, A.; Rage, J.C.; Jalil, N.-E.; Bouya, B.; Amaghzaz, M. Reptilian assemblages from the latest Cretaceous—Palaeogene phosphates of Morocco: From Arambourg to present time. *Hist. Biol.* **2010**, *22*, 186–199. [CrossRef]
12. Jouve, S.; Bardet, N.; Jalil, N.-E.; Pereda-Suberbiola, X.; Bouya, B.; Amaghzaz, M. The oldest African crocodylian: Phylogeny, paleobiogeography, and differential survivorship of marine reptiles through the Cretaceous-Tertiary boundary. *J. Vertebr. Paleontol.* **2008**, *28*, 409–421. [CrossRef]
13. Vincent, P.; Bardet, N.; Pereda-Suberbiola, X.; Bouya, B.; Amaghzaz, M.; Meslouh, S. *Zarafasaura oceanis*, a new elasmosaurid (Reptilia: Sauropterygia) from the Maastrichtian Phosphates of Morocco and the palaeobiogeography of latest Cretaceous plesiosaurs. *Gondwana Res.* **2011**, *19*, 1062–1073. [CrossRef]
14. Vincent, P.; Bardet, N.; Houssaye, A.; Amaghzaz, M.; Meslouh, S. New plesiosaur specimens from the Maastrichtian Phosphates of Morocco and their implications for the ecology of the latest Cretaceous marine apex predators. *Gondwana Res.* **2013**, *24*, 796–805. [CrossRef]
15. Bardet, N.; Jalil, N.-E.; de Lapparent de Broin, F.; Germain, D.; Lambert, O.; Amaghzaz, M. A Giant Chelonioid Turtle from the Late Cretaceous of Morocco with a Suction Feeding Apparatus Unique among Tetrapods. *PLoS ONE* **2013**, *8*, e63586. [CrossRef]
16. Lapparent de Broin, F.; Bardet, N.; Amaghzaz, M.; Meslouh, S. A strange new chelonioid turtle from the Latest Cretaceous phosphates of Morocco. *Comptes Rendus Palevol* **2013**, *13*, 87–95. [CrossRef]
17. Houssaye, A.; Bardet, N.; Rage, J.-C.; Pereda-Suberbiola, X.; Bouya, B.; Amaghzaz, M.; Amalik, M. A review of *Pachyvaranus crassispondylus* Arambourg, 1952, a pachyostotic marine squamate from the latest Cretaceous phosphates of the margin of the southern Tethys. *Geol. Mag.* **2011**, *148*, 237–249. [CrossRef]
18. Bardet, N.; Pereda-Suberbiola, X.; Iarochène, M.; Bouyahyaoui, F.; Bouya, B.; Amaghzaz, M. *Mosasaurus beaugei* Arambourg, 1952 (Squamata, Mosasauridae) from the Late Cretaceous phosphates of Morocco. *Geobios* **2004**, *37*, 315–324. [CrossRef]
19. Bardet, N.; Pereda-Suberbiola, X.; Iarochène, M.; Bouya, B.; Amaghzaz, M. New species of *Halisaurus* from the Late Cretaceous phosphates of Morocco, and the phylogenetical relationships of the Halisaurinae (Squamata: Mosasauridae). *Zool. J. Linn. Soc.* **2005**, *143*, 447–472. [CrossRef]
20. Bardet, N.; Pereda-Suberbiola, X.; Iarochène, M.; Amalik, M.; Bouya, B. Durophagous Mosasauridae (Squamata) from the Upper Cretaceous phosphates of Morocco, with description of a new species of *Globidens*. *Neth. J. Geosci.* **2005**, *84*, 167–175.
21. Bardet, N.; Pereda-Suberbiola, X.; Schulp, A.S.; Bouya, B. New material of *Carinodens* (Squamata, Mosasauridae) from the Maastrichtian (Late Cretaceous) phosphates of Morocco. *Fort Hays Stud. Spec. Pap. Sternberg Mus. Nat. Hist.* **2007**, *3*, 29–36.

22. Cappetta, H.; Bardet, N.; Pereda-Suberbiola, X.; Adnet, S.; Akkrim, D.; Amalik, M.; Benabdallah, A. Marine vertebrate faunas from the Maastrichtian Phosphates of the Benguérir area (Ganntour Basin, Morocco): Biostratigraphy, palaeobiogeography and palaeoecology. *Palaeogeogr. Palaeoclimatol. Palaeoecol.* **2014**, *409*, 217–238. [CrossRef]
23. LeBlanc, A.R.H.; Caldwell, M.W.; Bardet, N. A new mosasaurine from the Maastrichtian (Upper Cretaceous) phosphates of Morocco and its implications for mosasaurine systematics. *J. Vertebr. Paleontol.* **2012**, *32*, 82–104. [CrossRef]
24. LeBlanc, A.R.H.; Mohr, S.R.; Caldwell, M.W. Insights into the anatomy and functional morphology of durophagous mosasaurines (Squamata: Mosasauridae) from a new species of *Globidens* from Morocco. *Zool. J. Linn. Soc.* **2019**, *186*, 1026–1052. [CrossRef]
25. Longrich, N.R.; Bardet, N.; Schulp, A.S.; Jalil, N.-E. *Xenodens calminechari* gen. et sp. nov., a bizarre mosasaurid (Mosasauridae, Squamata) with shark-like cutting teeth from the upper Maastrichtian of Morocco, North Africa. *Cretac. Res.* **2021**, *123*, 104764. [CrossRef]
26. Longrich, N.R.; Bardet, N.; Khaldoune, F.; Yazami, O.K.; Jalil, N.-E. *Pluridens serpentis*, a new mosasaurid (Mosasauridae: Halisaurinae) from the Maastrichtian of Morocco and implications for mosasaur diversity. *Cretac. Res.* **2021**, *126*, 104882. [CrossRef]
27. Longrich, N.R.; Jalil, N.-E.; Khaldoune, F.; Yazami, O.K.; Pereda-Suberbiola, X.; Bardet, N. *Thalassotitan atrox*, a giant predatory mosasaurid (Squamata) from the Upper Maastrichtian Phosphates of Morocco. *Cretac. Res.* **2022**, *140*, e105315. [CrossRef]
28. Longrich, N.R.; Jalil, N.-E.; Pereda-Suberbiola, X.; Bardet, N. *Stelladens mysteriosus*: A strange new mosasaurid (Squamata) from the Maastrichtian (Late Cretaceous) of Morocco. *Foss. Stud.* **2023**, *1*, 2–14. [CrossRef]
29. Longrich, N.R.; Polcyn, M.J.; Jalil, N.-E.; Pereda-Suberbiola, X.; Bardet, N. A bizarre new plioplatecarpine mosasaurid from the Maastrichtian of Morocco. *Cretac. Res.* **2024**, *160*, e105870. [CrossRef]
30. Polcyn, M.J.; Lindgren, J.; Bardet, N.; Cornelissen, D.; Verding, L.; Schulp, A.S. Description of new specimens of *Halisaurus arambourgi* Bardet & Pereda-Suberbiola, 2005 and comments on the phylogeny of Halisaurinae. *Bull. Soc. Géolog. Fr.* **2012**, *183*, 121–134.
31. Rempert, T.H.; Martens, B.P.; Melchers, A.P.V. First record of a Tylosaurine mosasaur from the Latest Cretaceous Phosphates of Morocco. *Open J. Geol.* **2022**, *12*, 883–906. [CrossRef]
32. Strong, C.R.C.; Caldwell, M.W.; Konishi, T.; Palci, A. A new species of longirostrine plioplatecarpine mosasaur (Squamata: Mosasauridae) from the Late Cretaceous of Morocco, with a re-evaluation of the problematic taxon *'Platecarpus' ptychodon*. *J. Syst. Palaeontol.* **2020**, *18*, 1769–1804. [CrossRef]
33. Schulp, A.S.; Bardet, N.; Bouya, B. A new species of the durophagous mosasaur *Carinodens* (Squamata, Mosasauridae) and additional material of *Carinodens belgicus* from the Maastrichtian phosphates of Morocco. *Neth. J. Geosci.* **2009**, *88*, 161–167. [CrossRef]
34. Bardet, N. Maastrichtian marine reptiles of the Mediterranean Tethys: A palaeobiogeographical approach. *Bull. Soc. Géolog. Fr.* **2012**, *183*, 573–596. [CrossRef]
35. Crofts, S.B.; Smith, S.M.; Anderson, P.S.L. Beyond description: The many facets of dental biomechanics. *Integr. Comp. Biol.* **2020**, *60*, 594–607. [CrossRef] [PubMed]
36. Fischer, V.; Bennion, R.F.; Foffa, D.; MacLaren, J.A.; McCurry, M.R.; Melstrom, K.M.; Bardet, N. Ecological signal in the size and shape of marine amniote teeth. *Proc. R. Soc. B* **2022**, *289*, 20221214. [CrossRef]
37. Lindgren, J.; Siverson, M. *Tylosaurus ivoensis*: A giant mosasaur from the Early Campanian of Sweden. *Earth Environ. Sci. Trans. R. Soc. Edinb.* **2002**, *93*, 73–93. [CrossRef]
38. Massare, J.A. Tooth morphology and prey preference of Mesozoic marine reptiles. *J. Vertebr. Paleontol.* **1987**, *7*, 121–137. [CrossRef]
39. Schulp, A.S.; Vonhof, H.B.; Van Der Lubbe, J.H.J.L.; Janssen, R.; Van Baal, R.R. On diving and diet: Resource partitioning in type-Maastrichtian mosasaurs. *Neth. J. Geosci.* **2003**, *92*, 165e170. [CrossRef]
40. Hornung, J.J.; Reich, M. Tylosaurine mosasaurs (Squamata) from the Late Cretaceous of northern Germany. *Neth. J. Geosci.* **2015**, *94*, 55–71. [CrossRef]
41. McCurry, M.R.; Evans, A.R.; Fitzgerald, E.M.G.; McHenry, C.R.; Bevitt, J.; Pyenson, N.D. The repeated evolution of dental apicobasal ridges in aquatic-feeding mammals and reptiles. *Biol. J. Linn. Soc.* **2019**, *127*, 245–259. [CrossRef]
42. Holwerda, F.M.; Bestwick, J.; Purnell, M.A.; Jagt, J.W.M.; Schulp, A.S. Three-dimensional dental microwear in type-Maastrichtian mosasaur teeth (Reptilia, Squamata). *Sci. Rep.* **2023**, *13*, 18720. [CrossRef]
43. McKensie, A.S.; Brock, G.A.; McCurry, M.R. The impact of apicobasal ridges on dental load-bearing capacity in aquatic-feeding predatory amniotes. *Paleobiology* **2024**, *50*, 346–363. [CrossRef]
44. Sander, P.M. The microstructure of reptilian tooth enamel: Terminology, function, and phylogeny. *Münch. Geowiss. Abh.* **1999**, *A38*, 1–102.
45. Gren, J.A.; Lindgren, J. Dental histology of mosasaurs and a marine crocodylian from the Campanian (Upper Cretaceous) of southern Sweden: Incremental growth lines and dentine formation rates. *Geol. Mag.* **2013**, *151*, 134–143. [CrossRef]
46. Owocki, K.; Madzia, D. Predatory behaviour in mosasaurid squamates inferred from tooth microstructure and mineralogy. *Cretac. Res.* **2020**, *111*, 104430. [CrossRef]

47. Street, H.P.; LeBlanc, A.R.H.; Caldwell, M.W. A histological investigation of dental crown characters used in mosasaur phylogenetic analyses. *Vertebr. Anat. Morphol. Palaeontol.* **2021**, *9*, 82–104. [CrossRef]
48. Rieppel, O.; Kierney, M. Tooth Replacement in the Late Cretaceous Mosasaur *Clidastes*. *J. Herpetol.* **2005**, *39*, 688–692. [CrossRef]
49. Caldwell, M.W. Ontogeny, anatomy and attachment of the dentition in mosasaurs (Mosasauridae: Squamata). *Zool. J. Linn. Soc.* **2007**, *149*, 687–700. [CrossRef]
50. LeBlanc, A.R.H.; Lamoureux, D.O.; Caldwell, M.W. Mosasaurs and snakes have a periodontal ligament: Timing and extent of calcification, not tissue complexity, determines tooth attachment mode in reptiles. *J. Anat.* **2017**, *231*, 869–885. [CrossRef]
51. Notholt, A.J.G. Phosphorite Resources in the Mediterranean (Tethyan) Phosphogenic Province: A Progress Report. *Sci. Géolo. Bull. Mém.* **1985**, *77*, 9–17.
52. Lucas, J.; Prévôt-Lucas, L. Tethyan phosphates and bioproductites. In *The Tethys Ocean*; Springer: Berlin/Heidelberg, Germany, 1996; pp. 367–391.
53. Brives, A. Sur les terrains éocènes dans le Maroc occidental. *Comptes Rendus l'Acad. Sci.* **1905**, *140*, 395–397.
54. Brives, A. Sur le Sénonien et l'Eocène de la bordure nord de l'Atlas marocain. *Comptes Rendus l'Acad. Sci.* **1908**, *146*, 873–875.
55. Office Chérifien des Phosphates. The phosphates basins of Morocco. In *Phosphates Deposits of the World, Vol. 2—Phosphate Rock Resources*; Northolt, A.J.G., Sheldon, R.P., Davidson, D.F., Eds.; Cambridge University Press: Cambridge, UK, 1989; pp. 301–311.
56. El Bamiki, R.; Raji, O.; Ouabid, M.; Elghali, A.; Khadiri Yazami, O.; Bodinier, J.-L. Phosphate Rocks: A Review of Sedimentary and Igneous Occurrences in Morocco. *Minerals* **2021**, *11*, 1137. [CrossRef]
57. Charrière, A.; Haddoumi, H.; Mojon, P.-O.; Ferrière, J.; Cuche, D.; Zili, L. Mise en évidence par charophytes et ostracodes de l'âge Paléocène des dépôts discordants sur les rides anticlinales de la région d'Imilchil (Haut Atlas, Maroc): Conséquences paléogéographiques et structurales. *Comptes Rendus Palevol* **2009**, *8*, 9–19. [CrossRef]
58. Martin, J.E.; Vincent, P.; Tacail, T.; Khaldoune, F.; Jourani, E.; Bardet, N.; Balter, V. Calcium isotopic evidence for vulnerable marine ecosystem structure prior to the K/Pg extinction. *Curr. Biol.* **2017**, *27*, 1641–1644. [CrossRef]
59. Cosmidis, J.; Benzerara, K.; Gheerbrant, E.; Eestève, I.; Bouya, B.; Amaghzaz, M. Nanometer-scale characterization of exceptionally preserved bacterial fossils in Paleocene phosphorites from Ouled Abdoun (Morocco). *Geobiology* **2013**, *11*, 139–153. [CrossRef]
60. Kocsis, L.; Gheerbrant, E.; Mouflih, M.; Cappetta, H.; Yans, J.; Amaghzaz, M. Comprehensive stable isotope investigation of marine biogenic apatite from the late Cretaceous–early Eocene phosphate series of Morocco. *Palaeogeogr. Palaeoclimatol. Palaeoecol.* **2014**, *394*, 74–88. [CrossRef]
61. Kocsis, L.; Gheerbrant, E.; Mouflih, M.; Cappetta, H.; Ulianov, A.; Chiaradia, M.; Bardet, N. Gradual changes in upwelled seawater conditions (redox, pH) from the late Cretaceous through early Paleogene at the northwest coast of Africa: Negative Ce anomaly trend recorded in fossil bio-apatite. *Chem. Geol.* **2016**, *421*, 44–54. [CrossRef]
62. Russell, D.A. Systematics and morphology of American mosasaurs. *Bull. Peabody Mus. Nat. Hist.* **1967**, *23*, 1–240.
63. Woodward, A.S. Note on a tooth of an extinct Alligator (*Bottosaurus belgicus*, sp. nov.) from the Lower Danian of Ciply, Belgium. *Geol. Mag.* **1891**, *8*, 114–115. [CrossRef]
64. Christiansen, P.; Bonde, N. A new species of gigantic mosasaur from the Late Cretaceous of Israel. *J. Vertebr. Paleontol.* **2002**, *22*, 629–644. [CrossRef]
65. Bonhomme, V.; Picq, S.; Gaucherel, C.; Claude, J. Momocs: Outline Analysis Using R. *J. Stat. Softw.* **2014**, *56*, 1–24. [CrossRef]
66. Simões, T.R.; Vernygora, O.; Paparella, I.; Jiménez-Huidobro, P.; Caldwell, M.W. Mosasauroid phylogeny under multiple phylogenetic methods provides new insights on the evolution of aquatic adaptations in the group. *PLoS ONE* **2017**, *12*, e0176773. [CrossRef] [PubMed]
67. Madzia, D.; Cau, A. Inferring 'weak spots' in phylogenetic trees: Application to mosasauroid nomenclature. *PeerJ* **2017**, *5*, e3782. [CrossRef]
68. Bell, G.L., Jr. A phylogenetical revision of North American and Adriatic Mosasauroidea. In *Ancient Marine Reptiles*; Callaway, J.M., Nicholls, E.L., Eds.; Academic Press: San Diego, CA, USA, 1997; pp. 293–332.
69. Schulp, A.S.; Jagt, J.; Camp, W.M.; Fonken, F. New material of the mosasaur *Carinodens belgicus* from the Upper Cretaceous of The Netherlands. *J. Vertebr. Paleontol.* **2004**, *2*, 744–747. [CrossRef]
70. Bardet, N. The mosasaur collections of the Muséum National d'Histoire Naturelle of Paris. *Bull. Soc. Géolo. Fr.* **2012**, *183*, 35–53. [CrossRef]
71. Street, H.P.; Caldwell, M.W. Rediagnosis and redescription of *Mosasaurus hoffmannii* (Squamata: Mosasauridae) and an assessment of species assigned to the genus *Mosasaurus*. *Geol. Mag.* **2017**, *154*, 521e557. [CrossRef]
72. Agassiz, L. *Recherches sur les Poissons Fossiles*; Imprimerie Petitpierre: Neuchâtel, Switzerland, 1833–1843; Volume 5, 1420p.
73. Dollo, L. *Globidens fraasi*, mosasaurien mylodonte nouveau du Maestrichtien (Crétacé supérieur) du Limbourg, et l'Ethologie de la nutrition chez les mosasauriens. *Arch. Biol.* **1913**, *28*, 609–626.
74. Schulp, A.S.; Averianov, A.O.; Yarkov, A.A.; Trikolidi, F.A.; Jagt, J.W.M. First record of the Late Cretaceous durophagous mosasaur *Carinodens belgicus* (Squamata, Mosasauridae) from Volgograd region (Russia) and Crimea (Ukraine). *Russ. J. Herpetol.* **2006**, *13*, 175–180.

75. Thurmond, J.T. New name for the mosasaur *Compressidens* Dollo, 1924. *J. Paleontol.* **1969**, *43*, 1298.
76. Kaddumi, H.F. The first and most complete *Carinodens* (Squamata: Mosasauridae) skeleton yet with a description of a new species from the Harrana Fauna. In *Fossils of the Harrana Fauna and the Adjacent Areas*; Publications of the Eternal River Museum of Natural History: Amman, Jordan, 2009; pp. 49–64.
77. Schulp, A.S. Feeding the mechanical mosasaur: What did *Carinodens* eat? *Neth. J. Geosci.* **2005**, *84*, 345–357. [CrossRef]
78. Zdansky, O. The occurrence of mosasaurs in Egypt and in Africa in general. *Bull. l'Inst. d'Egypte* **1935**, *17*, 83–94. [CrossRef]
79. Bardet, N.; Pereda-Suberbiola, X.; Jalil, N.-E. A new mosasauroid (Squamata) from the Late Cretaceous (Turonian) of Morocco. *Comptes Rendus Palevol* **2003**, *2*, 607–616. [CrossRef]
80. Owen, R. *Odontography; or, a Treatise on the Comparative Anatomy of the Teeth; Their Physiological Relations, Mode of Development, and Micro-scopic Structure, in the Vertebrate Animals*; Hippolyte Bailliere: London, UK, 1840–1845; 655p.
81. Schulp, A.S.; Polcyn, M.J.; Mateus, O.; Jacobs, L.L.; Morais, M.L. A new species of Prognathodon (Squamata, Mosasauridae) from the Maastrichtian of Angola, and the affinities of the mosasaur genus *Liodon*. *Fort Hays Stud. Spec. Pap. Sternberg Mus. Nat. Hist.* **2007**, *3*, 1–12.
82. Conybeare, W.D. *Mosasaurus*. The saurus of the Meuse, the Maestricht animal of Cuvier. In *Outlines in Oryctology: An Introduction to the Study of Fossil Organic Remains*, 1st ed.; Parkinson, J., Ed.; Sherwood, Neely, Jones & Phillips: London, UK, 1822; pp. 298–301.
83. Camp, C.L. California mosasaurs. *Mem. Univ. Calif.* **1942**, *13*, 1–68.
84. Lindgren, J.; Jagt, J.W.M.; Caldwell, M.W. A fishy mosasaur: The axial skeleton of *Plotosaurus* (Reptilia, Squamata) reassessed. *Lethaia* **2007**, *40*, 153–160. [CrossRef]
85. Dortangs, R.W.; Schulp, A.S.; Mulder, E.W.A.; Jagt, J.W.M.; Peeters, H.H.G.; De Graaf, D.T. A large new mosasaur from the Upper Cretaceous of The Netherlands. *Neth. J. Geosci.* **2002**, *81*, 1–8. [CrossRef]
86. Lambert, O.; Bianucci, G.; de Muizon, C. Macroraptorial sperm whales (Cetacea, Odontoceti, Physeteroidea) from the Miocene of Peru. *Zool. J. Linn. Soc.* **2016**, *179*, 404–474. [CrossRef]
87. Osborn, H.F. *Tyrannosaurus* and other Cretaceous carnivorous dinosaurs. *Bull. Am. Mus. Nat. Hist.* **1905**, *21*, 259–265.
88. Lewy, Z.; Cappetta, H. Senonian elasmobranch teeth from Israel. Biostratigraphic and paleoenvironmental implications. *Neues Jahrb. Geol. Paläontol. Monatshefte* **1989**, *4*, 212–222. [CrossRef]
89. Meyer von, C.E.H. Mitteilungen, an Professor Bronn gerichtet. *Neues Jahrb. Mineral. Geol. Geogn. Petrefaktenkunde* **1837**, *4*, 413–418.
90. Fanti, F.; Miyashita, T.; Cantelli, L.; Mnasri, F.; Dridi, J.; Contessi, M.; Cau, A. The largest thalattosuchian (Crocodylomorpha) supports teleosaurid survival across the Jurassic-Cretaceous boundary. *Cretac. Res.* **2016**, *61*, 263–274. [CrossRef]
91. Tichy, G.; Karl, H.-V. The structure of fossil teeth of chelonophagous crocodiles (Diapsida: Crocodylia). *Stud. Geol. Salmant.* **2004**, *40*, 115–124.
92. Polcyn, M.J.; Jacobs, L.L.; Schulp, A.S.; Mateus, O. The North African Mosasaur Globidens phosphaticus from the Maastrichtian of Angola. *Hist. Biol.* **2010**, *22*, 175–185. [CrossRef]
93. Gilmore, C.W. A new mosasauroid reptile from the Cretaceous of Alabama. *Proc. U. S. Natl. Mus.* **1912**, *41*, 479–484. [CrossRef]
94. Luan, X.; Walker, C.; Dangaria, S.; Ito, Y.; Druzinsky, R.; Jarosius, K.; Lesot, H.; Rieppel, O. The mosasaur tooth attachment apparatus as paradigm for the evolution of the gnathostome periodontium. *Evol. Dev.* **2009**, *11*, 247–259. [CrossRef]
95. Konishi, T.; Caldwell, M.W.; Nishimura, T.; Sakurai, K.; Tanoue, K. A new halisaurine mosasaur (Squamata: Halisaurinae) from Japan: The first record in the western Pacific realm and the first documented insights into binocular vision in mosasaurs. *J. Syst. Palaeontol.* **2016**, *14*, 809–839. [CrossRef]
96. Shaker, A.A.; Longrich, N.R.; Strougo, A.; Asan, A.; Bardet, N.; Mousa, M.K.; Tantawy, A.A.; Abu El-Kheir, G.A. A new species of *Halisaurus* (Mosasauridae: Halisaurinae) from the lower Maastrichtian (Upper Cretaceous) of the Western Desert, Egypt. *Cretac. Res.* **2024**, *154*, 105719. [CrossRef]
97. Marsh, O.C. Notice of some new mosasauroid reptiles from the greensand of New Jersey. *Am. J. Sci.* **1869**, *48*, 392–397. [CrossRef]
98. Dollo, L. Note sur l'ostéologie des Mosasauridae. *Bull. Musée R. d'Hist. Nat. Belg.* **1882**, *1*, 55–80.
99. Brown, D.S. The English Upper Jurassic Plesiosauroidea (Reptilia) and a review of the phylogeny and classification of the Plesiosauria. *Bull. Br. Mus. Nat. Hist. Geol.* **1981**, *35*, 253–347.
100. Rothschild, B.; Martin, L. Mosasaur ascending: The phylogeny of bends. *Neth. J. Geosci.* **2005**, *84*, 341e344.
101. Dollo, L. Première note sur les mosasauriens de Mesvin. *Mém. Soc. Belg. Géol. Paléontol. d'Hydrol.* **1889**, *3*, 271–304.
102. Lingham-Soliar, T. A new mosasaur *Pluridens walkeri* from the Upper Cretaceous, Maastrichtian of the Iullemmeden Basin, Southwest Niger. *J. Vertebr. Paleontol.* **1998**, *18*, 709–717. [CrossRef]
103. Polcyn, M.J.; Bardet, N.; Albright, L.B., III; Titus, A. A new lower Turonian mosasaurid from the Western Interior Seaway and the antiquity of the unique basicranial circulation pattern in Plioplatecarpinae. *Cretac. Res.* **2023**, 105621. [CrossRef]
104. Plasse, M.; Valentin, X.; Garcia, G.; Guinot, G.; Bardet, N. New remains of Mosasauroidea (Reptilia, Squamata) from the Upper Cretaceous (Santonian) of Aude, southern France. *Cretac. Res.* **2024**, *157*, 105823. [CrossRef]
105. Polcyn, M.J.; Bell, G.L., Jr. *Russellosaurus coheni* n. gen., n. sp., a 92 million-year-old mosasaur from Texas (USA), and the definition of the parafamily Russellosaurina. *Neth. J. Geosci.* **2005**, *84*, 321–333. [CrossRef]

106. Polcyn, M.J.; Bardet, N.; Amaghzaz, M.; Gonçalves, O.A.; Houssaye, A.; Jourani, E.; Kaddumi, H.F.; Lindgren, J.; Mateus, O.; Meslouh, S.; et al. An extremely derived plioplatecarpine mosasaur from the Maastrichtian of Africa and the Middle East. In Proceedings of the 5th Triennal Mosasaur Meeting—A Global Perspective on Mesozoic Marine Amniotes, Uppsala, Sweden, 16–20 May 2016; pp. 32–33.
107. Wright, K.R.; Shannon, S.W. A new plioplatecarpine mosasaur (Squamata, Mosasauridae) from Alabama. *J. Vertebr. Paleontol.* **1998**, *8*, 102–107. [CrossRef]
108. Azzaroli, A.; De Giuli, C.; Ficcarelli, G.; Torre, D. An aberrant mosasaur from the Upper Cretaceous of north western Nigeria. *Atti Accad. Naz. Lincei. Cl. Sci. Fis. Mat. Naturali. Rend.* **1972**, *52*, 398–402.
109. Marx, F.G.; Lambert, O.; Uhen, M.D. Cetacean Palaeobiology. In *Topics in Palaeobiology*; Benton, M.J., Ed.; John Wiley & Sons: Chichester, UK, 2016; 346p.
110. Dollo, L. Notes sur les vertébrés récemment offerts au Musée de Bruxelles par M. Alfred Lemonnier. *Bull. Soc. Belg. Géol. Paléontol. d'Hydrol.* **1889**, *3*, 181–182.
111. Thévenin, A. Mosasauriens de la Craie Grise de Vaux-Eclusier près de Péronne (Somme). *Bull. Soc. Géolo. Fr.* **1896**, *24*, 900–916.
112. Persson, P.O. Studies on Mesozoic marine reptile faunas with particular regard to the Plesiosauria. *Inst. Mineral. Paleontol. Quat. Geol. Univ. Lund Swed.* **1963**, *118*, 1–15.
113. Hector, J. On the Fossil Reptilia of New Zealand. *Trans. Proc. N. Z. Inst.* **1874**, *6*, 333–358.
114. Bell, G.L., Jr.; Polcyn, M.J. *Dallasaurus turneri*, a new primitive mosasauroid from the Middle Turonian of Texas and comments on the phylogeny of Mosasauridae (Squamata). *Neth. J. Geosci.* **2005**, *84*, 177–194. [CrossRef]
115. Marsh, O.C. Note on *Rhinosaurus*. *Am. J. Sci.* **1872**, *4*, 147.
116. Dollo, L. Le Hainosaure. *Rev. Quest. Sci.* **1885**, *18*, 285–289.
117. Lindgren, J. The first record of *Hainosaurus* (Reptilia: Mosasauri dae) from Sweden. *J. Paleontol.* **2005**, *79*, 1157–1165. [CrossRef]
118. Fernandez, M.; Martin, J.E. Description and phylogenetic relationships of *Taniwhasaurus antarcticus* (Mosasauridae, Tylosaurinae) from the Upper Campanian (Cretaceous) of Antarctica. *Cretac. Res.* **2009**, *30*, 717–726. [CrossRef]
119. Polcyn, M.J.; Everhart, M.J. Description and phylogenetic analysis of a new species of *Selmasaurus* (Mosasauridae: Plioplatecarpinae) from the Niobrara Chalk of western Kansas. In Proceedings of the Second Mosasaur Meeting, Hays, KS, USA, 3–6 May 2007; pp. 13–28.
120. Azzaroli, A.; De Giuli, C.; Ficcarelli, G.; Torre, D. Late Cretaceous Mosasaurs from the Sokoto District, Nigeria. *Atti Della Accad. Naz. Lincei. Cl. Sci. Fis. Mat. Nat.* **1975**, *13*, 21–34.
121. Soliar, T. The mosasaur *Goronyosaurus* from the Upper Cretaceous of Sokoto State, Nigeria. *Palaeontology* **1998**, *31*, 747–762.
122. Lingham Soliar, T. Mosasaurs from the Upper Cretaceous of Niger. *Palaeontology* **1991**, *34*, 653–670.
123. Jiménez-Huidobro, P.; Caldwell, M.W. Reassessment and Reassignment of the Early Maastrichtian Mosasaur *Hainosaurus bernardi* Dollo, 1885, to *Tylosaurus* Marsh, 1872. *J. Vertebr. Paleontol.* **2016**, *36*, e1096275. [CrossRef]
124. Jiménez-Huidobro, P.; Caldwell, M.W. A new hypothesis of the phylogenetic relationships of the Tylosaurinae (Squamata: Mosasauroidea). *Front. Earth Sci.* **2019**, *7*, 47. [CrossRef]
125. Kiernan, C.R. Stratigraphic distribution and habitat segregation of mosasaurs in the Upper Cretaceous of western and central Alabama, with an historical review of Alabama mosasaur discoveries. *J. Vertebr. Paleontol.* **2002**, *22*, 91–103. [CrossRef]
126. Bardet, N.; Pereda-Suberbiola, X.; Corral, J.C. A tylosaurine Mosasauridae (Squamata) from the Late Cretaceous of the Basque-Cantabrian Region. *Estud. Geol.* **2006**, *62*, 213–218. [CrossRef]
127. Caldwell, M.W.; Konishi, T.; Obata, I.; Muramoto, K. New species of *Taniwhasaurus* (Mosasauridae, Tylosaurinae) from the upper Santonian-lower Campanian (Upper Cretaceous) of Hokkaido, Japan. *J. Vertebr. Paleontol.* **2008**, *28*, 339–348. [CrossRef]
128. Foffa, D.; Young, M.T.; Stubbs, T.L.; Dexter, K.G.; Brusatte, S.L. 2018 The long-term ecology and evolution of marine reptiles in a Jurassic seaway. *Nat. Ecol. Evol.* **2018**, *2*, 1548–1555. [CrossRef]
129. MacLaren, J.A.; Bennion, R.F.; Bardet, N.; Fischer, V. Global ecomorphological restructuring of dominant marine reptiles prior to the 2 K/Pg mass extinction. *Proc. R. Soc. B* **2022**, *289*, 0585. [CrossRef]
130. Chatterjee, S.; Small, B.J. New plesiosaurs from the Upper Cretaceous of Antarctica. *Geol. Soc. Spec. Publ.* **1989**, *47*, 197–215. [CrossRef]
131. O'Keefe, F.R.; Otero, R.A.; Soto-Acuña, S.; O'Gorman, J.P.; Godfrey, S.J.; Chatterjee, S. Cranial anatomy of *Morturneria seymourensis* from Antarctica, and the evolution of filter feeding in plesiosaurs of the Austral Late Cretaceous. *J. Vertebr. Paleontol.* **2017**, *37*, e1347570. [CrossRef]

diversity

MDPI

Article

Mosasaur Feeding Ecology from the Campanian Bearpaw Formation, Alberta, Canada: A Preliminary Multi-Proxy Approach

Femke M. Holwerda [1,2,*], Mark T. Mitchell [1], Madelon van de Kerk [3] and Anne S. Schulp [2,4]

1 Royal Tyrrell Museum of Palaeontology, P.O. Box 7500, Drumheller, AB T0J 0Y0, Canada
2 Department of Geosciences, Utrecht University, Princetonlaan 8a, 3584 CB Utrecht, The Netherlands; a.s.schulp@uu.nl
3 School of Environment & Sustainability, Western Colorado University, 1 Western Way, Gunnison, CO 81231, USA; mvandekerk@western.edu
4 Naturalis Biodiversity Center, P.O. Box 9517, 2300 RA Leiden, The Netherlands
* Correspondence: f.m.holwerda@uu.nl

Abstract: Mosasaur taxa recovered from the Bearpaw Formation, Alberta, Canada, generally show exceptional preservation after rapid burial. Since the mosasaur community consisted of two dominant taxa, *Mosasaurus missourensis Prognathodon overtoni*, and three less prevalent taxa *Tylosaurus proriger*, *Mosasaurus conodon*, and *Plioplatecarpus primaevus*, some form of habitat or dietary niche partitioning is to be expected. To test this, several approaches are tried. Two-dimensional microwear analysis is used as an exploratory method to quantify tooth abrasion by food items. The good preservation of skull material reveals complete tooth rows of the upper and lower jaws for *Mosasaurus missouriensis* and *Prognathodon overtoni*, as well as isolated teeth for all taxa. The teeth are also measured for tooth bending strength to test stress regarding usage of teeth per taxon. Energy-dispersive X-ray spectroscopy (EDX) measurements show trace elements which may be a preliminary test for relative trophic level positioning. Some mosasaurs have their stomach contents preserved, providing direct and unambiguous evidence of diet. The results show a relatively clear separation of two-dimensional microwear counts between *Mosasaurus*, *Prognathodon*, and *Plioplatecarpus*, which, to some extent, is reflected in the EDX analyses, too. Tooth bending strength measurements show a clear difference between the latter three mosasaurs but no difference between the lower and upper jaws in *Mosasaurus* or *Prognathodon*. The combination of these three techniques maps a clear dietary niche differentiation of Bearpaw mosasaurs, which lays the groundwork for future analytical or chemical palaeoecological studies.

Keywords: Mosasauridae; Bearpaw Formation; dental microwear; feeding ecology; Cretaceous; geochemical analysis

1. Introduction

The Campanian Bearpaw Formation of Southern Alberta, Canada, represents one of the northernmost extensions of the Western Interior Seaway [1,2]. In Southern Alberta, it consists of marine shale with nodular concretions [3]. Ammolite mining, a precious stone derived from fossilized ammonites [4,5] at the city of Lethbridge, as well as other excavations, have yielded a considerable amount of vertebrate and invertebrate marine fossils, representing a Campanian marine ecosystem (74.1 Ma) with average bathymetry of

~40 m (MM pers. obs.) Prominently and consistently present are mosasaurs (Reptilia: Squamata) [6]. During the Campanian, mosasaurs already showed a worldwide distribution [7]. Although the southern exposures of the Western Interior Seaway (e.g., Kansas, Alabama, and New Jersey, all present in the USA) show higher mosasaur diversity [8], the northern edge of the Western Interior Seaway in Alberta preserves a mosasaur community consisting of four to five taxa: *Mosasaurus missouriensis*, potentially *Mosasaurus conodon* (Amelia Zietlow, pers. comm.), *Prognathodon overtoni*, *Plioplatecarpus primaevus*, and *Tylosaurus* sp., likely *Tylosaurus proriger* (Amelia Zietlow, pers. comm. and pers. obs). *Mosasaurus missouriensis* and *Prognathodon overtoni* are found to be the most common taxa. Specimens from the ammolite mines around Lethbridge are especially well preserved, although mosasaur material is also found at Manyberries, Alberta [9]. The local mosasaur taxonomy has been extensively researched in previous studies [9,10]. The excellent preservation after the rapid burial of some of these specimens also revealed soft tissue and stomach contents, thus directly elucidating the animal's diet [10]. The latter study already hypothesized ecological niche differentiation between the most common mosasaurs *Prognathodon* and *Mosasaurus*, citing differences in tooth morphology and dental mesowear. Here, we continue with this mosasaur research and present a study on the feeding ecology of all Southern Alberta mosasaurs to map the highest levels of the Bearpaw foodweb.

Previous research already demonstrated the applicability of two-dimensional dental microwear and three-dimensional surface texture microwear to elucidate mosasaur feeding behavior [11,12]. Microwear is defined as microscopic damage to teeth, usually on wear facets, by tooth–tooth and tooth–food abrasion [13]. However, the preliminary results of these previous mosasaur microwear studies clearly show the need for additional lines of evidence beyond microwear to be explored, as microwear only indicates the use of the animal's teeth in the final weeks to months before its death. For isolated teeth, only the size, shape, and type of animal can be ascertained, but the well-preserved Bearpaw Fm specimens can provide information on differences in tooth morphology on the toothrow, bite force, skull kinetics, and size of the animal to reveal the trophic level. Finally, geochemical analysis can be employed on the well-preserved dental remains. Strontium concentration in relation to Calcium and Barium ratios in teeth has been successfully shown to indicate relative trophic position, following from studies conducted on extinct mammals [14,15]. In recent years, Strontium, Calcium, and Barium have been proven to function as dietary niche proxies in extinct marine as well as terrestrial organisms [16].

Here, a number of Bearpaw Fm. mosasaur remains, both isolated teeth and crania, are studied in order to determine a first trophic-level framework. Two-dimensional microwear, tooth bending strength, and Sr/Ba and Ca/Ba ratios via EDX analysis are all measured in order to provide a multi-proxy approach to determine local Bearpaw Fm feeding ecology.

2. Materials and Methods

2.1. Microwear

Isolated mosasaur teeth from the collections of the Royal Tyrrell Museum of Palaeontology (RTMP) collections were used as is; however, teeth from tooth-bearing bones were cast from direct molds using Moldstar 20T parts A and B mixed equally. See Table 1 for provenance and number of teeth used.

Table 1. Materials and number of teeth used for each analysis, including localities.

Taxon	*Mosasaurus*	*Prognathodon*	*Tylosaurus*	*Plioplatecarpus*	Elasmosaur	Sawfish	Shark
Type	Skull and teeth	Skull and teeth	Teeth	Teeth	Teeth	Teeth	Teeth
Locality	Lethbridge	Lethbridge	Manyberries	Lethbridge	DPP	Iddlesleigh	Iddlesleigh
Formation	Bearpaw	Bearpaw	Bearpaw	Bearpaw	Bearpaw	Bearpaw	Bearpaw
Microwear	10	5	2	3	-	-	-
EDX	16	16	9	7	5	4	15
Bending Strength	30	28	-	5	-	-	-

Two-dimensional microwear was measured using a Hitachi tabletop TM-1000 Scanning Electron Microscope (Hitachi High-Tech Canada Inc., Toronto, ON, Canada). Specimens were mounted on putty, and the apex was focused on, as this is usually where the bulk of the microwear is visible on the wear facet of the tooth [11]. Magnifications ranged between 200× to 400×. Four to six images of each tooth were taken, and all microwear visible in each image was counted (see Supplementary Information for microwear counts).

Microwear was counted by hand by one single observer (FH) to avoid inter-observer bias [11,17]. Microwear was categorized into gouges (large scratches), thin scratches, and pits, as in [11]. See Figure 1 for microwear types. After confirming that assumptions for parametric tests were met, data were analyzed using *t*-tests for comparisons between two groups using or ANOVAs for comparisons among more than two groups. We tested for interspecific differences in microwear on teeth from all jaw locations combined, for intraspecific differences between teeth from different jaw locations, and for interspecific differences between teeth from the same jaw location. Significant ANOVAs were performed, followed by a post hoc Tukey HSD test. All statistics were run in R version 4.3.1 [18] and are included in our Supplementary File S1. Due to the relative scarcity of isolated teeth of all taxa, both worn as well as new teeth were used for each analysis. Only teeth from skulls with an 'open mouth' where the dentition was readily available for casting were used, limiting the number of complete toothrows useful for this study. A study on differences between casts of teeth (using different casting and molding agents) and actual teeth in terms of microwear quality is in preparation by FH and Dr. Lorna O'Brien, former chief technician at the RTMP, and is therefore intended as a follow-up to this current study.

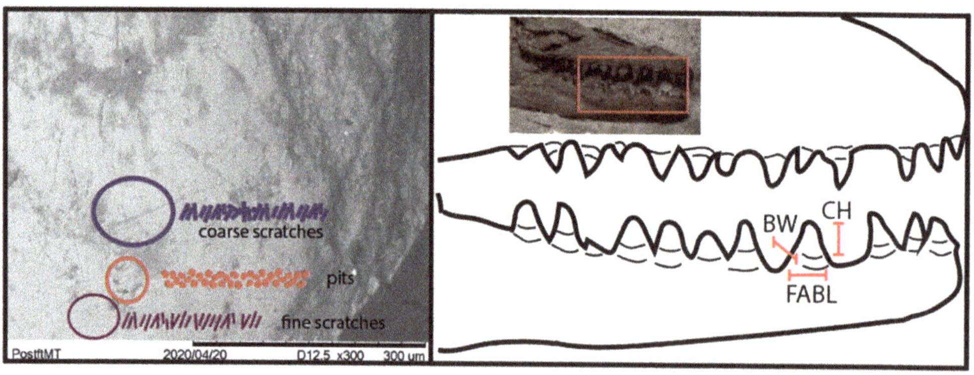

Figure 1. Selection of types of microwear and schematic of mosasaur tooth bending strength measurements.

2.2. EDX

Energy-dispersive X-ray spectroscopy (EDX) was used for elemental analysis. The Hitachi tabletop TM-1000 used for SEM at the Royal Tyrrell Museum of Palaeontology has a simplified EDX system which was used in this exploratory study. Measurements are displayed as discrete counts, which have been converted into ratios. Diagenetic noise was ruled out by checking for sulfur (S) spikes [19]. If no peak was found, the tooth could be measured for Sr, Ba, and Ca.

All trace elements are stored in bioapatite via food and water intake, but Ca is preferentially stored over Sr and Ba. This means that in every step up the food chain, Sr and Ba decrease until these are finally the lowest in terrestrial hypercarnivores. In a marine setting, Sr, Ba, and Ca are proven to distinguish different trophic levels as well. Therefore, all mosasaur taxa, the indeterminate elasmosaurs, and sawfish and shark teeth have been measured for these trace elements. Sr/Ca and Ba/Ca ratios were statistically tested using Anova and inverse log ratios displayed in a principal component analysis.

2.3. Tooth Bending Strength

Tooth bending strength, or the amount of stress a tooth can structurally withstandwhen in use, was calculated using the formula used in [20] for tooth measurements in theropods. The tooth crown height (CH) and foreaft (mesiodistal) basal length (FABL) and basal (anteroposterior) width (BW) at the base were measured in mm. The tooth bending strength formula is expressed as follows (see Figure 1):

Bending strength of anteroposterior side:

$$(AP) = (((\pi \times 0.5 \times FABL) \times ((0.5 \times BW)^2))/4)/(CH \times (0.5 \times FABL)))$$

Bending strength of mediolateral side:

$$(ML) = (((\pi \times (0.5 \times FABL) \times ((0.5 \times BW)^3))/4)/(CH \times (0.5 \times BW)))$$

The Log_{10} value of the crown height (Log_{10} CH) is plotted against the Log_{10} value of the anteroposterior bending strength and mediolateral bending strength in scatter plots. See [20] for the rationale behind the calculation.

3. Results

3.1. Microwear

The two-dimensional microwear counts are shown in the overview graph in Figure 2. Though the data quality differed between teeth and molds, some intraspecific patterns could be discerned. All mosasaurs have significant high numbers of pits dominated by *Prognathodon* and *Plioplatecarpus*. The latter taxon also has the highest number of fine scratches on average, followed by *Mosasaurus*. These two are the only taxa with higher numbers of fine scratches than gouges and pits in this dataset. *Prognathodon* and Tylosaurus, in contrast, display a higher number of pits than gouges and fine scratches in their microwear. *Tylosaurus* has more fine scratches than gouges, whereas *Prognathodon* has the lowest number of fine scratches compared to its gouges and pits. Intraspecific differences were found to be statistically significant by an ANOVA for several taxa and for all microwear types (fine scratches, gouges, and pits). For all microwear types, significant differences between *Plioplatecarpus* and *Mosasaurus* were found, as well as between *Prognathodon* and *Plioplatecarpus* (see Statistics Section in Supplementary File S1).

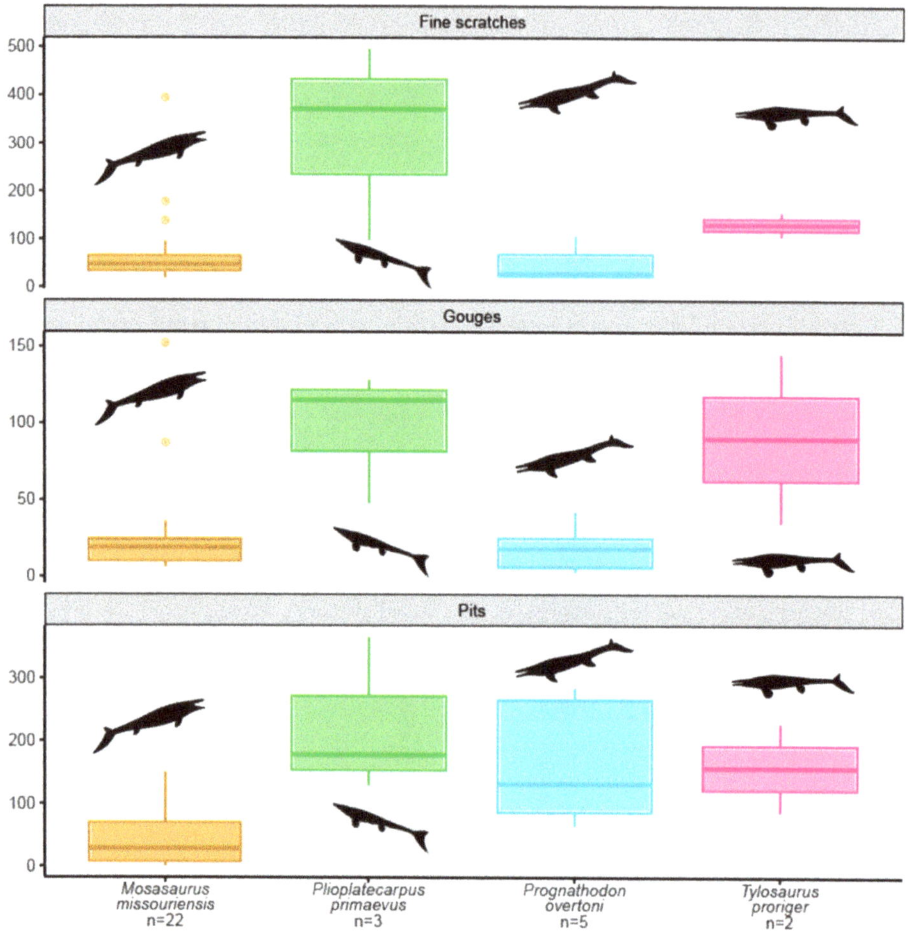

Figure 2. Microwear between all mosasaur taxa. Mosasaur silhouettes from Phylopic; see Data Availability Statement for all creators and CC licenses.

When looking at differences between premaxillary, maxillary, and dentary microwear counts (only between *Mosasaurus* and *Prognathodon* of which complete skulls and measurable dentition were available), some patterns in the counts are visible (see Figure 3). *Mosasaurus* has high counts of fine scratches and pits in its dentary, compared to lower counts of scratches in both the maxilla and premaxilla. The premaxilla has higher counts of pits than the maxilla. For *Prognathodon*, the maxilla displays high counts of pits and fine scratches, with much less microwear displayed in the dentary and premaxilla, hinting at the most tooth–tooth or tooth–food abrasion occuring in the lower jaws.

Between the larger and smaller *Mosasaurus* specimens, there are some differences as well. The large *Mosasaurus missouriensis* TMP 2012.010.0001 shows high counts of fine scratches both in the premaxilla and the maxilla, as well as a high number of pits in the premaxilla. The smaller *Mosasaurus missouriensis* TMP 2008.036.0001 shows the highest numbers of fine scratches and pits in the dentary, as well as a high number of fine scratches in the maxilla. This shows that there is a similar microwear pattern for fine scratches in the maxillae of both larger and smaller *Mosasaurus missouriensis* and, therefore, tooth–tooth or tooth–food abrasion, as well as high individual or ontogenetic variation. However, when tested with an ANOVA for statistically significant microwear differences between jaw locations between *Mosasaurus* and *Prognathodon*, no significance was found (see Statistics Section in Supplementary File S1). Similarly, interspecific microwear differences between jaw locations were not statistically significant for *Prognathodon* nor for *Mosasaurus*.

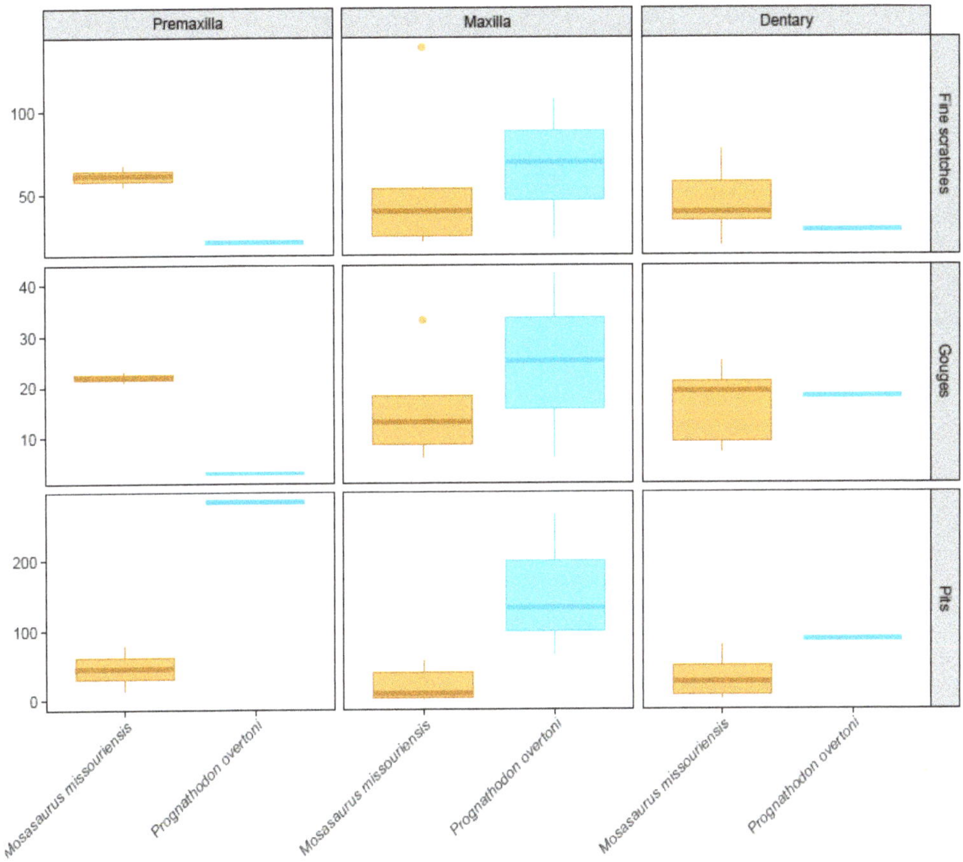

Figure 3. Microwear counts between upper and lower jaws between *Mosasaurus* and *Prognathodon*.

3.2. EDX Analysis

The lowest rates of Sr/Ca are seen in *Prognathodon* (and sawfish), followed closely by *Mosasaurus*; see Figure 4. Whilst the range for *Mosasaurus missouriensis* in terms of Sr/Ca is quite high, the rates of Sr/Ca are the highest in Tylosaurus and the indeterminate elasmosaur teeth. Finally, the hybodont shark teeth show a high variation in their Sr/Ca ratio. *Plioplatecarpus* has the most intermediate Sr/Ca ratio.

Prognathodon and *Mosasaurus missouriensis* have a similarly large range for their Ba/Ca ratio; see Figure 5. However, *Prognathodon* and *Plioplatecarpus* have the lowest Ba/Ca rates, with *Mosasaurus missouriensis*, possibly *Mosasaurus conodon*, and hybodont sharks following with an intermediate level of Ba/Ca. *Tylosaurus*, the indeterminate elasmosaurs, and sawfish have the highest average Ba/Ca ratios, with the elasmosaurs displaying the narrowest range. While both Sr/Ca and Ba/Ca rates meet the assumptions for ANOVA (normality, homoscedasticity, and independence), only the Sr/Ca ratio showed significant differences between taxa, whereas the Ba/Ca ratio did not. However, the Ba/Ca ratio showed a p value on the lower side, indicating that with more data, the p value might be statistically significant.

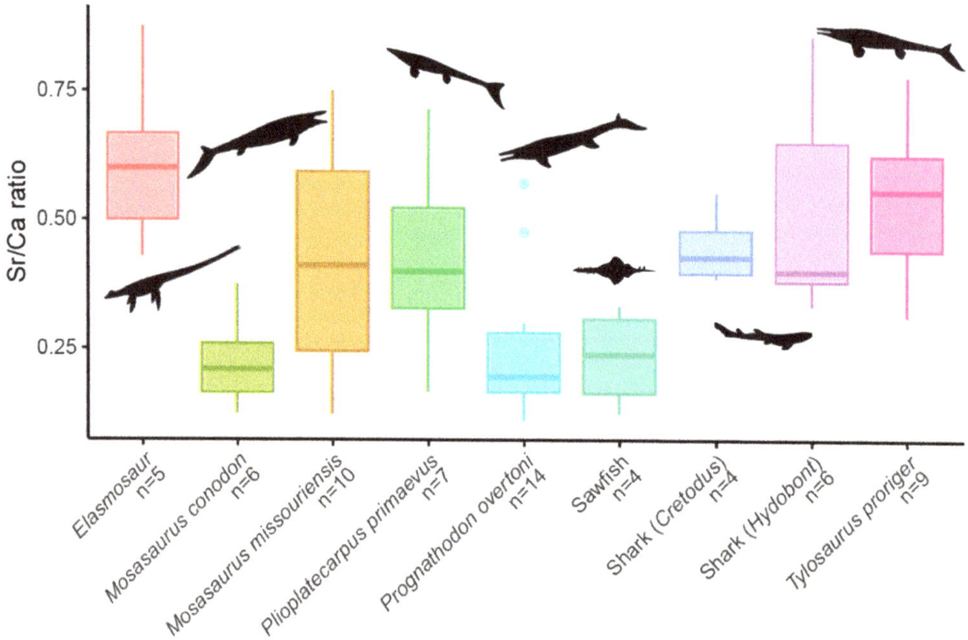

Figure 4. Sr/Ca ratios between Bearpaw sea vertebrates.

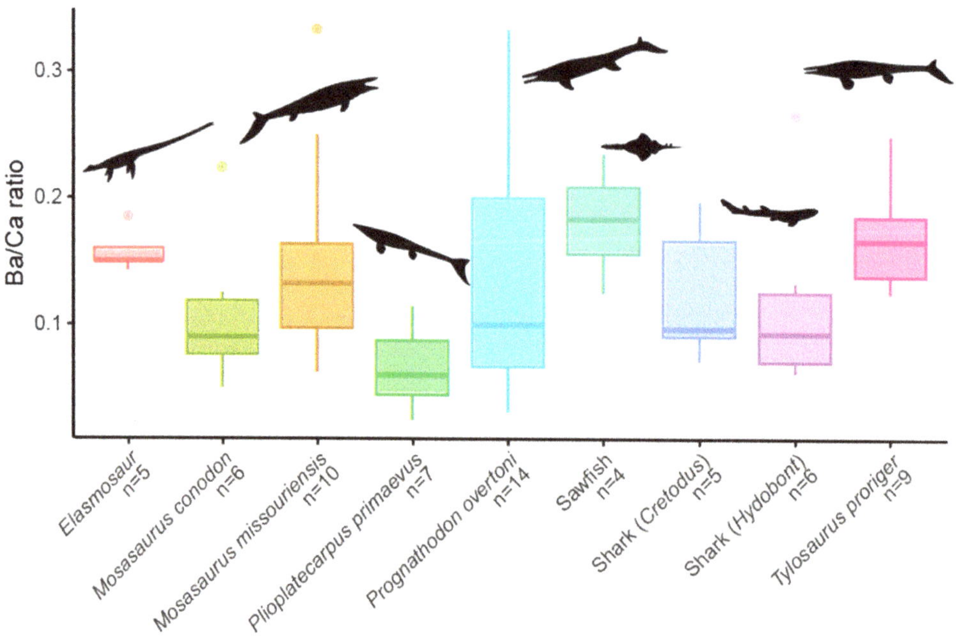

Figure 5. Ba/Ca ratios for all Bearpaw vertebrates.

3.3. Tooth Bending Strength

Following [20], a Log_{10} value of the bending strength of the anteroposterior side (AP, x-axis) is plotted against the Log_{10} value of the bending strength of the mediolateral side (ML, y-axis). The resulting scatterplot (Figure 6) roughly follows the trend seen in [20] for the tooth bending strength of several theropods, extant and extinct felids, and extant and extinct canids. Indeterminate mosasaurs were also plotted. Therefore, the general trend of the scatterplot is perhaps not surprising (Figure 6). However, there are differences between the mosasaurs measured here: *Mosasaurus missouriensis* scatters with lower bending strength ranges, *Prognathodon overtoni* scores the higher/highest bending strength, and *Plioplatecarpus* takes an intermediate position, though it is still higher than *Mosasaurus* on average. There is not much of a difference between the larger and smaller *Mosasaurus* specimens in terms of the overall bending strength; however, there is a difference between the upper and lower jaws in larger specimens overall. The tooth bending strength meets the assumptions for ANOVA.

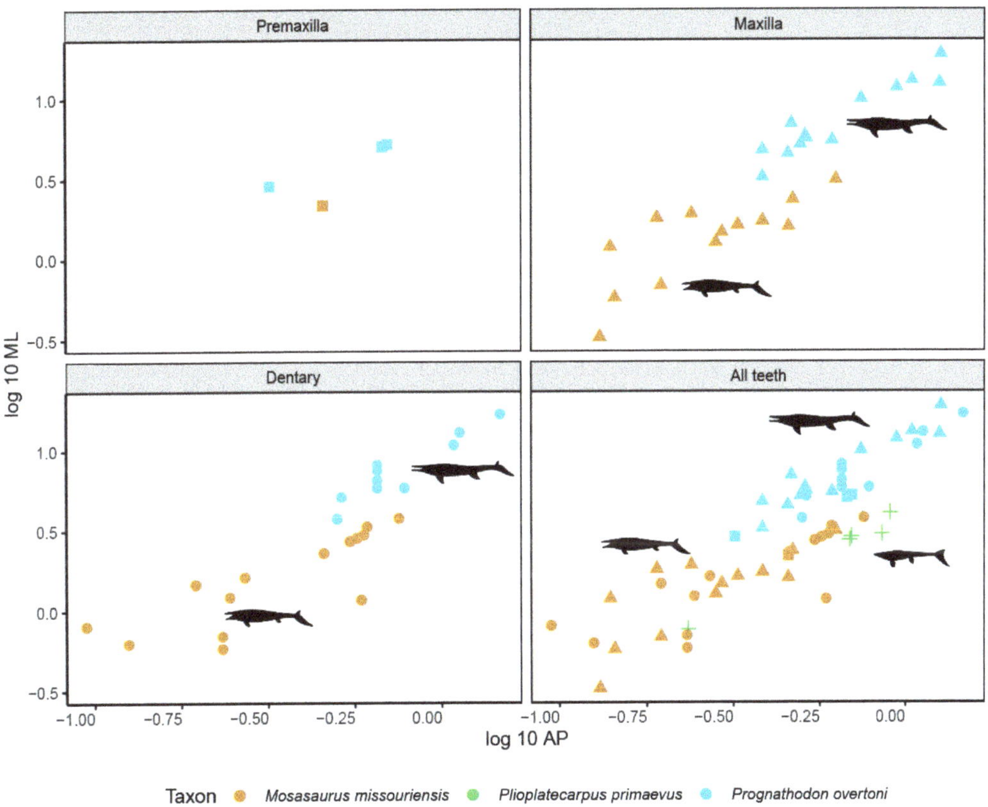

Figure 6. Tooth bending strength for *Mosasaurus* (orange), *Prognathodon* (blue), and *Plioplatecarpus* (green).

4. Discussion

4.1. EDX-Based Sr/Ca and Ba/Ca Differences

A PCA of the EDX measurements is shown in Figure 7.

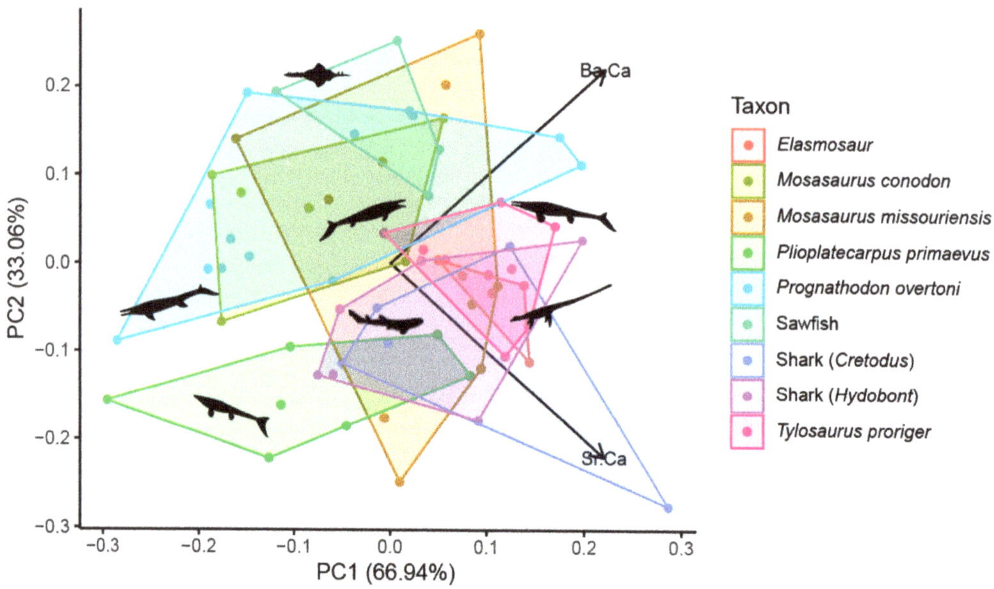

Figure 7. PCA of all Bearpaw vertebrates' Sr/Ca and Ba/Ca ratios.

High Ba values are observed in marine fauna feeding near the base of the marine foodweb, which is then reflected in high Ba levels in higher trophic levels. The highest Ba/Ca ratio is found in plants and invertebrates that bio-accumulate it in high doses; higher trophic level feeders that directly feed from this source will have higher rates of Ba/Ca. This is observed in conodonts (but the Ba-rich food source in question remains elusive) [16]. In fish, Ba is seen as indicative of diet, as Sr remains constant in fish [21]. That said, Ba/Ca ratios in fish are highly variable, whilst the Sr/Ca ratio is conservative, so if a mosasaur is mostly piscivorous, their dentition should indicate variable Ba/Ca ratios as well. We indeed see higher Ba/Ca ratios in presumably piscivorous elasmosaurs, but not the highest levels of this sample of marine vertebrates. *Tylosaurus* seems to closely follow the ranges of Sr/Ca and Ba/Ca of elasmosaurs, only with slightly lower values of Sr (likely confirming a higher trophic level) and slightly wider-ranging Ba levels (likely confirming a wider array of diet than elasmosaurs, the latter which seem to be specialized into feeding/predation on one category). Tylosaurs from the Campanian of Alabama are thought to have occupied shallower, near-shore waters [22], and in the Bearpaw sea, they might have occupied a similar niche as elasmosaurs. However, as *Tylosaurus saskatchewanensis* from the Bearpaw sea of Saskatchewan (more into the northern middle section of the Western Interior Seaway) and *Tylosaurus proriger* from the Niobrara Fm of Kansas (the southern section of the Western Interior Seaway) have small mosasaurs and elasmosaurs in their stomach content, respectively, it is also not impossible for the EDX analysis to reveal a direct predator–prey relationship [23]. Elasmosaurs show a narrow range in Sr, with the highest Sr values of this marine vertebrate assemblage, and a very narrow range in Ba values. This places them at a lower trophic level than the mosasaurs, with the aforementioned narrow Ba range indicating a specific food source, likely fish and/or squid. Elasmosaurs from elsewhere are known to have benthic diets as well as more sarcophagous diets [24,25]. However, the Bearpaw elasmosaurs seem to have been more restrictive and specialized in their dietary niche, which is also observed in Western Interior Seaway elasmosaurs from the Pierre Shale as well as Upper Cretaceous plesiosaurs from South Australia [26,27].

Hybodont sharks from Cretaceous of Thailand [28] have a diverse diet, interpreted as opportunist feeding, with potentially some shell-crushing durophagy of hybodontiform sharks with low cusped teeth similar to some Bearpaw sharks. In other hybodont sharks, direct evidence of predation on ammonites is shown [29,30]. Both predation on shellfish and ammonites, as filterfeeding organisms of plankton, would account for high rates of Ba/Ca in the upper ranges of *Prognathodon*, *Mosasaurus*, and hybodont sharks. Furthermore, similarly to mosasaurs, hybodontid and cretodid sharks have 'clutching or piercing' dentitions, thought to make them active predators of large prey (such as fish, other sharks, and other marine vertebrates) [31]. Therefore, some overlap in Sr/Ca and Ba/Ca ratios, especially with the more piscivorous mosasaurs, is explained here.

4.2. Inferred Diet and Niche Partitioning

See Table 2 for an overview of the inferred diets of the mosasaurs in this study. The authors of [9] already hypothesized about ecological niche differentiation for the Alberta Bearpaw *Mosasaurus* and *Prognathodon* based on differences in tooth morphology and differences observed in dental mesowear. Together with the two other less common mosasaurs from the area, this study confirms the above hypothesis. The authors of [10] placed *Mosasaurus missouriensis* firmly in the 'cut' guild sensu [32] based on their recurved teeth, no exposed horizontal wear facets, and—perhaps most compellingly—mostly fish-based contents, as well as squid, in the gut. This is different in the Bearpaw *Prognathodon overtoni* TMP 2007.034.0001, which was found not only with turtle remains in its stomach, but also with large horizontal wear facets on its dentition. The authors of [33] demonstrated that the dental spalling or large, oval, longitudinal wear facets on the metriorhynchid *Dakosaurus* points to macrophagy or eating large prey by shearing them in pieces before consumption. The authors of [33] observed this type of spalling in indeterminate mosasaurs from the American Museum of Natural History. Both *Tylosaurus* and *Plioplatecarpus* were also placed in this macrophagous group of mosasaurs by the authors of [10]. Although this study did not find significant spalling as mesowear and used microwear instead, the two-dimensional microwear largely confirms the *Mosasaurus*-*Prognathodon* niche differentiation hypothesis. Empirically, both intraspecific and interspecific differences between the upper and lower toothrows in terms of microwear were also found. The statistical analysis did not find these differences to be significant, though. It is possible that more data would be needed to explore this further, as the authors of [34] found intraspecific microwear differences along the toothrows of crocodylians and varanids, both of whom could be argued to be extant equivalents of extinct marine reptiles. Moreover, the highly kinetic skulls of mosasaurs would infer a similar outcome, as with skull kinesis, different parts of the dentition would be expected to generate different microwear.

Coupling back to the EDX, although there is some overlap between *Prognathodon* and *Mosasaurus*, there is a stark difference between both types provided by the Sr/Ca and Ba/Ca ratios. *Mosasaurus* seems to be an 'all-rounder', overlapping with almost all other groups. Indeed, a Dental Microwear Texture Analysis (DMTA) of type-Maastrichtian *Mosasaurus hoffmanni* displays a generalist diet for this large taxon [12]. Though not as large as *M. hoffmanni*, *M. missouriensis* was definitely on the larger side of the Bearpaw mosasaur spectrum (reaching up to 8 m in length). Its two-dimensional microwear shows a general high distribution of fine scratches, as well as pits, indeed confirming usage of its teeth for both softer and harder food items. Finally, the tooth bending strength between the two taxa clearly separates *Mosasaurus* from *Prognathodon*, with the latter showing higher tooth bending strength and therefore teeth that could withstand higher stresses, which concurs with the microwear and EDX.

Table 2. Sr/Ca and Ba/Ca ranges of all mosasaurs from this study, with dominant microwear and inferred prey.

Taxon	*Mosasaurus*	*Prognathodon*	*Tylosaurus*	*Plioplatecarpus*
Sr/Ca range	0.2–0.73	0.11–0.57	0.31–0.77	0.17–0.71
Ba/Ca range	0.06–0.33	0.03–0.33	0.13–0.21	0.03–0.11
Dominant microwear	Fine scratches	Pits	Gouges	Fine scratches
Inferred prey	Large range of vertebrate prey items Large range of invertebrate prey items	Harder vertebrate prey items (turtles and large fish) Harder invertebrate prey items (shellfish and ammonites)	Fish and larger vertebrates	Softer prey items (squid and fish) Harder prey items (large fish and ammonites)

Prognathodon is represented at one end of the spectrum (Figure 7), closer to sawfish than sharks, and is clearly separated from *Plioplatecarpus*. Dental microwear texture analysis of type-Maastrichtian mosasaurs revealed that prognathodons surprisingly not only foraged for harder prey items, but softer invertebrates as well, likely displaying foraging behavior on benthic invertebrates [12]. Modern sawfish show a propensity to bottom-feeding durophagy, which is another link to the microwear in this current study, the DMTA [12], and the *Prognathodon* stomach content of TMP 2007.034.0001 and its mesowear. Finally, a partial lobster was recovered as putative stomach content of Bearpaw *Prognathodon* TMP 2007.034.0001, but this was never confirmed as the position of the lobster was not as undeniable as stomach content as the turtle and fish were [9]. If it was stomach content, this would again confirm both *Prognathodon*'s position in the PCA diagram as overlapping with bottom-feeding sawfish, as well as being a consumer of both harder prey and invertebrates. The tooth bending strength shows higher levels for *Prognathodon* compared to the other measured mosasaurs (Figure 6). *Prognathodon* teeth are higher in FABL and BW than *Mosasaurus* and *Plioplatecarpus*. This emphasizes the fact that *Prognathodon*'s teeth were suitable to process much harder prey. Together with a high overall count of pits, especially compared to fine scratches and gouges, this shows a handling of prey items through more oral food processing than *Mosasaurus*.

Plioplatecarpus also occupies a distinct position in the Sr/Ca Ba/Ca spectrum with little overlap with sharks and elasmosaurs. Traditionally, plioplatecarpids, by virtue of their presumed piscivorous tooth morphology, are placed in the piscivorous 'cut' end of the guild sensu [32]. However, type-Maastrichtian *Plioplatecarpus marshi* displays a surprising tendency for harder invertebrate consumption [12]. Two-dimensional microwear counts from the Bearpaw *Plioplatecarpus primaevus* equally show microwear pointing to harder prey items besides fine scratches pointing to softer prey. The authors of [22] already questioned plioplatecarpids exclusively feeding on softer prey items such as belemnites, and with its EDX results showing an offset from the other Bearpaw mosasaurs, away from sharks and piscivorous elasmosaurs, this smaller type of mosasaur presents the most enigmatic feeding ecology. A clue, perhaps, comes from another Bearpaw *Plioplatecarpus*: TMP 2022.043.0001 [35]. This particular specimen has very large orbits, which, according to [35] indicates visual foraging at the deepest reaches of the photic zone. The large parietal is similarly proposed by the authors of [35] to have been of use for fast movement and chasing agile prey, as well as dodging larger predators (such as *Prognathodon* and *Mosasaurus*). This hypothesis would suit an existence in a different feeding niche from the other Bearpaw mosasaurs, as indicated by the EDX results, and to a lesser extent, the microwear of this study. Another clue could come from a unique basicranial circulation pattern discovered by the authors of [36]. Here, the type of circulation to the cranium that plioplatecarpines display is likened to those of arboreal snakes and helps with being upside down whilst hunting for prey. Perhaps *Plioplatecarpus* dove for prey out of reach of other mosasaurs. This does not, however, answer the question of why *Plioplatecarpus* demonstrates such 'rough' microwear. Even with this limited sample, *Plioplatecarpus* shows an intermediate to high tooth bending strength (Figure 6). Harder prey such as large fish with hard ganoid scales or ammonites, could have been on the *Plioplatecarpus* menu, but there is no unequivocal evidence of this [37–39]. Moreover, an isotopic analysis of *Plioplatecarpus* dentition from the type Maastrichtian does not support a deep diving lifestyle [40].

An additional future line of evidence to answer some of these standing questions would involve using DMTA on these Bearpaw mosasaurs as well as an isotope analysis of dental enamel. Both are currently in preparation for future studies. What seems to become more clear, however, is that Cretaceous marine ecosystems were high in productivity, supporting rich fauna, and were able to support multiple predators, as seen in [12,40,41].

5. Conclusions

A combination of two-dimensional microwear analysis, EDX analysis using Sr/Ca and Ba/Ca ratios, and calculated tooth bending strengths for the Alberta Bearpaw mosasaurs *Mosasaurus missouriensis*, *Prognathodon overtoni*, *Plioplatecarpus primaevus*, *Tylosaurus proriger*, and possibly *Mosasaurus conodon*, shows a distinct niche differentiation pattern. There is a clear separation between *Mosasaurus* and *Prognathodon*, with the former showing generalist dietary preferences and the latter showing a tendency to feed on harder prey items and a possible tendency for macrophagy. *Plioplatecarpus* remains an enigmatic smaller mosasaur in terms of dietary preferences, although it also shows a distinct dietary area according to EDX and the tooth bending strength analysis, with a mainly piscivorous note from its microwear, and additional foraging on harder prey that is still unidentified. *Tylosaurus* seems to follow a similar diet to the Bearpaw elasmosaurs or even predated on them. Future DMTA and isotopic studies will further clarify the dietary niches of the Bearpaw mosasaurs.

Supplementary Materials: The following supporting information can be downloaded at: https://www.mdpi.com/article/10.3390/d17030205/s1, Supplementary File S1, and at Figshare. https://doi.org/10.6084/m9.figshare.28009238.v1.

Author Contributions: Conceptualization, F.M.H. and A.S.S.; methodology, F.M.H. and M.v.d.K.; software, M.v.d.K.; writing—original draft preparation, F.M.H., M.T.M., M.v.d.K. and A.S.S.; writing—review and editing, F.M.H., M.T.M., M.v.d.K. and A.S.S. All authors have read and agreed to the published version of the manuscript.

Funding: F.M.H. received funding from the Dr. Elizabeth Nicholls Postdoctoral Research Fellowship provided by the Royal Tyrrell Museum of Palaeontology Cooperating Society (RTMPCS).

Institutional Review Board Statement: Not applicable.

Data Availability Statement: All data are available at FigShare (https://doi.org/10.6084/m9.figshare.28009238.v1) under CC license 4.0 at Mosasaur feeding ecology from the Campanian Bearpaw Formation, Alberta, Canada: a preliminary multi-proxy approach—Supplementary File S1. *Mosasaurus* from Phylopic by T.K. Robinson, 2022, https://creativecommons.org/licenses/by-nc/3.0/; *Plioplatecarpus* from Phylopic by Matt Crooke, 2014, https://creativecommons.org/licenses/by-sa/3.0/; *Tylosaurus* from Phylopic by Scott Hartman, 2017, https://creativecommons.org/licenses/by/3.0/; *Prognathodon* from Phylopic by Ian Reid, 2017, https://creativecommons.org/licenses/by/3.0/; *Albertonectes*/elasmosaur from Phylopic by F. Danota, 2013, https://creativecommons.org/publicdomain/zero/1.0/; Sawfish from Phylopic by Taylor Greenfield, 2021, https://creativecommons.org/licenses/by/3.0/; Shark from Phylopic by B. Lang, 2016, https://creativecommons.org/publicdomain/zero/1.0/. All Phylopic.org sites accessed on 2 March 2024.

Acknowledgments: We thank the RTMP technical staff for all their help and hard work: Marilyn LaFramboise, Joe Sanchez, Becky Sanchez, Brian Strilisky, Tom Courtenay, Rhian Russell, and Lorna O'Brien. We thank Don Henderson, Steve Cumbaa, and Don Brinkman for their technical discussions on tooth bending strength and Bearpaw shark and fish taxonomy. We thank Jeff Liston and Darren Naish for their fruitful discussions on marine reptile niche differentiation. The authors would like to thank reviewer 1 (Emily Bamforth and reviewer 2 (anonymous) for providing kind and collegiate reviews and very helpful comments and recommendations, which improved this manuscript. F.M.H. received funding from the Elizabeth Nicholls Postdoctoral Research Fellowship, generously provided by the Royal Tyrrell Museum of Palaeontology Cooperating Society (RTMPCS). This paper is dedicated to the memory of Knaagt Liston-Holwerda, who showed us that the world is there to explore.

Conflicts of Interest: The authors declare no conflicts of interest.

References

1. Eberth, D.A. Review and comparison of Belly River Group and Edmonton Group stratigraphy and stratigraphic architecture in the southern Alberta Plains. In Conference of the Canadian Society of Petroleum Geology, Calgary, Alberta, Abstracts 117; 2002. Available online: https://www.researchgate.net/publication/237349569_Review_and_Comparison_of_Belly_River_Group_and_Edmonton_Group_Stratigraphy_and_Stratigraphic_Architecture_in_the_Southern_Alberta_Plains (accessed on 18 October 2024).
2. Zubalich, R.; Capozzi, R.; Fanti, F.; Catuneanu, O. Evolution of the Western Interior Seaway in west-central Alberta (late Campanian, Canada): Implications for hydrocarbon exploration. *Mar. Pet. Geol.* **2020**, *124*, 104779. [CrossRef]
3. Wall, J.H.; Sweet, A.R.; Hills, L.V. Paleoecology of the Bearpaw and contiguous Upper Cretaceous formations in the C.P.O.G. Strathmore well, southern Alberta. *Bull. Can. Pet. Geol.* **1971**, *19*, 691–702. [CrossRef]
4. Mychaluk, K.A. Update on Ammolite Production from Southern Alberta, Canada. *Gems Gemol.* **2009**, *45*, 192–196. [CrossRef]
5. Tanke, D. Personal, historical and technical perspectives on the growing role of light and heavy industry on ver-tebrate palaeontology in Alberta, Canada. In *Eighteenth Annual Symposium, Abstracts*; Alberta Palaeontological Society: Calgary, AB, Canada, 2014; pp. 49–50.
6. Russell, D.A. *Systematics and Morphology of American Mosasaurs (Reptilia, Sauria)*; Yale University Press: New Haven, CT, USA, 1967.
7. Polcyn, M.J.; Jacobs, L.L.; Araújo, R.; Schulp, A.S.; Mateus, O. Physical drivers of mosasaur evolution. *Palaeogeogr. Palaeoclimatol. Palaeoecol.* **2014**, *400*, 17–27. [CrossRef]
8. Nicholls, E.L.; Russell, A.P. Paleobiogeography of the Cretaceous Western Interior Seaway of North America: The vertebrate evidence. *Palaeogeogr. Palaeoclim. Palaeoecol.* **1990**, *79*, 149–169. [CrossRef]
9. Konishi, T.; Brinkman, D.; Massare, J.A.; Caldwell, M.W. New exceptional specimens of *Prognathodon overtoni* (Squamata, Mosasauridae) from the upper Campanian of Alberta, Canada, and the systematics and ecology of the genus. *J. Vertebr. Paléontol.* **2011**, *31*, 1026–1046. [CrossRef]
10. Konishi, T.; Newbrey, M.G.; Caldwell, M.W. A small, exquisitely preserved specimen of *Mosasaurus missouriensis* (Squamata, Mosasauridae) from the upper Campanian of the Bearpaw Formation, western Canada, and the first stomach contents for the genus. *J. Vertebr. Paléontol.* **2014**, *34*, 802–819. [CrossRef]
11. Holwerda, F.M.; Beatty, B.L.; Schulp, A.S. Dental macro-and microwear in *Carinodens belgicus*, a small mosasaur from the type Maastrichtian. *Neth. J. Geosci.* **2013**, *92*, 267–274. [CrossRef]
12. Holwerda, F.M.; Bestwick, J.; Purnell, M.A.; Jagt, J.W.M.; Schulp, A.S. Three-dimensional dental microwear in type-Maastrichtian mosasaur teeth (Reptilia, Squamata). *Sci. Rep.* **2023**, *13*, 18720. [CrossRef]
13. Teaford, M.F.; Walker, A. Quantitative differences in dental microwear between primate species with different diets and a comment on the presumed diet of *Sivapithecus*. *Am. J. Phys. Anthr.* **1984**, *64*, 191–200. [CrossRef]
14. Sillen, A. Biogenic and diagenetic Sr/Ca in Plio-Pleistocene fossils of the Omo Shungura Formation. *Paleobiology* **1986**, *12*, 311–323. [CrossRef]
15. Sillen, A. Elemental and isotopic analyses of mammalian fauna from Southern Africa and their implications for paleodietary research. *Am. J. Phys. Anthr.* **1988**, *76*, 49–60. [CrossRef]
16. Terrill, D.F.; Jarochowska, E.; Henderson, C.M.; Shirley, B.; Bremer, O. Sr/Ca and Ba/Ca ratios support trophic partitioning within a Silurian conodont community from Gotland, Sweden. *Paleobiology* **2022**, *48*, 601–621. [CrossRef]
17. Mihlbachler, M.C.; Beatty, B.L.; Caldera-Siu, A.; Chan, D.; Lee, R. Error rates and observer bias in dental microwear analysis using light microscopy. *Palaeontol. Electron.* **2012**, *15*, 1–22. [CrossRef]
18. R Development Core Team. *R: A Language and Environment for Statistical Computing*; Version 3.6.3; R Foundation for Statistical Computing: Vienna, Austria, 2020; Available online: http://www.r-project.org/ (accessed on 18 October 2024).

19. Shirley, B.; Jarochowska, E. Chemical characterisation is rough: The impact of topography and measurement parameters on energy-dispersive X-ray spectroscopy in biominerals. *Facies* **2022**, *68*, 7. [CrossRef]
20. Farlow, J.O.; Brinkman, D.L.; Abler, W.L.; Currie, P.J. Size, shape, and serration density of theropod dinosaur lateral teeth. *Mod. Geol.* **1991**, *16*, 161–198.
21. Peek, S.; Clementz, M.T. Sr/Ca and Ba/Ca variations in environmental and biological sources: A survey of marine and terrestrial systems. *Geochim. Cosmochim. Acta* **2012**, *95*, 36–52. [CrossRef]
22. Kiernan, C.R. Stratigraphic distribution and habitat segregation of mosasaurs in the Upper Cretaceous of western and central Alabama, with an historical review of alabama mosasaur discoveries. *J. Vertebr. Paléontol.* **2002**, *22*, 91–103. [CrossRef]
23. Everhart, M.J. Plesiosaurs as the food of mosasaurs; new data on the stomach contents of a Tylosaurus proriger (Squamata; Mosasauridae) from the Niobrara Formation of western Kansas. *Mosasaur* **2004**, *7*, 41–46.
24. McHenry, C.R.; Cook, A.G.; Wroe, S. Bottom-Feeding Plesiosaurs. *Science* **2005**, *310*, 75. [CrossRef]
25. Sørensen, A.M.; Surlyk, F.; Lindgren, J. Food resources and habitat selection of a diverse vertebrate fauna from the upper lower Campanian of the Kristianstad Basin, southern Sweden. *Cretac. Res.* **2013**, *42*, 85–92. [CrossRef]
26. Cicimurri, D.J.; Everhart, M.J. An Elasmosaur with Stomach Contents and Gastroliths from the Pierre Shale (Late Cretaceous) of Kansas. *Trans. Kans. Acad. Sci.* **2001**, *104*, 129–143. [CrossRef]
27. White, J.M.; Barron, A.; McCurry, M.R.; Denham, T. Investigating gut contents of the leptocleidian plesiosaur *Umoonasaurus demoscyllus* using micro-CT imaging. *Alcheringa Australas. J. Palaeontol.* **2023**, *47*, 206–210. [CrossRef]
28. Cuny, G.; Suteethorn, V.; Kamha, S.; Buffetaut, E. Hybodont sharks from the lower Cretaceous Khok Kruat Formation of Thailand, and hybodont diversity during the Early Cretaceous. *Geol. Soc. Lond. Spéc. Publ.* **2008**, *295*, 93–107. [CrossRef]
29. Vullo, R. Direct evidence of hybodont shark predation on Late Jurassic ammonites. *Naturwissenschaften* **2011**, *98*, 545–549. [CrossRef]
30. Wright, J.K.; Bassett-Butt, L.; Collinson, M. Fatally bitten ammonites from the Lower Calcareous Grit Formation (Upper Jurassic) of NE Yorkshire, UK. *Proc. Yorks. Geol. Soc.* **2014**, *60*, 1–8. [CrossRef]
31. Cumbaa, S.L.; Underwood, C.J.; Schröder-Adams, C.J.; Arratia, G. Paleoenvironments and paleoecology of the vertebrate fauna from a Late Cretaceous marine bonebed, Canada. *Mesoz. Fishes* **2013**, *5*, 509–524.
32. Massare, J.A. Tooth morphology and prey preference of Mesozoic marine reptiles. *J. Vertebr. Paléontol.* **1987**, *7*, 121–137. [CrossRef]
33. Young, M.T.; Brusatte, S.L.; Beatty, B.L.; De Andrade, M.B.; Desojo, J.B. Tooth-On-Tooth Interlocking Occlusion Suggests Macrophagy in the Mesozoic Marine Crocodylomorph *Dakosaurus*. *Anat. Rec.* **2012**, *295*, 1147–1158. [CrossRef]
34. Bestwick, J.; Unwin, D.M.; Henderson, D.M.; Purnell, M.A. Dental microwear texture analysis along reptile tooth rows: Complex variation with non-dietary variables. *R. Soc. Open Sci.* **2021**, *8*, 201754. [CrossRef]
35. Holmes, R.B. Evaluation of the photosensory characteristics of the lateral and pineal eyes of *Plioplatecarpus* (Squamata, Mosasauridae) based on an exceptionally preserved specimen from the Bearpaw Shale (Campanian, Upper Cretaceous) of southern Alberta. *J. Vertebr. Paléontol.* **2024**, *43*, e2335174. [CrossRef]
36. Polcyn, M.J.; Bardet, N.; Albright, L.B.; Titus, A. A new lower Turonian mosasaurid from the Western Interior Seaway and the antiquity of the unique basicranial circulation pattern in Plioplatecarpinae. *Cretac. Res.* **2023**, *151*, 105621. [CrossRef]
37. Kauffman, E.G. Mosasaur Predation on Upper Cretaceous Nautiloids and Ammonites from the United States Pacific Coast. *Palaios* **2004**, *19*, 96–100. [CrossRef]
38. Klompmaker, A.A.; Waljaard, N.A.; Fraaije, R.H. Ventral bite marks in Mesozoic ammonoids. *Palaeogeogr. Palaeoclimatol. Palaeoecol.* **2009**, *280*, 245–257. [CrossRef]
39. Tsujita, C.J.; Westermann, G.E. Were limpets or mosasaurs responsible for the perforations in the ammonite Placenticeras? *Palaeogeogr. Palaeoclimatol. Palaeoecol.* **2001**, *169*, 245–270. [CrossRef]
40. Schulp, A.S.; Vonhof, H.B.; van der Lubbe, J.H.J.L.; Janssen, R.; van Baal, R.R. On diving and diet: Resource partitioning in type-Maastrichtian mosasaurs. *Neth. J. Geosci.* **2013**, *92*, 165–170. [CrossRef]
41. Cortés, D.; Larsson, H.C. Top of the food chains: An ecological network of the marine Paja Formation biota from the Early Cretaceous of Colombia reveals the highest trophic levels ever estimated. *Zool. J. Linn. Soc.* **2024**, *202*, zlad092. [CrossRef]

diversity

MDPI

Article

The Evolution of Mosasaurid Foraging Behavior Through the Lens of Stable Carbon Isotopes

Michael J. Polcyn [1,2,*], John A. Robbins [2], Anne S. Schulp [1,3], Johan Lindgren [4] and Louis L. Jacobs [2]

1 Faculty of Geosciences, Utrecht University, Princetonlaan 8a, 3584CB Utrecht, The Netherlands; a.s.schulp@uu.nl
2 Huffington Department of Earth Sciences, Southern Methodist University, Dallas, TX 75275, USA
3 Naturalis Biodiversity Center, Darwinweg 2, 2333CR Leiden, The Netherlands
4 Department of Geology, Lund University, Sölvegatan 12, 223 62 Lund, Sweden; johan.lindgren@geol.lu.se
* Correspondence: mpolcyn@smu.edu

Abstract: A large data set of new and previously published measurements of $\delta^{13}C$ values derived from tooth enamel (*n* = 223, of which 93 are new) are compiled to explore patterns of foraging area preferences of Late Cretaceous mosasaurid squamates over evolutionary time scales (~93–66 Ma). Our results indicate that small-bodied halisaurines are restricted to a relatively nearshore range, overlapping the lower end of the range of plioplatecarpines and some mosasaurine taxa. Most moderately sized plioplatecarpines occupy a relatively narrow foraging area in much of the nearshore and proximal offshore marine foraging area for the majority of their existence. Tylosaurines exhibit a greater offshore marine range than plioplatecarpines, consistent with their large body size and the robustness of their feeding apparatus. The largest tylosaurine taxa are replaced by *Mosasaurus* in the Late Campanian–Maastrichtian in the offshore foraging range. Mosasaurine taxa are found to occupy the broadest range of foraging areas, but their ranges are taxonomically segregated, consistent with adult body size and the diversity of feeding adaptations such as tooth morphologies and skull architecture seen in that subfamily. Where foraging areas of multiple taxa overlap, differences are typically in tooth form, reflecting prey preference or feeding niche. Foraging area occupation by multiple taxa with similar tooth forms suggests that other factors such as body size and prey acquisition style may have allowed for the finer partitioning of resources. Deep diving and long submergence may have also contributed to the depleted signals recovered for some of the large-bodied durophages and the largest of the macrophagous apex predators.

Keywords: paleoecology; foraging area; Mosasauridae; $\delta^{13}C$; stable carbon isotopes; Late Cretaceous

1. Introduction

In this contribution, we report measurements of $\delta^{13}C$ derived from tooth enamel as a proxy for foraging area for a large temporal, geographic, and taxonomic sample of mosasaurs. Mosasaurs are an extinct clade of lizards that entered the marine realm at about 98 Ma and went extinct at the Cretaceous/Paleogene (K/Pg) boundary event at 66 Ma [1] (Figure 1). Early mosasaurs were morphologically similar to extant monitor lizards, and Mosasauria was recovered in recent phylogenetic analyses as the sister taxon to Varanoidea [2]. During their 32-million-year history, they adapted to life in water, undergoing significant morphological changes, most apparent in the development of flippers, fluked sculling tails, and a fusiform body shape [3]. Despite our increasing knowledge of the morphology and phylogenetic relationships of mosasaurs, comparatively little is known of their feeding ecology. Most studies of this nature have used dental morphology and gut content to infer prey preference or the suitability of dentition to a particular prey type, but do not address the spatial segregation of their foraging area [4–20].

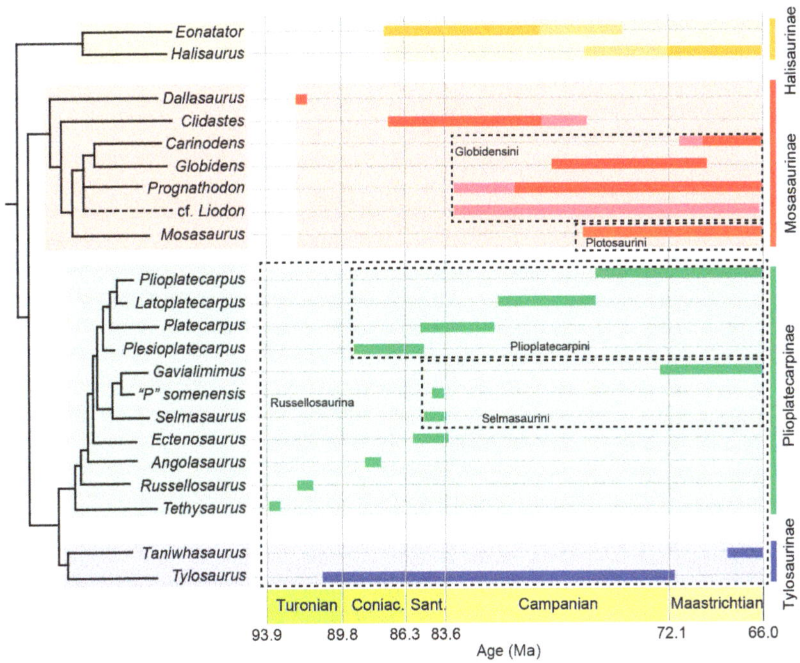

Figure 1. Simplified cladogram of mosasaurid relationships. See text for details.

Other studies have addressed the habitat preference of mosasaurs by inferring shoreline proximity and paleobathymetry from sedimentary analyses of the deposits in which mosasaur remains have been found [21–26]. Analyses of rare earth elements (REEs) have also been applied to mosasaur bones to infer the paleobathymetry of their habitats [27–29]. Although facies and REE analyses do provide clues as to the depositional environment in which the fossils are preserved, they do not necessarily reflect where the animals lived and foraged because of the possible postmortem transport of carcasses (e.g., [30]).

Conversely, carbon isotope measurements ($\delta^{13}C$) of marine vertebrates relate more directly to the diet [31] and foraging area [32] of the animal while alive. This is because marine primary producers have a wide range of $\delta^{13}C$ compositions reflecting the spatial gradient (nearshore to offshore) of particulate organic matter (POM; [33–35]). The ^{13}C-depleted nature of offshore primary producers propagates through the food web, causing larger, offshore-dwelling predators to have depleted $\delta^{13}C$ values compared to their nearshore counterparts. Clementz and Koch [32] used this phenomenon to identify differences in foraging zones between multiple marine, freshwater, and terrestrial mammals by analyzing the $\delta^{13}C$ values of tooth enamel.

Robbins et al. [36] were the first to explore mosasaur diet and foraging segregation through the examination of $\delta^{13}C$ values from tooth enamel, employing the methods of Clementz and Koch [32]. That study analyzed a small number of mosasaurine mosasaurids, a dolichosaurid, and the extant Galápagos marine iguana, *Amblyrhynchus cristatus*, and found that larger animals generally had lower $\delta^{13}C$ compositions, consistent with foraging farther from shore. Comparison with the isotopic ranges of tooth enamel $\delta^{13}C$ in modern pinnipeds grouped by foraging zone [32] revealed that smaller mosasaurs such as *Dallasaurus, Clidastes,* and juvenile *Globidens* had δ^{13}C values similar to modern kelp-bed-dwelling animals, with some specimens of *Clidastes* just bordering on the nearshore marine range. Larger mosasaurs, such as *Mosasaurus* and an adult *Globidens* from north Texas, fell within the nearshore marine range. Specimens of *Globidens* and *Prognathodon* from the Maastrichtian of Angola fell into the offshore range, which Robbins et al. [36] argued may have been the result of offshore foraging, deep diving (see below), or a combination of the two. Other studies [37–40] employed geographically and temporally restricted samples to infer resource partitioning in contemporaneous assemblages. In those studies, more-depleted δ^{13}C values were generally correlated with increased body size. Most recently, Leuzinger et al. [41] explored high-latitude stable isotopes, interpreting the composition of their mosasaur sample as an indication of a deeper, more offshore foraging area compared to the values recovered for their fish sample.

The simplified cladogram presented in Figure 1 is a synthesis of recent phylogenetic analyses [42–45]. Although there are a small number of basal branching, plesiopedal mosasaurids present in the Cenomanian and Turonian (~98 Ma–~90 Ma), most mosasaurids belong to one of four broadly recognized subfamilies (Halisaurinae, Tylosaurinae, Plioplatecarpinae, and Mosasaurinae) which appear to diverge in the Late Cenomanian or Early Turonian [43,46]. The compositions of the subfamily-ranked clades are relatively consistent among various workers, but the ingroup relationships of those subfamilies are somewhat more labile (e.g., [42–45,47–53]). Two additional subfamilies were erected by Palci and Caldwell [54] but not recovered in recent analyses [43,45], and the constituent taxa of those, along with *Ectenosaurus, Yaguarasaurus,* and *Angolasaurus,* are shown here as basal branching plioplatecarpines, but topologies may differ among analyses [42–44,51,55].

Later branching plioplatecarpines are included in the tribes Selmasaurini and Plioplatecarpini. Longrich et al. [45] erected the tribe Selmasaurini to include the genera *Selmasaurus, Gavialimimus, Goronyosaurus,* and *Khinjaria,* to which we provisionally add, here, *"Platecarpus" somenensis* Thevenin 1896. The tribe *Plioplatecarpini* was originally erected by Russell [21] to include the genera *Platecarpus, Plioplatecarpus,* and *Ectenosaurus,* but was revised by Longrich et al. [45] to include the genera *Plesioplatecarpus, Platecarpus, Latoplatecarpus,* and *Plioplatecarpus,* but not *Ectenosaurus,* as that taxon is more basal branching in their analysis (see also [52]). Species currently referred to *Ectenosaurus* are united by a number of characters unique among mosasaurids and likely form a monophyletic clade [55]; however, monophyly was not recovered in the strict consensus of recent analyses [55,56]. Our sample includes a single specimen (SMU76350) from Texas referred to the genus *Ectenosaurus* [57].

Mosasaurinae is a large and diverse clade that originates in the Turonian, represented by *Dallasaurus,* a small, plesiopedal taxon, recovered as a basal branching member of that clade [58]. The genus *Clidastes,* which we here consider monophyletic (but see also [59]), appears in the Coniacian and ranges into the middle Campanian [60]. In the Campanian, mosasaurines develop taxonomically disparate dental morphologies [1,61,62]. We use the tribe name Plotosaurini (sensu [47]) to include the genera *Mosasaurus, Plotosaurus, Jormungandr, Bentiabasaurus, Moanasaurus,* and *Rikisaurus. Plotosaurini* is the sister clade to the poorly resolved tribe Globidensini (sensu [47]), which includes the genera *Globidens, Prognathodon, Plesiotylosaurus, Thalassotitan,* certain taxa previously referred to the genus *Liodon* (fide [63]), and provisionally, *Carinodens.*

Halisaurinae has long been considered to have branched prior to the Mosasaurinae–Russellosaurina split [47,64]; however, some analyses recover Halisaurinae more closely related to Russellosaurina [50]. Halisaurines share a plesiomorphic configuration of the temporal arcade and quadrate suspensorium, a novel path for the internal carotid artery in the parabasisphenoid, and complete paired nasals separating the premaxilla from the frontal, overlying the latter [65,66]. Given the plesiomorphic nature of this subfamily, we illustrate it in Figure 1 as branching basal to the Mosasaurinae–Russellosaurina split.

This study explores patterns of the foraging area preference of mosasaurs through time using $\delta^{13}C$ measurements from their tooth enamel. The data are compiled from both published and previously unpublished sources [36,37,39–41]. The sample includes species of four subfamilies, ranges in age from the lower middle Turonian (~92.5 Ma) to the Late Maastrichtian (~66 Ma), and includes specimens from North America (USA: Texas, Kansas, Arkansas), Israel, the Netherlands, Denmark, Sweden, Angola, Argentina, and Antarctica. We present the results by taxon and discuss patterns of foraging area segregation through time.

2. Materials and Methods

2.1. Sample Selection

A total of 93 new samples were analyzed for this contribution (see Appendix A for sample and replicate details). Teeth with good surface preservation and few cracks were preferentially selected to avoid breakage during the mechanical removal of the enamel, although incomplete teeth with broken bases or tips were also used, as these breaks made the identification of the dentine and enamel layers easier. One sample (*P. currii*) was recovered as enamel fragments left in a silicone mold of a single tooth of the holotype specimen. Where practical, molds of the teeth prepared for this study were made prior to sampling, and casts were deposited in the Shuler Museum of Paleontology at Southern Methodist University, Dallas, TX, USA. The mold of the *P. currii* tooth is housed at Lund University.

2.2. Sample Pretreatment

Teeth with adhesives on their exterior were first subjected to a 24 h suspension in acetone followed by vigorous rinsing and scrubbing with a soft brush before placement in a 0.1 M acetic acid solution for 72 h to remove diagenetic carbonate adhered to the surface of the tooth. This pretreatment method has been shown to remove diagenetic carbonate without affecting the original isotopic signature of the bioapatite [67]. The teeth were thoroughly washed again with deionized water, followed by two 20 min suspensions in an ultrasonic methanol bath to remove organic matter, and a third suspension in deionized water.

2.3. Sample Enamel Removal

After thorough drying, enamel was selectively removed using a low-speed rotary tool with a 0.5 mm tungsten carbide burr bit. The enamel was removed in passes and collected on sterile weigh paper and regularly transferred to a vial. The low-speed nature of the drill and regular collection of material assured that the specimen was not heated and that only enamel was collected. Any material that was present on the weigh paper when the drill struck dentin was discarded. One sample (*P. currii*) was only available as crushed material. Enamel was picked from this sample mechanically under a microscope to avoid the presence of any dentin.

2.4. Sample Analysis

For samples analyzed in the Stable Isotope Laboratory at Southern Methodist University, Dallas, TX, USA, between 0.7 and 5 mg of powdered enamel were reacted under vacuum with 102% phosphoric acid for at least 4 h at 25 °C before extraction on a vacuum line. Purified CO_2 gas was measured for isotopic composition on a Finnigan MAT 252 mass spectrometer and calibrated against a tank of carbon dioxide gas from Oztech Trading

Corporation (Houston, TX, USA). This tank was periodically calibrated using a suite of international (NBS-18 [δ^{13}C −5.01‰, δ^{18}O −23.2‰], NBS-19 [δ^{13}C 1.95‰, δ^{18}O −2.2‰]) and in-house (Carrara marble [δ^{13}C −2.2‰, δ^{18}O −5.9‰]) carbonate standards.

Samples analyzed in the Stable Isotope Geosciences Facility at Texas A&M University, College Station, TX, USA, were weighed between 0.5 and 2 mg and reacted under vacuum with 102% phosphoric acid for ~15 min at 75 °C in a Kiel IV carbonate device coupled to a ThermoFinnigan MAT 253 mass spectrometer. These samples were run alongside multiple NBS-19 standards for calibration and quality control. Isotopic compositions are reported in δ-notation and expressed in part per thousand (‰) with respect to V-PDB. Analytical uncertainty from both labs was less than 0.1‰.

2.5. Compiled Data

The previously published data sets (n = 130 samples total) were run in different labs in the USA and Europe [36,37,39–41] using slightly different methods. For instance, samples were reacted at higher temperatures according to the methods of [37,40] than in the US labs [34,37, this study]. While there is concern about the high variability in oxygen isotopic compositions with different reaction temperatures and methods [68–70], the carbon isotopes, which are the focus of this study, show little variation when analyzed at different temperatures or through different techniques (see [70]) if the calibration methods are similar (i.e., one-point calibration; [71]). Each lab that measured the teeth reported their calibrated carbon isotopic compositions using international standards (SMU: NBS-18 [δ^{13}C −5.01‰, δ^{18}O −23.2‰], NBS-19 [δ^{13}C 1.95‰, δ^{18}O −2.2‰]; TAMU: NBS-19; UNIL: NBS-19; Vrije Universiteit (Amsterdam) Earth Science Stable Isotope Laboratory: IAEA-603 [δ^{13}C 2.46‰, δ^{18}O −2.37‰]). This calibration step, along with the consistent behavior of carbon isotopic fractionation using different reaction temperatures and techniques, allowed us to compare results between labs. Isotopic compositions from each lab are reported in δ-notation and expressed in part per thousand (‰) with respect to V-PDB. Analytical uncertainty based on the standards run alongside the samples was 0.14‰ or better.

2.6. Taxonomic Identifications

The teeth in our new data set were identified to taxon by sampling from published specimens (e.g., [51,72]) or by comparison with specimens in institutional collections and descriptions and illustrations in the literature (e.g., [17,63,73–75]). The taxonomic identifications of samples from previous studies were maintained, with the exception of two samples referred to *Plioplatecarpus* [41], here considered *Taniwhasaurus* sp. based on comparisons with figured specimens [76]. We referred the teeth from Sweden to the taxon *"Platecarpus"* cf. *"P." somenensis* because it compares well with the European holotype material of *"Platecarpus" somenensis* Thévenin 1896, which we consider a selmasaurin plioplatecarpine, and is not in reference to large North American plioplatecarpin plioplatecarpines that have been erroneously assigned to that taxon [47,52]. Species referred to the genus *Liodon* are considered to be closely related to or in some cases included in the genus *Prognathodon* [63]; however, we maintained the assignments to *Liodon* for the published samples [41] in our compiled data set.

2.7. Institutional Abbreviations

HUJ, Hebrew University of Jerusalem, Israel; LO, Paleontological Collection, Department of Geology, Lund University, Lund, Sweden; MGUAN PA, Museum of Geology, Universidade Agostinho Neto, Luanda, Angola; NHMM, Natuurhistorisch Museum Maastricht, Maastricht, the Netherlands; SMU, Shuler Museum of Paleontology, Southern Methodist University, Dallas, TX, USA.

3. Results

The detailed results are presented in Appendix A and graphically summarized in Figure 2. In the following sections, we present the findings by taxonomic group, placing them within foraging areas following the zonation scheme of Clementz and Koch [32]. We use the general terms "more nearshore" and "more offshore" to encompass the range of values from 0‰ to −16‰. We use the term "Nearshore Zone" (NSZ) to include the combined ranges of the "Kelp Bed Zone" (KBZ) and that portion of the "Estuarine Zone" (EZ) of Clementz and Koch [32] that falls below −8.0‰. We use "Nearshore Marine Zone" (NSMZ) and "Offshore Marine Zone" (OSMZ) for those partially overlapping zones defined by Clementz and Koch [32].

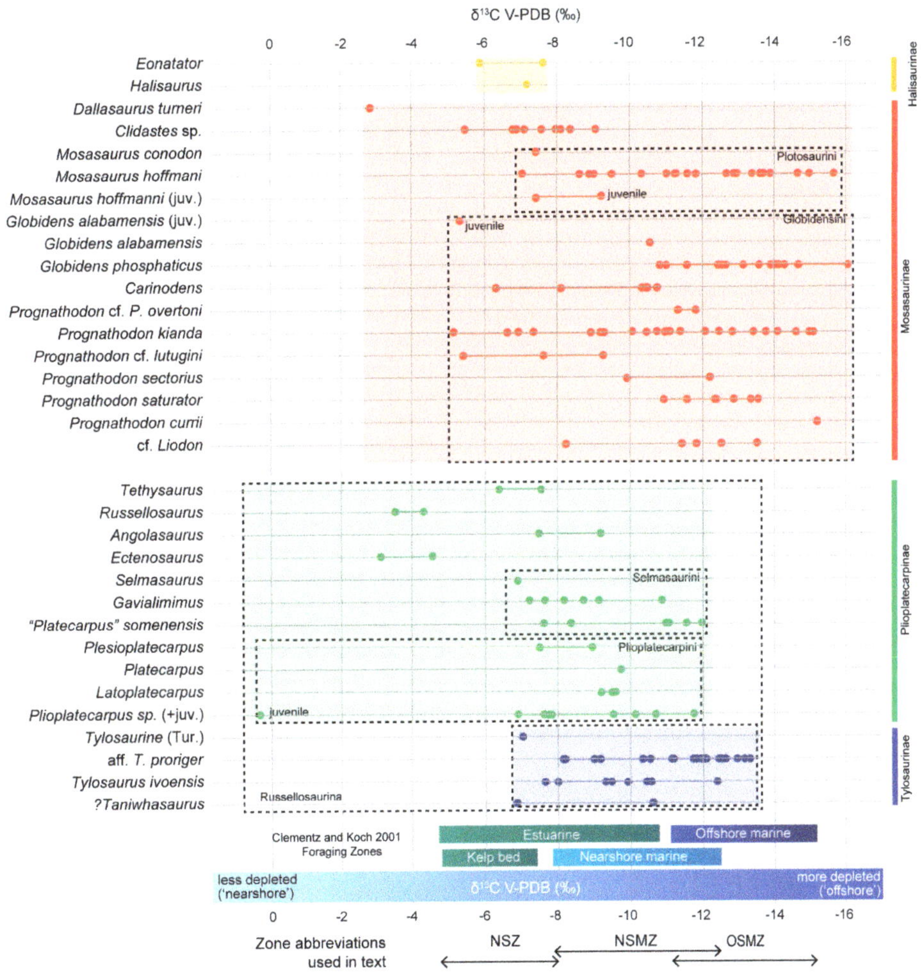

Figure 2. Mosasaurid $\delta^{13}C$ values by taxon. Color coding denotes subfamily membership: yellow = Halisaurinae; red = Mosasaurinae; green = Plioplatecarpinae; blue = Tylosaurinae. See Appendix A for details.

3.1. $\delta^{13}C$ by Taxon

3.1.1. Halisaurinae

Samples of *Eonatator* sp. from the latest early Campanian of Åsen, Sweden, yielded $\delta^{13}C$ values of −5.9 and −7.7‰, and *Halisaurus* sp., from the Lower Maastrichtian of Angola [39], yielded a value of −7.2‰, suggesting foraging in the NSZ.

3.1.2. Early-Diverging Plioplatecarpinae

Tethysaurus nopscai, from the lower Turonian of Morocco, with an average $\delta^{13}C$ value of −6.9‰, suggests a NSZ preference. *Russellosaurus coheni,* from the middle Turonian of Texas ranged from −3.5‰ to −4.3‰, suggesting foraging in the least depleted NSZ, exceeding the ranges given by [32]. *Ectenosaurus* sp. from the upper Santonian of west Texas [57] had a range of −3.1‰ to −4.6‰, indicating foraging in the most enriched NSZ and similar to the results for *Russellosaurus. Angolasaurus bocagei* from the Coniacian of Angola [74] yielded values of −7.4‰ to −9.2‰, within the range of later branching plioplatecarpins and within the NSMZ.

3.1.3. Plioplatecarpinae: Plioplatecarpini

Three North American plioplatecarpins yielded values between −7.5‰ and −9.6‰, falling within the lower half of the NSMZ. *Plesioplatecarpus* sp. from Kansas and Texas yielded −8.8‰ and −7.5‰, respectively. *Latoplatecarpus* sp. from Kansas yielded −7.6‰ and the Texas sample yielded a range of values from −9.1‰ to −9.6‰. *Plioplatecarpus* sp. from north central Texas and western Arkansas yielded −7.6‰ and −7.7‰, respectively. The northern European *Plioplatecarpus* samples yielded a range from about −7.0‰ to −11.1‰. One Scandinavian *Plioplatecarpus* sample yielded −9.5‰, while a juvenile sample yielded a value of −0.2‰, which appears to be an outlier and is likely diagenetically altered. Apart from the aforementioned outlier, most *Plioplatecarpus* samples fall within the NSMZ, but the European–Scandinavian sample's upper range is more depleted than the North American sample.

3.1.4. Plioplatecarpinae: Selmasaurini

Three selmasaurin taxa yielded a range of values from −6.9‰ to −11.5‰. *Selmasaurus johnsoni* from the lower Santonian of western Kansas yielded −6.9‰, falling within the NSZ. Three specimens of the selmasaurin "*Platecarpus*" cf. "*P.*" cf. *somenensis* from the lower Campanian of Åsen, Sweden, had average $\delta^{13}C$ ratios of −10.9‰, −11.3‰, and −11.5‰, within the overlapping part of the NSMZ and the OSMZ, the most depleted (offshore) values of any plioplatecarpine. An individual from Ugnsmunnarna, Sweden, also lower Campanian, had an average value of −7.7‰, within the distal part of the NSZ. *Gavialimimus* sp. [20] from the Lower Maastrichtian of Bentiaba, Angola, yielded values ranging from −7.2‰ to −10.9‰, suggesting that it was foraging in much of the NSMZ.

3.1.5. Tylosaurinae

An undescribed basal branching tylosaurine from the Turonian of North Texas (LO7786) yielded a $\delta^{13}C$ ratio of −7.0‰, indicating foraging in the depleted portion of the NSMZ. Eight specimens of *Tylosaurus* aff. *T. proriger* yielded values ranging from −8.2‰ to −13.3‰, occupying most of the NSMZ but extending into the more depleted portion of the OSMZ. Nine samples of *Tylosaurus ivoensis* from Åsen, Sweden, yielded average $\delta^{13}C$ ratios between −7.6 and −12.4‰, suggesting foraging primarily in the NSMZ, but also ranging into the OSMZ.

3.1.6. Early-Diverging Mosasaurinae

The basal branching mosasaurine *Dallasaurus turneri* from the middle Turonian of Texas yielded a $\delta^{13}C$ value of −2.8‰, suggesting foraging in the most enriched NSZ, similar to the results for *Russellosaurus* and *Ectenosaurus*. *Clidastes* from Texas yielded values ranging from −6.7 ‰ and −8.3‰, and *Clidastes* from Åsen, Sweden, ranged from −5.4 ‰ to −9.1‰, indicating that *Clidastes* was foraging in the NSZ and the less depleted portion of the NSMZ.

3.1.7. Mosasaurinae: Plotosaurini

A single specimen of *Mosasaurus conodon* yielded a value of −7.4‰, foraging in the most depleted part of the NSZ. Adult specimens of *Mosasaurus hoffmanni* ranged from −7.1‰ to −14.9‰, with specimens from Texas yielding values from −7.4‰ to −9.5‰, the Netherlands −7.1‰ to −14.9‰, and Angola from −8.9‰ to −12.9‰, ranging through all three foraging zones and beyond the OSMZ. Juvenile specimens of *M. hoffmanni* from northern Europe yielded values of −7.1‰ to −9.2‰, ranging from the most depleted part of the NSZ to the less depleted quarter of the NSMZ.

3.1.8. Mosasaurinae: Globidensini

A juvenile *Globidens alabamaensis* yielded an average $\delta^{13}C$ value of −5.3‰, at the most enriched end of the NSZ. A co-occurring adult specimen of *G. alabamaensis* yielded an average value of −10.5‰, in the lower part of the more depleted half of the NSMZ. Samples of *Globidens phosphaticus* teeth from the Lower Maastrichtian of Bentiaba, Angola, yielded average $\delta^{13}C$ values ranging from −10.8‰ and −16.1‰, falling mostly within, but also exceeding, the OSMZ. *Carinodens* from the Upper Maastrichtian of northern Europe yielded values ranging from −10.3‰ to −10.8‰; a single sample from Denmark yielded −6.3‰ and a single Angolan sample yielded −8.1‰. Taken together, these suggest that *Carinodens* foraged in the upper half of the NSZ and much of the NSMZ. A single specimen of *P. currii* from the latest Campanian of Israel gave a single measurement of −15.1‰, falling within the most depleted portion of the OSMZ.

A *Prognathodon* cf. *P. overtoni* specimen from the Lower Maastrichtian of north Texas had a $\delta^{13}C$ average value of −11.5‰ [36], falling within the overlapping portions of the NSMZ and OSMZ. Two specimens of *Prognathodon* cf. *P. lutugini* from the middle Campanian of Åsen, Sweden, yielded $\delta^{13}C$ values between −5.4‰ and −9.3‰, ranging from the NSZ into the lower half of the NSMZ. *Prognathodon sectorius* from the Maastrichtian of northern Europe yielded $\delta^{13}C$ values of −9.9‰ to −12.2‰, approximately in the upper half of the NSMZ and the overlapping, less depleted part of the OSMZ. *Prognathodon saturator* from the Maastrichtian of northern Europe yielded values ranging from −10.9‰ to −13.6‰, falling within the more depleted part of the NSMZ and ranging into the upper half of the OSMZ. *Prognathodon kianda* from the Lower Maastrichtian of Bentiaba, Angola, ranged from −5.2‰ to −15‰, suggesting foraging across all three defined zones. Specimens tentatively referred to *Liodon* from Antarctica ranged between −8.2‰ and −13.5‰, encompassing the NSMZ and ranging into the upper half of the OSMZ.

3.2. *$\delta^{13}C$ Through Time*

Figure 3 illustrates the geographic distribution of mosasaurs and feeding guild coverage [8,17,37,77] for five time slices and $\delta^{13}C$ values by taxon through time, ranging from the early middle Turonian (~93 Ma) though to the latest Maastrichtian (66 Ma), for the taxa analyzed in this contribution.

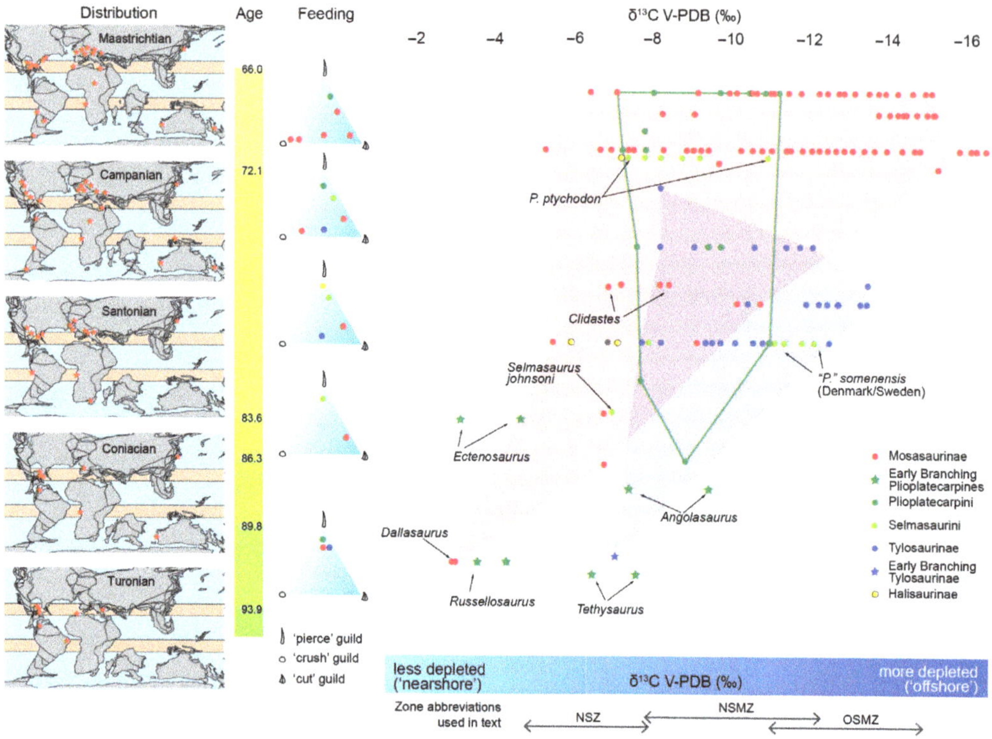

Figure 3. Temporal distribution of mosasaurs, tooth guild occupation, and isotope values by taxon. Geographic distribution of mosasaurs through time after [1]; horizontal tan bands span 15° to 30° north and south latitude. Tooth-morphology-based guild membership is subjectively assessed following previous research [8,17,77]. Color denotes subfamily membership: yellow = Halisaurinae; red = Mosasaurinae; green = Plioplatecarpinae; blue = Tylosaurinae.

3.2.1. Early Radiations

Foraging area segregation at the subfamily level appears to be established by the lower middle Turonian. Early-diverging plioplatecarpines, such as *Tethysaurus* and *Russellosaurus*, range farther offshore than *Dallasaurus,* which is also differentiated by its small size (~1 m) and tooth form [58]. *Tethysaurus* has the most depleted values, falling within the NSMZ (but see discussion below). A three-to-four-meter-long basal branching tylosaurine from the upper part of the middle Turonian of Texas occupies the proximal part of the NSMZ, but also maintains a conservative dentition, similar to *Russellosaurus* in size and shape.

3.2.2. Coniacian–Santonian

By the Coniacian–Santonian, most moderately sized plioplatecarpines, such as *Angolasaurus*, fall within the NSMZ, a foraging range they occupy through the remainder of the Cretaceous. Exceptions to this include the rare taxa *Ectenosaurus*, *Selmasaurus*, and "*P.*" cf. *somenensis*, all of which deviate from contemporaneous plioplatecarpins in dental morphology, skull architecture, or both. Within the Coniacian–Santonian sample, *Clidastes* is the first mosasaur bearing blade-like teeth. Although *Clidastes* is approximately the same adult size as contemporary plioplatecarpines, it appears to prefer a more nearshore foraging area. However, there are only a couple of contemporary plioplatecarpine samples and no tylosaurine samples for this interval, limiting direct comparisons.

3.2.3. Campanian

In the middle Campanian, most plioplatecarpin plioplatecarpines fall within the NSMZ, as in the prior and following stages. The exceptions to this are the early occurrences of *Plioplatecarpus* which appear in the upper Campanian of the North American Gulf Coast, with our samples from Texas and Arkansas falling within the most distal part of the NSZ. Also in the Campanian, tylosaurines achieve gigantic proportions [1] and range from the NSMZ into the proximal two-thirds of the OSMZ. *Clidastes* maintains a relatively nearshore habitat and is present into the middle Campanian, but is not known from the upper Campanian or thereafter. In the lower and middle Campanian, mosasaurines show a marked increase in taxonomic diversity and morphological disparity (Figure 1; [1]), with the first appearance of globidensin mosasaurines represented by *Prognathodon*, *Liodon*, and the durophagous *Globidens*. The first appearance of a plotosaurin mosasaurine is *Mosasaurus conodon*, with our sample falling within the foraging range of *Clidastes*. The body size and dentition of *Mosasaurus conodon* is similar to that of the large middle Campanian *Clidastes*, which it likely replaced in that feeding niche in the later part of the Campanian. *Prognathodon* cf. *P. overtoni* falls within the proximal part of the OSMZ and bears a robust crush–cut guild dentition, while *Prognathodon* cf. *P. lutugini* falls within the NS and NSMZ, has a sectorial dentition, and falls within the cut guild. *Globidens* is found in the NSZ and NSMZ and bears a nearly hemispherical crushing dentition in the middle tooth row.

3.2.4. Maastrichtian

In the Maastrichtian, plioplatecarpin plioplatecarpines are represented by *Plioplatecarpus* from northern Europe in our sample set. They fall within a slightly expanded range compared to previous stages, in the distal part of the NSZ and spanning the entire NSMZ. This is also the case for the selmasaurin *Gavialimimus* sp. from the Maastrichtian of Angola. The large plotosaurin mosasaurine *Mosasaurus hoffmanni* also appears in the Maastrichtian and occupies the distal NSZ to beyond the defined limit of the OSMZ. This taxon is similar to *Tylosaurus* in size and dental morphology, completely overlaps with its foraging range, and thus appears to be the ecological replacement for large tylosaurines, which are not known after the Early Maastrichtian.

A remarkable increase in disparity in tooth morphology occurs in Globidensini. There is increased taxonomic diversity in durophagous taxa with the appearance of large-bodied, low-crowned durophages such as *Globidens phosphaticus*, the high-crowned durophage *Prognathodon currii*, and small-bodied taxa such as *Carinodens*. There is also significantly expanded taxonomic diversity and tooth disparity in large macrophagous taxa such as *Prognathodon saturator* and a range of other forms bearing more or less sectorial dentitions but varying in their degree of lateral compression, such as in the taxa *Prognathodon sectorius* and *P. kianda*. The Maastrichtian durophagous taxa in our sample are both relatively depleted. *Globidens phosphaticus* ranges from the distal part of the NSMZ to beyond the OSMZ, the most depleted (offshore) values for any taxon in our sample. Our single sample of *P. currii* falls in the most depleted part of the OSMZ. *P. saturator* and *P. sectorius* (which overlap the more enriched part of the range of *P. saturator*) fall within approximately the middle of the foraging range recovered for the contemporaneous *M. hoffmanni*. However, those three co-occurring taxa differ from one another in tooth morphology [37,78] and, in the case of *P. saturator*, in skull architecture as well [79].

4. Discussion

4.1. Trophic Level and Foraging Habit

Based on tooth form and known gut content, the mosasaur taxa discussed here were carnivorous [8,17,20,77]. To facilitate comparisons between taxa, we assume dietary fractionation was the same for all taxa. Fractionation factors from diet to enamel ($\alpha_{diet\text{-}enamel}$) vary less among carnivores compared to herbivorous taxa (see [80,81]), typically centering around 9‰ for terrestrial carnivores [82,83] and ~10‰ for modern odontocetes [84–87]. While enrichment in whole-body ^{13}C occurs from one trophic level to the next, its relevance is relatively minor (up to 1‰ per trophic step, [31]). The opposite is true for $\delta^{13}C_{POM}$ values, which become more depleted (increasingly negative) with distance from shore and with depth, which propagates depletion from primary producers to top predator $\delta^{13}C$ enamel values through the trophic chain [32].

Foraging depth may also be a factor in interpreting $\delta^{13}C$ values. Benner et al. [88] demonstrated a decrease in primary producer $\delta^{13}C$ of about 3‰ between waters 100 to 200 m deep. This decrease in primary producer $\delta^{13}C$ with increasing depth propagates through the food web, resulting in carnivores that consistently feed at depth having depleted $\delta^{13}C$ ratios compared to shallow-foraging individuals. For example, Clementz and Koch [32] argued that the depleted nature of elephant seal tooth enamel $\delta^{13}C$ (~−13‰) relative to other pinnipeds may be in part related to their habit of deep diving (>300 m, with some dives exceeding 1000 m [89]). Thus, we interpret $\delta^{13}C$ values primarily as a proxy for foraging area, which may indicate shoreline proximity, water depth, or some combination of the two.

4.2. Physiological Factors

Air-breathing marine amniotes accumulate CO_2 in their bodies during oxygen consumption while breath-holding [90–92]. This increased amount of respired CO_2 in the blood among deep divers (or long-duration breath-holding) leads to a more depleted $\delta^{13}C$ value compared to shallow-diving or surface-dwelling individuals. During long-term dives, marine reptiles may increase the accumulation of CO_2 in their body fluids, including the blood, as a result of the increased resistance to blood flow [90,93,94]. The CO_2 present in the blood then reacts with water, forming carbonic acid. This acid dissociates into carbonate and bicarbonate (HCO_3), the latter being the predominant form of CO_2 in the blood [91]. The majority of bone and enamel carbonate is derived from the respired CO_2 in the blood. Thus, the ^{13}C-depleted nature of respired CO_2 results in a negative correlation between the amounts of respired CO_2 incorporated in bone carbonate and the $\delta^{13}C$ of that carbonate [95].

Physiological differences between shallow and deep divers may influence the fractionation captured in tooth enamel. Deep-diving marine amniotes typically store more oxygen in their blood and tissues than their lungs [94]. During long, deep dives, oxygen release is favored over oxygen uptake, which efficiently uses stored oxygen and increases the amount of respired CO_2 in the blood. The ^{13}C-depleted respired CO_2 is then incorporated into the tissue carbonate. Shallow divers lack the deep-diving adaptation of increased oxygen storage in the blood and tissues, instead storing more oxygen in the lungs [94]. During long dive periods, blood pH decreases. This pH decrease occurs in tandem with the decrease in pO_2 in the lungs, a situation which favors oxygen uptake rather than release. The Bohr effect, a decrease in oxygen affinity with decreasing pH, is disadvantageous under these long, shallow dive conditions. The Bohr effect decreases over the course of a prolonged dive to compensate for the drop in pH, which allows for the release of the last traces of O_2 in the lungs [94]. This method is a less efficient oxygen transport system than that of deep-divers, resulting in lower amounts of respired CO_2 in the blood compared to deep-diving species. The carbonate precipitated in equilibrium with the blood of a shallow-diving animal should therefore be enriched in ^{13}C compared to deep-divers.

Biasatti [96] examined modern sea turtle remains for evidence of the effects of respiratory physiology, diet, and latitude on the δ^{13}C ratios of sea turtle bone carbonate. Samples from the Green sea turtle, *Chelonia mydas*, a shallow-diving species, were similar to those of barnacles found on their carapaces and the δ^{13}C ratios of marine dissolved inorganic carbonate (DIC; 0–1‰). Biasatti [96] also argued that the deep-diving adaptations and behavior of Leatherback (*Dermochelys coriacea*) and Olive Ridley (*Lepidochelys olivacea*) sea turtles cause a greater accumulation of CO_2 in the blood compared to species not adapted for deep diving. The depleted values from the Leatherback turtle (−6 to −13‰) and Olive Ridley turtle (−7 to −9‰) were similar to the ranges found among deep-diving cetaceans [96,97].

Although carbon isotope ratios may be influenced by depth [88] and dive duration [95,96], when viewed in the context of depositional environment or other clues as to feeding style, these factors may provide more nuanced interpretations of the raw δ^{13}C values in certain taxa. One example is the large Maastrichtian durophage *Globidens phosphaticus* and the locality of Bentiaba, Angola [73]. Skeletal remains and numerous shed teeth are found within a small geographic area at Bentiaba, associated with the remains of other taxa which yielded much more enriched δ^{13}C values [39]. Thus, shoreline proximity alone cannot account for the depleted values seen in *Globidens phosphaticus* (~−11~−16‰). The depositional setting in the locality is relatively nearshore and the local paleobathymetry is estimated to be 50–150 m [39]. The taxon possesses large, irregularly hemispherical teeth, well suited for crushing hard-shelled prey. The locality is also rich in the remains of a large inoceramid bivalve, reported to be at least one of the prey items of the genus [12]. Taken together, it is reasonable to infer that at least part of the significantly depleted values seen in this taxon is due to the increased dive time (and depth) associated with bottom foraging. Additionally, feeding low on the trophic chain, *Globidens* would not manifest the 1‰ per trophic step enrichment of the apex predators in the ecosystem, given comparable shoreline proximity.

4.3. Other Factors

Local ecological setting and latitude are known to affect δ^{13}C values. However, patterns of mosasaur assemblages in disparate geographies (and time slices) of our data are broadly similar (Figure 4), with larger taxa generally exhibiting more depleted values (Figure 5). This body-size partitioning correlates well with tooth forms, where small-bodied piscivorous taxa occupy a more nearshore range, large-bodied macrophages forage in a more offshore range, and generalists with more or less sectorial dentitions foraging in a broad range irrespective of body size (e.g., *Prognathodon kianda*). Where two taxa are found to occupy the same isotopic range, there is little or no overlap in tooth form. Where differences between similar taxa occur between localities, these may be an artifact of noise in the sample, local ecological effects related to shoreline proximity, latitude, or some combination of these. Nonetheless, the broader patterns of foraging area partitioning are relatively consistent, although comparisons among geographies reveal some differences in detail.

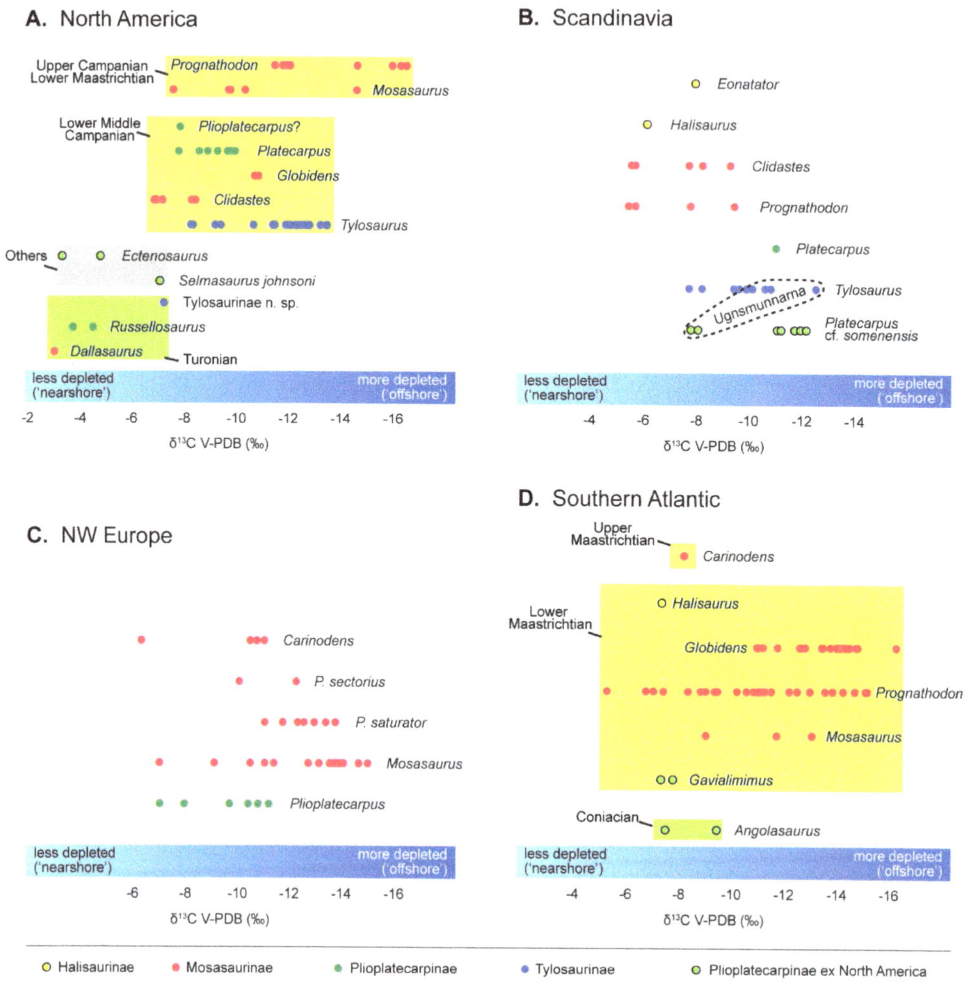

Figure 4. $\delta^{13}C$ by select geographic region. Color coding denotes subfamily membership: yellow = Halisaurinae; red = Mosasaurinae; green = Plioplatecarpinae; blue = Tylosaurinae.

For instance, the two localities from Scandinavia yielded taxonomic patterns of range occupation that differ somewhat from one another. The samples were collected from sediments at Åsen and Ugnsmunnarna, deposited at about 50 °N paleolatitude (Figure 4B). They are latest early Campanian in age (middle Campanian in the North American tripartite Campanian system) and fall within the informal *Belemnellocamax mammillatus* biozone, which is roughly 300 kyr in duration [98]. The marine succession at Åsen was deposited in a nearshore environment. The presence of floodplain sediments, hybodont shark teeth, and the topography of the adjacent basement rock (i.e., the Höljeån valley), suggest that a fluvial system was located close nearby. At Ugnsmunnarna the fauna preserves a more ontogenetically mature sample, and was likely more distal or a deeper water deposition than Åsen during the latest early Campanian [99]. The recovered $\delta^{13}C$ values range from −5.4‰ to −12.4‰, with significant overlap among taxa; however, segregation patterns do exist. Halisaurines, with a range of values from −5.9‰ to −7.7‰, overlap with the lowest part of the range of *Tylosaurus* (−7.6‰ to −12.4 ‰) and "*P.*" cf. *somenensis* (−7.6‰ to −12‰) and occupy roughly the upper half of the range of both *Clidastes* (−5.4‰ to −9.1‰)

and *Prognathodon* (−5.4‰ to −9.3‰). The relatively enriched values for the halisaurine *Eonatator* and its small body size (~3 m), suggest that this taxon was foraging in the depleted part of the Nearshore Zone. The range of values of *Clidastes* and *Prognathodon*, co-occurring in the Åsen sample, nearly overlap, suggesting that they probably occupied the same nearshore and nearshore marine foraging area, but may have been partitioning by prey size, with *Clidastes* being significantly smaller (~4 m) than the larger (~8 m) *Prognathodon* species present in the locality (e.g., [75,99]). The aggregate range of values for *Tylosaurus* and "*P.*" cf. *somenensis* largely overlap; however, the samples from Ugnsmunnarna occupy opposite parts of the range compared to those at Åsen (Figure 4B). It is unclear if the influence of meteoric water via river discharge at the Åsen locality had any effect on the recovered values; however, one would expect a more uniform skewing of all the taxa in the locality if this were the case. Instead, the aggregate *Tylosaurus* and *Clidastes* values are roughly comparable to the North American sample, but the nearshore Åsen sample does not contain the more depleted values for *Tylosaurus*; they are present in the more distal, offshore sample from Ugnsmunnarna, suggesting instead that this may be an artifact of sampling. If this is the case, the genera *Tylosaurus* and "*P.*" cf. *somenensis* co-occur in the mid-shelf to most offshore part of the range, but differences in tooth forms in those taxa (e.g., [98]) suggest possible partitioning by prey type and size.

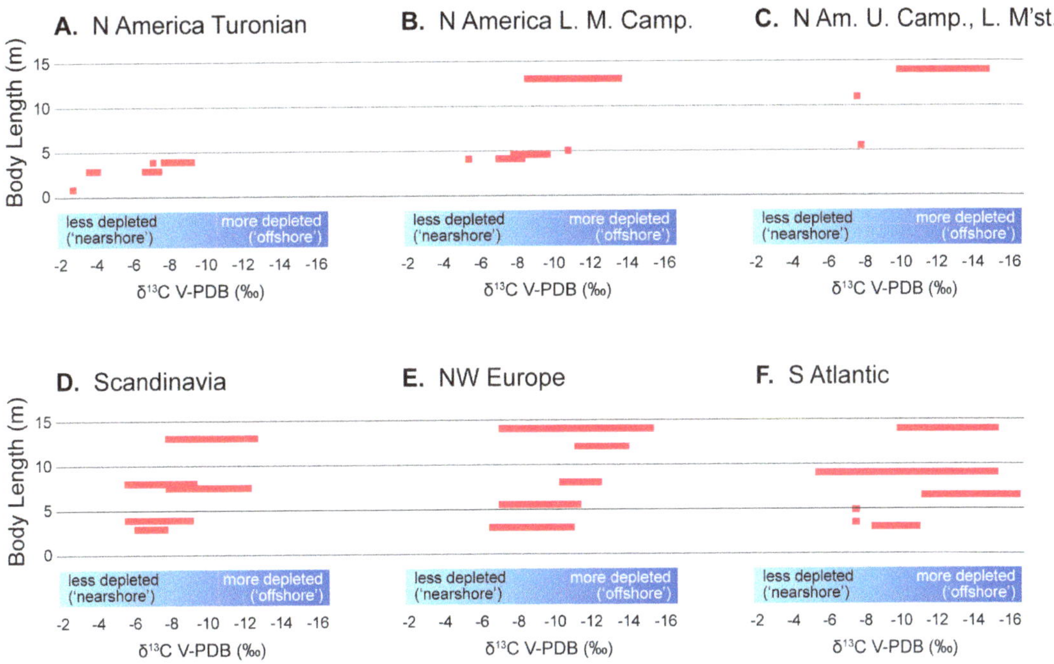

Figure 5. δ^{13}C by select geographic region and body length. Size data taken from [1].

The northeastern European sample is from the Maastrichtian type area, deposited at about 45 °N paleolatitude. Although the values at the depleted end of the range are similar to the other localities, the most enriched values (Figure 4C) are about ~2–3‰ more depleted than the other localities. Schulp et al. [37] reported isotope values for five mosasaur taxa (Figure 4C), finding that depleted δ^{13}C values largely correlate with increased body size, as previously noted by Robbins et al. [36]. With the exception of *P. saturator*, the taxa reported by [37] range through much of the Late Maastrichtian [100] and are found on a relatively shallow carbonate shelf, with water depths ranging from 80 m in the uppermost Gulpen

Formation to between 2 and 15 m in the uppermost part of the Maastricht Formation [101]. The majority of the sample (80%) is from the uppermost part of the Maastricht Formation, from relatively shallow water. As pointed out by Wendler [102], myriad variables influence the absolute values of $\delta^{13}C$ at specific localities. Thus, the more depleted values seen in the northeastern European sample, may possibly be due to local shoreline proximity and water depth, latitudinal factors, or other factors that negatively influence primary producers. This may also be the case for the *Tethysaurus* from the lower Turonian of the North African margin, which yielded relatively depleted values, incongruent with its size and level of marine adaptation. The locality that produced *Tethysaurus* was deposited in an open-platform environment [49] but under unusual and harsh conditions, indicated by a depauperate fauna [103].

Notwithstanding the large number of variables influencing primary producers [102] and the exceptions noted above, recovered $\delta^{13}C$ values with respect to body size, taxonomy, and tooth form (Figures 2–5) are largely consistent across geographies.

4.4. Evolution of Foraging Behavior in Mosasaurs

As shown above (Figure 3), early in their evolution, mosasaurs generally occupied a relatively nearshore range. Some taxonomic segregation of foraging area is already apparent between the small-bodied mosasaurines (e.g., *Dallasaurus*) and the moderately sized basal branching plioplatecarpines and tylosaurines (Figure 4). In the Coniacian–Maastrichtian, moderately sized plioplatecarpines consistently occupy the NSMZ, while both tylosaurines and mosasaurines expand their range farther offshore. Small-bodied halisaurines appear to favor a nearshore foraging area.

Dental morphology (see, e.g., [17,37,77]) broadly corresponds with the recovered $\delta^{13}C$ values. All plioplatecarpine and tylosaurine mosasaurs possess simple slightly recurved conical tooth crowns, only varying in ornamentation and robustness. A laterally compressed sectorial tooth form is first seen in early-branching mosasaurines, such as *Clidastes* in the Coniacian, and in the Campanian two new clades of mosasaurines appear: the Plotsaurini and the Globidensini (Figure 1). Globidensin mosasaurines display the most disparate range of tooth forms, suggesting a high degree of segregation by prey type and differences in feeding style, consistent with the broad range of $\delta^{13}C$ values recovered for the clade. Globidensini originates in the Campanian, evidenced by novel and diverse tooth forms, such as *Prognathodon overtoni,* which possess a robust bicarinate but squat dentition; the robust but more laterally compressed dentition of *Prognathodon* (*Dollosaurus*) *lutugini* [75,104]; and the broadly domed, durophagous dentition of *Globidens alabamaensis* [105]. By the Late Campanian, diversity in the Globidensini increases, with additional species of *Globidens* [62,106] and *Prognathodon* [107]. In the Maastrichtian, there is continued diversification, with an apparent radiation of sectorial toothed forms such as *Eremiasaurus*, *P. sectorius*, *P. kianda*, and other species previously referred to *Liodon* [63] (see also [15]). Large macrophages such as *P. saturator*, *Thalassotitan* from Morocco [17,18], and a large *Prognathodon* sp. from Angola [108] also appear in the Maastrichtian, along with durophagous forms such as the small-bodied *Carinodens* [109,110], new species of *Globidens* [73,111], and high-crowned durophages such as *Igdamanosaurus* [112] and *P. currii* [72]. Body size does not appear to correlate with foraging range in the globidensin taxa *Prognathodon kianda* and *P. lutugini*.

The range of tooth forms in globidensins stands in contrast to that of plotosaurins. A sectorial dentition is present in *Clidastes* and *Mosasaurus conodon*, while large-bodied forms such as *Mosasaurus hoffmanni* evolved a large robust macrophagous dentition and *Plotosaurus* evolved a high-aspect-ratio piscivorous (piercing) dentition. Among the plotosaurin taxa in our sample, evidence of ecological replacement occurs with the appearance of *Mosasaurus conodon* and coincident absence of a large *Clidastes* in the upper Campanian, with which it shares a similar dentition, body size, and isotopically determined foraging range. In the Maastrichtian, the derived species *M. hoffmanni* possesses a robust tooth form and attains a large body size, both traits similar to derived *Tylosaurus* species, a taxon absent from the OSMZ in the Maastrichtian. Where foraging areas of multiple taxa overlap, apparent differences in tooth form, body size, or both occur, reflecting prey preference, suitability, or feeding niche. Foraging area occupation by multiple taxa with similar tooth forms suggests greater competition and the finer partitioning of resources.

5. Conclusions

In this contribution, we presented a large taxonomic and temporal sample of $\delta^{13}C$ values derived from tooth enamel, revealing patterns of foraging area preferences of Late Cretaceous mosasaurid squamates over evolutionary time scales (~93–66 Ma). Our results for individual localities are broadly consistent with previous studies and their predictions, but new insights were gained from the addition of the temporal axis. Notwithstanding the relatively small sample size, sparse temporal sampling for certain timeframes, and other sources of uncertainty, broad patterns of foraging segregation are apparent. Inference of foraging area preference in conjunction with other data, such as body size, tooth form, skull architecture, and swimming ability, provides the basis to evaluate niche occupation and competition over evolutionary time scales.

Patterns of foraging area segregation at the subfamily level are established early in the record, and there is a trend of increasing offshore range extension through time, broadly mirroring increases in body size in certain taxa. The nearshore foraging range, occupied by more basal forms in the Turonian, appears to be increasingly under-represented through time. Derived russellosaurian mosasaurs are found to occupy nearshore marine and the more enriched parts of offshore marine environments, although tylosaurines achieve a large body size early and also are found to range somewhat farther offshore than plioplatecarpines. Early plioplatecarpines range from nearshore to nearshore marine areas, and the taxa with the most depleted values overlap with the earliest tylosaurines. From the Coniacian through the Maastrichtian, plioplatecarpines are largely restricted to the Nearshore Marine Zone, with a few outliers in the tribe Selmasaurini (e.g., *Selmasaurus johnsoni*, *"P." somenensis*) and the plioplatecarpin genus *Ectenosaurus*. Early small-bodied mosasaurines occupy a relatively nearshore foraging area but diversify in the Campanian, and thereafter occupy the widest range of foraging areas of all the subfamilies. One mosasaurine tribe, the Globidensini, shows the greatest taxonomic diversity, plasticity of tooth form, and range of body size, and also exhibits the broadest range of isotope values of any clade. *Mosasaurus conodon* appears to replace the superficially similar, large middle Campanian *Clidastes*, and large mosasaurines (e.g., *Mosasaurus hoffmanni* and some species of *Prognathodon*) replace the large tylosaurines (*Tylosaurus*) in the upper Campanian or Lower Maastrichtian. These may be two examples of competitive exclusion as a source of extinction.

Author Contributions: Conceptualization, M.J.P., L.L.J. and J.A.R.; methodology, M.J.P. and J.A.R.; validation, M.J.P. and J.A.R.; formal analysis, M.J.P. and J.A.R.; resources, M.J.P., L.L.J., J.L. and A.S.S.; data curation, M.J.P.; writing—original draft preparation, M.J.P. and J.A.R.; visualization, M.J.P. and A.S.S.; writing—review and editing, M.J.P., L.L.J., J.L. and A.S.S. All authors have read and agreed to the published version of the manuscript.

Funding: This research received no external funding.

Data Availability Statement: Supporting data are available on request to the corresponding author.

Acknowledgments: We thank Kurt Ferguson for useful discussions and Dale Winkler for assistance with specimens in the Shuler Museum of Paleontology at SMU. We thank Chris Strganac for access to unpublished isotope data from Bentiaba, Angola, samples and Richard Zakrzewski for the permission to sample tooth enamel in the Fort Hays Sternberg Museum collection. We thank the ISEM at SMU and The Saurus Institute for their ongoing support.

Conflicts of Interest: The authors declare no conflicts of interest.

Appendix A

Table A1. The $\delta^{13}C$ values, rounded to one decimal place and listed by specimen and sample. Data sources are as follows: * = this study; [36] = Robbins et al. (2008); [37] = Schulp et al. (2013); [39] = Strganac et al. (2015); [40] = Giltaij et al. (2021); and [41] = Leuzinger et al. (2023). The Scandinavian specimens processed for this study were un-numbered specimens from the Kristianstad Basin (Sweden) collected by one of us (J.L.) and referenced in the table for this contribution as Asen1–Asen8, Kris1, Kris2, Ugn0, and Ugn1. The cell colors denote subfamily membership: yellow = Halisaurinae; red = Mosasaurinae; green = Plioplatecarpinae; and blue = Tylosaurinae. The values for $\delta^{13}C$ and $\delta^{18}O$ are represented in ‰ with respect to V-PDB, and wt.% carbonate and $\delta^{18}O$ were included for the samples for which they were available.

Specimen # (Sample #)	Taxon	Region	$\delta^{13}C$	$\delta^{18}O$	wt.% Carb *	Data Src.
Asen1 (1)	*Eonatator* sp.	Scandinavia	−7.7	−3.7	7.8	*
Asen1 (2)	*Eonatator* sp.	Scandinavia	−5.9	−2.8	7.1	*
Asen2 (1)	*Clidastes* cf. *C. propython*	Scandinavia	−9.1	−3.3	8.2	*
Asen3 (1)	*Clidastes* cf. *C. propython*	Scandinavia	−5.4	−3.1	6.4	*
Asen3 (2)	*Clidastes* cf. *C. propython*	Scandinavia	−5.4	−2.1	4.2	*
Asen3 (3)	*Clidastes* cf. *C. propython*	Scandinavia	−7.5	−2.3	4.5	*
Asen4 (1)	*Clidastes* cf. *C. propython*	Scandinavia	−8.0	-	7.8	*
SMU76391 (1)	*Clidastes* sp.	North Texas	−7.1	−3.0	5.4	[36]
SMU76391 (2)	*Clidastes* sp.	North Texas	−8.1	−3.5	4.9	[36]
SMU76404 (1)	*Clidastes* sp.	North Texas	−8.3	−3.1	3.2	[36]
SMU76281 (1)	*Clidastes* sp.	North Texas	−6.8	−3.1	5.1	[36]
SMU72184 (1)	*Clidastes* cf. *C. propython*	North Texas	−6.7	−4.3	4.6	*
SMU62504 (1)	*Clidastes* sp.	North Texas	−6.7	−3.9	4.4	*
TMM43209-1 (1)	*Dallasaurus turneri*	North Texas	−2.8	−4.4	4.8	*
MGUAN PA149 (1)	*Globidens phosphaticus*	West Africa	−13.8	−2.5	4.2	[36]
MGUAN PA149 (2)	*Globidens phosphaticus*	West Africa	−14.6	−2.3	4.4	[36]
MGUAN PA149 (3)	*Globidens phosphaticus*	West Africa	−14.0		4.5	[36]
MGUAN PA149 (4)	*Globidens phosphaticus*	West Africa	−14.0		4.8	[36]
MGUAN PA149 (5)	*Globidens phosphaticus*	West Africa	−13.9		3.2	[36]
MGUAN PA149 (6)	*Globidens phosphaticus*	West Africa	−14.1		5.0	[36]
MGUAN PA149 (7)	*Globidens phosphaticus*	West Africa	−14.0		5.9	[36]
MGUAN PA149 (8)	*Globidens phosphaticus*	West Africa	−14.1		6.0	[36]
MGUAN PA149 (9)	*Globidens phosphaticus*	West Africa	−14.2		4.0	[36]
MGUAN PA149 (10)	*Globidens phosphaticus*	West Africa	−14.1		5.8	[36]
MGUAN PA149 (11)	*Globidens phosphaticus*	West Africa	−14.0		3.0	[36]
MGUAN PA149 (12)	*Globidens phosphaticus*	West Africa	−14.0		4.4	[36]

Table A1. *Cont.*

Specimen # (Sample #)	Taxon	Region	$\delta^{13}C$	$\delta^{18}O$	wt.% Carb *	Data Src.
MGUAN PA149 (13)	*Globidens phosphaticus*	West Africa	−13.6		5.0	[36]
MGUAN PA149 (14)	*Globidens phosphaticus*	West Africa	−14.1		5.7	[36]
MGUAN PA149 (15)	*Globidens phosphaticus*	West Africa	−14.1		2.8	[36]
MGUAN PA149 (16)	*Globidens phosphaticus*	West Africa	−13.9		4.4	[36]
MGUAN PA149 (17)	*Globidens phosphaticus*	West Africa	−14.1		4.0	[36]
MGUAN PA149 (18)	*Globidens phosphaticus*	West Africa	−14.2		3.9	[36]
MGUAN PA149 (19)	*Globidens phosphaticus*	West Africa	−13.9		3.3	[36]
MGUAN PA149 (20)	*Globidens phosphaticus*	West Africa	−13.9		5.0	[36]
SMU76241 (1)	*Globidens alabamaensis* (juv.)	North Texas	−5.2	−2.6	4.1	[36]
SMU76241 (2)	*Globidens alabamaensis* (juv.)	North Texas	−5.3	−2.5	3.4	[36]
SMU76241 (3)	*Globidens alabamaensis* (juv.)	North Texas	−5.4	−2.6	3.7	[36]
SMU76280 (1)	*Globidens alabamaensis* (adult)	North Texas	−10.5	−3.4	3.6	[36]
SMU76280 (2)	*Globidens alabamaensis* (adult)	North Texas	−10.5	−2.2	4.3	[36]
SMU76348 (1)	*Mosasaurus* cf. *M. conodon*	North Texas	−7.4	−1.3	5.0	*
SMU76242 (1)	*Mosasaurus hoffmanni*	North Texas	−9.5	−5.3	1.0	[36]
SMU62079 (1)	*Prognathodon* sp.	North Texas	−10.1	−3.2	2.9	*
SMU76393 (1)	*Prognathodon kianda*	South Atlantic	−14.5	−2.1	4.3	[36]
SMU76393 (2)	*Prognathodon kianda*	South Atlantic	−14.9	−2.1	4.2	[36]
SMU76393 (3)	*Prognathodon kianda*	South Atlantic	−15.0	−2.0	5.0	[36]
HUJ OR 100 (1)	*Prognathodon currii*	Israel	−15.1	−2.4	3.5	*
SMU76504 (1)	*Prognathodon* cf. *P. overtoni*	North Texas	−11.3	−4.7	7.5	*
SMU76504 (2)	*Prognathodon* cf. *P. overtoni*	North Texas	−11.7	−4.5	7.5	*
Asen5 (1)	*?Prognathodon* cf. *P. lutugini*	Scandinavia	−5.4	−2.9	9.2	*
Asen5 (2)	*?Prognathodon* cf. *P. lutugini*	Scandinavia	−5.4	−2.9	8.6	*
Asen5 (3)	*?Prognathodon* cf. *P. lutugini*	Scandinavia	−7.6	−2.6	7.9	*
Asen5 (4)	*?Prognathodon* cf. *P. lutugini*	Scandinavia	−9.3	−2.4	7.6	*
SMU76503 (1)	*?Prognathodon* sp.	North Texas	−11.6	−2.3	5.4	*
SMU76503 (2)	*?Prognathodon* sp.	North Texas	−14.4	−2.2	8.1	*
SMU76503 (3)	*?Prognathodon* sp.	North Texas	−11.8	−3.4	9.1	*
SMU76503 (4)	*?Prognathodon* sp.	North Texas	−16.0	−2.6	8.4	*
SMU76503 (5)	*?Prognathodon* sp.	North Texas	−16.2	−2.7	6.6	*
SMU76503 (6)	*?Prognathodon* sp.	North Texas	−15.7	−2.6	8.4	*
SMU76334 (1)	*Tethysaurusnopscai*	North Africa	−7.5	−4.8	3.0	*
SMU76334 (2)	*Tethysaurusnopscai*	North Africa	−6.4	−3.9	2.0	*
MGUAN PA1 (1)	*Angolasaurus bocagei*	West Africa	−7.4	−0.6	3.7	*
MGUAN PA65 (1)	*Angolasaurus bocagei*	West Africa	−9.2	−2.8	4.5	*
SMU76350 (1)	*Ectenosaurus* sp.	West Texas	−3.1	−4.4	4.0	*
SMU76350 (2)	*Ectenosaurus* sp.	West Texas	−4.6	−3.0	4.0	*
FHSM VP16582 (1)	*Latoplatecarpus* sp.	Western Kansas	−7.6	−5.3	4.7	*
SMU76501 (1)	*Plesioplatecarpus* sp.	Western Kansas	−8.8	−5.8	4.8	*
Asen6 (1)	*Platecarpus* sp.	Scandinavia	−10.9		7.4	*

Table A1. *Cont.*

Specimen # (Sample #)	Taxon	Region	$\delta^{13}C$	$\delta^{18}O$	wt.% Carb *	Data Src.
Asen7 (1)	*"Platecarpus"* cf. *P. somenensis*	Scandinavia	−12.0	−4.6	8.2	*
Asen7 (2)	*"Platecarpus"* cf. *P. somenensis*	Scandinavia	−11.1	−4.3	8.3	*
Asen7 (3)	*"Platecarpus"* cf. *P. somenensis*	Scandinavia	−11.0	−2.9	8.4	*
Asen7 (4)	*"Platecarpus"* cf. *P. somenensis*	Scandinavia	−11.6	−4.1	7.8	*
Ugn0 (1)	*"Platecarpus"* cf. *P. somenensis*	Scandinavia	−7.8	−4.2	8.4	*
Ugn0 (2)	*"Platecarpus"* cf. *P. somenensis*	Scandinavia	−7.6	−4.1	8.4	*
SMU61617 (1)	*Latoplatecarpus* sp.	North Texas	−9.6	−4.7	6.1	*
SMU61617 (2)	*Latoplatecarpus* sp.	North Texas	−9.6	−4.4	4.8	*
SMU61617 (3)	*Latoplatecarpus* sp.	North Texas	−9.6	−3.2	5.5	*
SMU61617 (4)	*Latoplatecarpus* sp.	North Texas	−9.1	−2.8	5.4	*
SMU61617 (5)	*Latoplatecarpus* sp.	North Texas	−9.1	−4.1	5.0	*
SMU61617 (6)	*Latoplatecarpus* sp.	North Texas	−9.5	−5.1	4.9	*
SMU61617 (7)	*Latoplatecarpus* sp.	North Texas	−9.1	−3.9	4.5	*
SMU61617 (8)	*Latoplatecarpus* sp.	North Texas	−9.5	−3.6	4.9	*
SMU76381 (1)	*Indet. sp.*	North Texas	−8.4	−4.2	3.7	*
SMU76497 (1)	*Plesioplatecarpus* sp.	Texas	−7.5	−3.4	4.6	*
SMU76393 (1)	*Plioplatecarpus* sp.	Western Arkansas	−7.7	−2.8	5.4	*
SMU76498 (1)	*?Plioplatecarpus* sp.	North Texas	−7.6	−5.9	4.1	*
SMU73056 (1)	*Russellosaurus coheni*	North Texas	−4.3	−5.8	4.5	*
SMU73056 (2)	*Russellosaurus coheni*	North Texas	−3.5	−4.5	5.2	*
FHSM VP 13910 (1)	*Selmasaurus johnsoni*	Western Kansas	−6.9	−5.9	4.1	*
Asen8 (1)	*Tylosaurus ivoensis*	Scandinavia	−9.3	−3.6	6.1	*
Asen8 (2)	*Tylosaurus ivoensis*	Scandinavia	−7.6	−3.7	8.9	*
Kris1 (1)	*Tylosaurus ivoensis*	Scandinavia	−9.5	−2.5	5.9	*
Kris2 (1)	*Tylosaurus ivoensis*	Scandinavia	−8.0	−2.6	6.3	*
Ugn1 (1)	*Tylosaurus ivoensis*	Scandinavia	−12.4	−5.8	5.8	*
Ugn1 (2)	*Tylosaurus ivoensis*	Scandinavia	−10.4	−3.5	5.8	*
Ugn1 (3)	*Tylosaurus ivoensis*	Scandinavia	−9.5	−3.3	5.7	*
Ugn1 (4)	*Tylosaurus ivoensis*	Scandinavia	−9.9	−3.3	5.7	*
Ugn1 (5)	*Tylosaurus ivoensis*	Scandinavia	−10.6	−3.8	4.4	*
SMU76502 (1)	*Tylosaurus* sp.	North Texas	−8.2	−2.8	2.9	*
SMU61667 (1)	*Tylosaurus* sp.	North Texas	−8.1	−3.7	6.9	*
SMU61667 (2)	*Tylosaurus* sp.	North Texas	−9.0	−3.5	4.9	*
SMU61667 (3)	*Tylosaurus* sp.	North Texas	−10.5	−3.7	5.3	*
SMU61667 (4)	*Tylosaurus* sp.	North Texas	−11.2	−3.3	5.8	*
SMU61667 (5)	*Tylosaurus* sp.	North Texas	−11.7	−3.4	5.7	*

Table A1. *Cont.*

Specimen # (Sample #)	Taxon	Region	$\delta^{13}C$	$\delta^{18}O$	wt.% Carb *	Data Src.
SMU61667 (6)	*Tylosaurus* sp.	North Texas	−11.9	−4.5	6.5	*
SMU76500 (1)	*Tylosaurus* sp.	North Texas	−13.3	−6.7	3.8	*
SMU61113 (1)	*Tylosaurus* sp.	North Texas	−11.8	−3.6	5.1	*
SMU61113 (2)	*Tylosaurus* sp.	North Texas	−12.4	−3.7	6.2	*
SMU61113 (3)	*Tylosaurus* sp.	North Texas	−13.2	−4.2	5.7	*
SMU61113 (4)	*Tylosaurus* sp.	North Texas	−12.5	−4.2	6.3	*
SMU61113 (5)	*Tylosaurus* sp.	North Texas	−12.1	−3.9	5.5	*
SMU61113 (6)	*Tylosaurus* sp.	North Texas	−12.1	−4.9	6.3	*
SMU61113 (7)	*Tylosaurus* sp.	North Texas	−12.1	−3.5	4.4	*
SMU61113 (8)	*Tylosaurus* sp.	North Texas	−12.6	−3.1	5.1	*
SMU61113 (9)	*Tylosaurus* sp.	North Texas	−12.1	−2.9	4.9	*
SMU76392 (1)	*Tylosaurus* sp.	North Texas	−13.0	−5.8	5.4	*
SMU76392 (2)	*Tylosaurus* sp.	North Texas	−12.6	−4.5	4.1	*
SMU76392 (3)	*Tylosaurus* sp.	North Texas	−12.5	−6.1	5.6	*
SMU76392 (4)	*Tylosaurus* sp.	North Texas	−12.5	−4.8	4.3	*
SMU76392 (5)	*Tylosaurus* sp.	North Texas	−12.2	−4.2	4.5	*
LO7786 (1)	*Tylosaurus* sp.	North Texas	−7.0	−4.8	6.7	*
SMU75374 (1)	*Tylosaurus* sp.	North Texas	−11.7	−6.2	6.3	*
SMU75374 (2)	*Tylosaurus* sp.	North Texas	−11.2	−6.3	5.9	*
SMU75374 (3)	*Tylosaurus* sp.	North Texas	−11.8	−6.9	6.8	*
SMU75374 (4)	*Tylosaurus* sp.	North Texas	−12.0	−6.2	5.3	*
SMU76382 (1)	*Tylosaurus* sp.	North Texas	−9.2	−4.2	2.	*
SMU75586 (1)	*Tylosaurus* sp.	North Texas	−10.3		8.2	*
NHMM 1980.6 (1)	*Carinodensbelgicus*	Northern Europe	−10.8	−4.0		[37]
NHMM 1980.7 (1)	*Carinodensbelgicus*	Northern Europe	−10.5	−2.3		[37]
NHMM 1980.7 (2)	*Carinodensbelgicus*	Northern Europe	-10.5	−2.3		[37]
NHMM 7354 (1)	*Carinodensbelgicus*	Northern Europe	-10.3	−3.6		[37]
NHMM 1984.88.1A (1)	*Plioplatecarpusmarshi*	Northern Europe	−10.7	−4.5		[37]
NHMM 1984.88.1A (2)	*Plioplatecarpusmarshi*	Northern Europe	-11.1	−4.5		[37]
NHMM 1984.88.1B	*Plioplatecarpusmarshi*	Northern Europe	-10.3			[37]
NHMM 1995 031 (1)	*Prognathodon sectorius*	Northern Europe	−9.9	−3.4		[37]
NHMM LV 150 (1)	*Prognathodon sectorius*	Northern Europe	−12.2	−3.0		[37]
NHMM 1998 141 (1)	*Prognathodonsaturator*	Northern Europe	−12.8			[37]
NHMM 1998 141 (2)	*Prognathodonsaturator*	Northern Europe	-13.3	−3.4		[37]
NHMM 1998 141 (3)	*Prognathodonsaturator*	Northern Europe	-13.6	−3.0		[37]
NHMM 1998 141 (4)	*Prognathodonsaturator*	Northern Europe	-12.3	−2.8		[37]
NHMM 1998 141 (5)	*Prognathodonsaturator*	Northern Europe	-12.3	−3.1		[37]
NHMM 1317.02 (1)	*Mosasaurushoffmanni*	Northern Europe	−9.3	−3.2		[37]
NHMM MK 591 (1)	*Mosasaurushoffmanni*	Northern Europe	−7.1			[37]
NHMM 4560 (1)	*Mosasaurushoffmanni*	Northern Europe	−11.3	−3.8		[37]
NHMM 1446 (1)	*Mosasaurushoffmanni*	Northern Europe	−13.7			[37]

Table A1. *Cont.*

Specimen # (Sample #)	Taxon	Region	$\delta^{13}C$	$\delta^{18}O$	wt.% Carb *	Data Src.
NHMM 1446 (2)	*Mosasaurushoffmanni*	Northern Europe	−13.0	−4.4		[37]
NHMM 1446 (3)	*Mosasaurushoffmanni*	Northern Europe	−12.6	−4.0		[37]
NHMM 1446 (4)	*Mosasaurushoffmanni*	Northern Europe	−13.8	−6.2		[37]
NHMM 1446 (5)	*Mosasaurushoffmanni*	Northern Europe	−14.0	−4.4		[37]
NHMM 1446 (6)	*Mosasaurushoffmanni*	Northern Europe	−13.4	−5.3		[37]
NHMM 1446 (7)	*Mosasaurushoffmanni*	Northern Europe	−13.6	−4.0		[37]
NHMM 1446 (8)	*Mosasaurushoffmanni*	Northern Europe	−14.9	−3.9		[37]
NHMM 1446 (9)	*Mosasaurushoffmanni*	Northern Europe	−14.5	−3.9		[37]
NHMM 1446 (10)	*Mosasaurushoffmanni*	Northern Europe	−13.7	−3.8		[37]
NHMM 1446 (11)	*Mosasaurushoffmanni*	Northern Europe	−13.6	−6.2		[37]
MGUAN PA 171 (1)	*Carinodens* sp.	West Africa	−8.1	−1.5		[39]
MGUAN PA 04 (1)	*Globidens phosphaticus*	West Africa	−14.6	−2.1		[39]
MGUAN PA 05 (1)	*Globidens phosphaticus*	West Africa	−13.2	−1.7		[39]
MGUAN PA 29 (1)	*Globidens phosphaticus*	West Africa	−14.3	−1.2		[39]
MGUAN PA 30 (1)	*Globidens phosphaticus*	West Africa	−10.8	−1.3		[39]
MGUAN PA 301 (1)	*Globidens phosphaticus*	West Africa	−12.6	−1.1		[39]
MGUAN PA 307 (1)	*Globidens phosphaticus*	West Africa	−11.6	−1.7		[39]
MGUAN PA 31 (1)	*Globidens phosphaticus*	West Africa	−12.4	−1.1		[39]
MGUAN PA 313 (1)	*Globidens phosphaticus*	West Africa	−12.5	−1.5		[39]
MGUAN PA 33 (1)	*Globidens phosphaticus*	West Africa	−11.0	−2.4		[39]
MGUAN PA 500 (1)	*Globidens phosphaticus*	West Africa	−16.1	−2.1		[39]
MGUAN PA 61 (1)	*Globidens phosphaticus*	West Africa	−14.3	−1.6		[39]
MGUAN PA 314 (1)	*Halisaurus* sp.	West Africa	−7.2	−2.2		[39]
MGUAN PA 309 (1)	*Mosasaurus* sp.	West Africa	−11.6	−1.4		[39]
MGUAN PA 309 (2)	*Mosasaurus* sp.	West Africa	−8.9	−1.4		[39]
MGUAN PA 44 (1)	*Mosasaurus* sp.	West Africa	−11.6	−2.6		[39]
MGUAN PA 46 (1)	*Mosasaurus* sp.	West Africa	−12.9	−1.9		[39]
MGUAN PA 177 (1)	*Gavialimimus* sp.	West Africa	−7.2	−2.1		[39]
MGUAN PA 312 (1)	*Gavialimimus* sp.	West Africa	−7.2	−2.1		[39]
MGUAN PA 312 (2)	*Gavialimimus* sp.	West Africa	−7.6	−2.4		[39]
MGUAN PA 525 (1)	*Gavialimimus* sp.	West Africa	−9.2	−2.6		[39]
MGUAN PA 525 (2)	*Gavialimimus* sp.	West Africa	−9.2	−3.3		[39]
MGUAN PA 321 (1)	*Gavialimimus* sp.	West Africa	−8.2	−2.3		[39]
MGUAN PA 38 (1)	*Gavialimimus* sp.	West Africa	−10.9	−1.5		[39]
MGUAN PA 55 (1)	*Gavialimimus* sp.	West Africa	−8.7	−3.6		[39]
MGUAN PA528 (1)	*Prognathodon kianda*	West Africa	−8.9	−1.4		[39]
MGUAN PA 526 (1)	*Prognathodon kianda*	West Africa	−9.3	−1.9		[39]
MGUAN PA 527 (1)	*Prognathodon kianda*	West Africa	−9.2	−2.3		[39]
MGUAN PA 28 (2)	*Prognathodon kianda*	West Africa	−11.1	−1.6		[39]
MGUAN PA 28 (1)	*Prognathodon kianda*	West Africa	−11.4	−2.3		[39]
MGUAN PA 304 (1)	*Prognathodon kianda*	West Africa	−7.3	−1.5		[39]
MGUAN PA 304 (2)	*Prognathodon kianda*	West Africa	−6.9	−1.6		[39]

Table A1. *Cont.*

Specimen # (Sample #)	Taxon	Region	$\delta^{13}C$	$\delta^{18}O$	wt.% Carb *	Data Src.
MGUAN PA 306 (1)	*Prognathodon kianda*	West Africa	−12.1	−2.4		[39]
MGUAN PA 315 (1)	*Prognathodon kianda*	West Africa	−10.1	−1.8		[39]
MGUAN PA 318 (1)	*Prognathodon kianda*	West Africa	−12.4	−2.4		[39]
MGUAN PA 35 (1)	*Prognathodon kianda*	West Africa	−10.7	−1.5		[39]
MGUAN PA 40 (1)	*Prognathodon kianda*	West Africa	−11.0	−1.7		[39]
MGUAN PA 41 (1)	*Prognathodon kianda*	West Africa	−5.2	−1.9		[39]
MGUAN PA 45 (1)	*Prognathodon kianda*	West Africa	−10.4	−3.3		[39]
MGUAN PA 47 (1)	*Prognathodon kianda*	West Africa	−13.7	−2.4		[39]
MGUAN PA 48 (1)	*Prognathodon kianda*	West Africa	−10.9	−1.6		[39]
MGUAN PA 50 (1)	*Prognathodon kianda*	West Africa	−14.1	−3.7		[39]
MGUAN PA 50 (2)	*Prognathodon kianda*	West Africa	−13.4	−1.9		[39]
MGUAN PA 53 (1)	*Prognathodon kianda*	West Africa	−12.8	−3.5		[39]
MGUAN PA 56 (1)	*Prognathodon kianda*	West Africa	−6.6	−2.0		[39]
NHMD 157504 (1)	*Carinodensminalmamar*	Scandinavia	−6.3	−3.1		[40]
NHMD 227349 (1)	*Mosasaurus* sp.	Scandinavia	−11.1	−2.5		[40]
OESM 8783 (1)	*Mosasaurus* sp.	Scandinavia	−8.6	−2.6		[40]
NHMM 1984089-1 (1)	*Mosasaurushoffmanni*	Northern Europe	−10.3	−2.7		[40]
NHMM 1984089-1 (2)	*Mosasaurushoffmanni*	Northern Europe	−9.0	−3.4		[40]
NHMD 226499 (1)	*Mosasaurus* cf. *M. hoffmanni*	Scandinavia	−6.9	−2.8		[40]
NHMD 226499 (2)	*Mosasaurus* cf. *M. hoffmanni*	Scandinavia	−11.0	−2.7		[40]
NHMD 227350 (1)	*Plioplatecarpus* sp.	Scandinavia	0.2	−3.4		[40]
NHMD 189763 (1)	*Plioplatecarpus* sp.	Scandinavia	−9.5	−2.8		[40]
NHMM 1997289 (1)	*Plioplatecarpusmarshi*	Northern Europe	−7.0	−2.4		[40]
NHMM 1997289 (2)	*Plioplatecarpusmarshi*	Northern Europe	−7.9	−4.0		[40]
NHMM 1998141-11 (1)	*Prognathodonsaturator*	Northern Europe	−10.9	−2.7		[40]
NHMM 1998141-7 (1)	*Prognathodonsaturator*	Northern Europe	−11.6	−2.5		[40]
MLP 15-I-24-41a: ARG-1 (1)	Mosasauridae indet.	Antarctica	−8.2	−4.0		[41]
MLP 15-I-24-44: ARG-3 (1)	Globidensini indet.	Antarctica	−11.2	−4.6		[41]
MLP 15-I-24-48: ARG-4 (1)	?*Mosasaurus* indet.	Antarctica	−11.8	−4.4		[41]
MLP 15-I-24-33a: ARG-5 (1)	cf. *Liodon* indet.	Antarctica	−13.5	−3.9		[41]
MLP 15-I-24-33a: ARG-5 (2)	cf. *Liodon* indet.	Antarctica	−11.4	−3.6		[41]
MLP 15-I-24-29: ARG-8 (1)	?*Taniwhasaurus* indet.	Antarctica	−10.6	−4.8		[41]
MLP 15-I-24-55a: ARG-10 (1)	?*Taniwhasaurus* indet.	Antarctica	−6.8	−4.8		[41]
MLP 15-I-24-25: ARG-13 (1)	cf. *Liodon* indet.	Antarctica	−11.8	−3.5		[41]

Table A1. *Cont.*

Specimen # (Sample #)	Taxon	Region	$\delta^{13}C$	$\delta^{18}O$	wt.% Carb *	Data Src.
MLP 15-I-24-33b: ARG-14 (1)	cf. *Liodon* indet.	Antarctica	−12.5	−2.6		[41]
MLP 13-XI-19-38: ARG-18 (1)	?*Mosasaurinae* indet.	Argentina	−11.1	−5.1		[41]
MLP 13-XI-19-38: ARG-19	?*Mosasaurinae* indet.	Argentina	−11.1	−4.2		[41]

References

1. Polcyn, M.J.; Jacobs, L.L.; Araújo, R.; Schulp, A.S.; Mateus, O. Physical Drivers of Mosasaur Evolution. *Palaeogeogr. Palaeoclimatol. Palaeoecol.* **2014**, *400*, 17–27. [CrossRef]
2. Augusta, B.G.; Zaher, H.; Polcyn, M.J.; Fiorillo, A.R.; Jacobs, L.L. A Review of Non-Mosasaurid (Dolichosaur and Aigialosaur) Mosasaurians and Their Relationships to Snakes. In *The Origin and Early Evolutionary History of Snakes*; Gower, D.J., Zaher, H., Eds.; Systematics Association Special Volume Series; Cambridge University Press: Cambridge, UK, 2022; pp. 157–179, ISBN 978-1-108-93889-1.
3. Lindgren, J.; Polcyn, M.J.; Young, B.A. Landlubbers to Leviathans: Evolution of Swimming in Mosasaurine Mosasaurs. *Paleobiology* **2011**, *37*, 445–469. [CrossRef]
4. Dollo, L. Le Hainosaure et Les Nouveaux Vertébrés Fossiles Du Musée de Bruxelles. *Rev. Des. Quest. Sci.* **1887**, *21*, 504–539.
5. Williston, S. Some Additional Characters of the Mosasaurs. *Kans. Univ. Q.* **1899**, *8*, 39–41.
6. Sternberg, C.H. Explorations of the Permian of Texas and the Chalk of Kansas, 1918. *Trans. Kans. Acad. Sci.* **1919**, *30*, 119–120. [CrossRef]
7. Camp, C.L. California Mosasaurs. *Mem. Univ. Calif.* **1942**, *13*, 1.
8. Massare, J.A. Tooth Morphology and Prey Preference of Mesozoic Marine Reptiles. *J. Vertebr. Paleontol.* **1987**, *7*, 121–137. [CrossRef]
9. Martin, J.E.; Bjork, P.P. Gastric Residues Associated with a Mosasaur from the Late Cretaceous (Campanian) Pierre Shale in South Dakota. In *Papers in Vertebrate Paleontology in Honor of Morton Green*; Martin, J.E., Ostrander, G.E., Eds.; Dakoterra; South Dakota School of Mines & Technology: Rapid City, SD, USA, 1987.
10. Everhart, M. Plesiosaurs as the Food of Mosasaurs; New Data on the Stomach Contents of a *Tylosaurus proriger* (Squamata; Mosasauridae) from the Niobrara Formation of Western Kansas. *Mosasaur* **2004**, *7*, 41–46.
11. Schulp, A.S. Feeding the Mechanical Mosasaur: What Did *Carinodens* Eat? *Neth. J. Geosci.* **2005**, *84*, 345–357. [CrossRef]
12. Martin, J.E.; Fox, J.E. Stomach Contents of *Globidens*, a Shell-Crushing Mosasaur (Squamata), from the Late Cretaceous Pierre Shale Group, Big Bend Area of the Missouri River, Central South Dakota. In *The Geology and Paleontology of the Late Cretaceous Marine Deposits of the Dakotas*; Martin, J.E., Paris, D., Eds.; Geological Society of America Special Papers; Geological Society of America: Boulder, CO, USA, 2007; Volume 427, pp. 167–176.
13. Einarsson, E.; Lindgren, J.; Kear, B.P.; Siverson, M. Mosasaur Bite Marks on a Plesiosaur Propodial from the Campanian (Late Cretaceous) of Southern Sweden. *GFF* **2010**, *132*, 123–128. [CrossRef]
14. Poynter, J.M. Using Dental Microwear Analysis to Predict Feeding Types in Mesozoic Marine Reptiles. Ph.D. Thesis, Northwest Missouri State University, Maryville, MO, USA, 2011.
15. Konishi, T.; Brinkman, D.; Massare, J.A.; Caldwell, M.W. New Exceptional Specimens of *Prognathodon overtoni* (Squamata, Mosasauridae) from the Upper Campanian of Alberta, Canada, and the Systematics and Ecology of the Genus. *J. Vertebr. Paleontol.* **2011**, *31*, 1026–1046. [CrossRef]
16. Konishi, T.; Newbrey, M.G.; Caldwell, M.W. A Small, Exquisitely Preserved Specimen of *Mosasaurus missouriensis* (Squamata, Mosasauridae) from the Upper Campanian of the Bearpaw Formation, Western Canada, and the First Stomach Contents for the Genus. *J. Vertebr. Paleontol.* **2014**, *34*, 802–819. [CrossRef]
17. Bardet, N.; Houssaye, A.; Vincent, P.; Pereda Suberbiola, X.; Amaghzaz, M.; Jourani, E.; Meslouh, S. Mosasaurids (Squamata) from the Maastrichtian Phosphates of Morocco: Biodiversity, Palaeobiogeography and Palaeoecology Based on Tooth Morphoguilds. *Gondwana Res.* **2015**, *27*, 1068–1078. [CrossRef]
18. Longrich, N.R.; Jalil, N.-E.; Khaldoune, F.; Yazami, O.K.; Pereda-Suberbiola, X.; Bardet, N. *Thalassotitan atrox*, a Giant Predatory Mosasaurid (Squamata) from the Upper Maastrichtian Phosphates of Morocco. *Cretac. Res.* **2022**, *140*, 105315. [CrossRef]
19. Holwerda, F.M.; Bestwick, J.; Purnell, M.A.; Jagt, J.W.M.; Schulp, A.S. Three-Dimensional Dental Microwear in Type-Maastrichtian Mosasaur Teeth (Reptilia, Squamata). *Sci. Rep.* **2023**, *13*, 18720. [CrossRef]

20. Polcyn, M.J.; Schulp, A.S.; Goncalves, A.O. Remarkably Well-Preserved in-Situ Gut-Content in a Specimen of *Prognathodon kianda* (Squamata: Mosasauridae) Reveals Multispecies Intrafamilial Predation, Cannibalism, and a New Mosasaurine Taxon. In *Windows into Sauropsid and Synapsid Evolution*; Lee, Y.-N., Ed.; Dinosaur Science Center Press: Hwaseong City, Republic of Korea, 2023; pp. 66–98.
21. Russell, D.A. *Systematics and Morphology of American Mosasaurs*; Yale University Press: New Haven, CT, USA, 1967; ISBN 978-1-933789-45-3.
22. Holmes, R.; Caldwell, M.W.; Cumbaa, S.L. A New Specimen of *Plioplatecarpus* (Mosasauridae) from the Lower Maastrichtian of Alberta: Comments on Allometry, Functional Morphology, and Paleoecology. *Can. J. Earth Sci.* **1999**, *36*, 363–369. [CrossRef]
23. Kiernan, C.R. Stratigraphic Distribution and Habitat Segregation of Mosasaurs in the Upper Cretaceous of Western and Central Alabama, with an Historical Review of Alabama Mosasaur Discoveries. *J. Vertebr. Paleontol.* **2002**, *22*, 91–103. [CrossRef]
24. Nicholls, E.L.; Meckert, D. Marine Reptiles from the Nanaimo Group (Upper Cretaceous) of Vancouver Island. *Can. J. Earth Sci.* **2002**, *39*, 1591–1603. [CrossRef]
25. Jacobs, L.L.; Polcyn, M.J.; Taylor, L.H.; Ferguson, K. Sea-Surface Temperatures and Palaeoenvironments of Dolichosaurs and Early Mosasaurs. *Geol. Mijnb.* **2005**, *84*, 269–281. [CrossRef]
26. Ifrim, C.; Stinnesbeck, W.; Frey, E. Upper Cretaceous (Cenomanian-Turonian and Turonian-Coniacian) Open Marine Plattenkalk Deposits in NE Mexico. *Neues Jahrb. Für Geol. Und Paläontologie Abh.* **2007**, *245*, 71–81. [CrossRef]
27. Patrick, D.; Martin, J.; Parris, D.; Grandstaff, D. Paleoenvironmental Interpretations of Rare Earth Element Signatures in Mosasaurs (Reptilia) from the Upper Cretaceous Pierre Shale, Central South Dakota, USA. *Palaeogeogr. Palaeoclimatol. Palaeoecol.* **2004**, *212*, 277–294. [CrossRef]
28. Kocsis, L.; Ősi, A.; Vennemann, T.; Trueman, C.N.; Palmer, M.R. Geochemical Study of Vertebrate Fossils from the Upper Cretaceous (Santonian) Csehbánya Formation (Hungary): Evidence for a Freshwater Habitat of Mosasaurs and Pycnodont Fish. *Palaeogeogr. Palaeoclimatol. Palaeoecol.* **2009**, *280*, 532–542. [CrossRef]
29. Harrell, T.L.; Pérez-Huerta, A. Habitat Preference of Mosasaurs Indicated by Rare Earth Element (REE) Content of Fossils from the Upper Cretaceous Marine Deposits of Alabama, New Jersey, and South Dakota (USA). *Neth. J. Geosci.* **2015**, *94*, 145–154. [CrossRef]
30. Reisdorf, A.G.; Bux, R.; Wyler, D.; Benecke, M.; Klug, C.; Maisch, M.W.; Fornaro, P.; Wetzel, A. Float, Explode or Sink: Postmortem Fate of Lung-Breathing Marine Vertebrates. *Palaeobiodivers. Palaeoenviron.* **2012**, *92*, 67–81. [CrossRef]
31. DeNiro, M.J.; Epstein, S. Influence of Diet on the Distribution of Carbon Isotopes in Animals. *Geochim. Cosmochim. Acta* **1978**, *42*, 495–506. [CrossRef]
32. Clementz, M.T.; Koch, P.L. Differentiating Aquatic Mammal Habitat and Foraging Ecology with Stable Isotopes in Tooth Enamel. *Oecologia* **2001**, *129*, 461–472. [CrossRef]
33. Fry, B.; Wainright, S.C. Diatom Sources of ^{13}C-Rich Carbon in Marine Food Webs. *Mar. Ecol. Prog. Ser.* **1991**, *76*, 149–157. [CrossRef]
34. Rau, G.H.; Takahashi, T.; Des Marais, D.J.; Repeta, D.J.; Martin, J.H. The Relationship between δ^{13}C of Organic Matter and [CO_2(Aq)] in Ocean Surface Water: Data from a JGOFS Site in the Northeast Atlantic Ocean and a Model. *Geochim. Cosmochim. Acta* **1992**, *56*, 1413–1419. [CrossRef]
35. Hemminga, M.; Mateo, M. Stable Carbon Isotopes in Seagrasses:Variability in Ratios and Use in Ecological Studies. *Mar. Ecol. Prog. Ser.* **1996**, *140*, 285–298. [CrossRef]
36. Robbins, J.; Ferguson, K.M.; Polcyn, M.J.; Jacobs, L.L. Application of Stable Carbon Isotope Analysis to Mosasaur Ecology. In *Proceedings of the Second Mosasaur Meeting*; Everhart, M.J., Ed.; Sternberg Museum of Natural History, Fort Hays State University: Hays, KS, USA, 2008; Volume Special Volume; pp. 123–130, ISBN 978-0-615-23109-9.
37. Schulp, A.S.; Vonhof, H.B.; Van Der Lubbe, J.H.J.L.; Janssen, R.; Van Baal, R.R. On Diving and Diet: Resource Partitioning in Type-Maastrichtian Mosasaurs. *Neth. J. Geosci.* **2013**, *92*, 165–170. [CrossRef]
38. Schulp, A.S.; Janssen, R.; Van Baal, R.R.; Jagt, J.W.M.; Mulder, E.W.A.; Vonhof, H.B. Stable Isotopes, Niche Partitioning and the Paucity of Elasmosaur Remains in the Maastrichtian Type Area. *Neth. J. Geosci.* **2017**, *96*, 29–33. [CrossRef]
39. Strganac, C.; Jacobs, L.L.; Polcyn, M.J.; Mateus, O.; Myers, T.S.; Salminen, J.; May, S.R.; Araújo, R.; Ferguson, K.M.; Gonçalves, A.O.; et al. Geological Setting and Paleoecology of the Upper Cretaceous Bench 19 Marine Vertebrate Bonebed at Bentiaba, Angola. *Neth. J. Geosci.* **2015**, *94*, 121–136. [CrossRef]
40. Giltaij, T.J.; Jeroen, V.D.L.; Lindow, B.; Schulp, A.S.; Jagt, J.W.M. Carbon Isotope Trends in North-West European Mosasaurs (Squamata; Late Cretaceous). *Bull. Geol. Soc. Den.* **2021**, *69*, 59–70. [CrossRef]
41. Leuzinger, L.; Kocsis, L.; Luz, Z.; Vennemann, T.; Ulyanov, A.; Fernández, M. Latest Maastrichtian Middle- and High-Latitude Mosasaurs and Fish Isotopic Composition: Carbon Source, Thermoregulation Strategy, and Thermal Latitudinal Gradient. *Paleobiology* **2023**, *49*, 353–373. [CrossRef]

42. Strong, C.R.C.; Caldwell, M.W.; Konishi, T.; Palci, A. A New Species of Longirostrine Plioplatecarpine Mosasaur (Squamata: Mosasauridae) from the Late Cretaceous of Morocco, with a Re-Evaluation of the Problematic Taxon *'Platecarpus' ptychodon*. *J. Syst. Palaeontol.* **2020**, *18*, 1769–1804. [CrossRef]
43. Polcyn, M.J.; Bardet, N.; Albright, L.B.; Titus, A. A New Lower Turonian Mosasaurid from the Western Interior Seaway and the Antiquity of the Unique Basicranial Circulation Pattern in Plioplatecarpinae. *Cretac. Res.* **2023**, *151*, 105621. [CrossRef]
44. Zietlow, A.R.; Boyd, C.A.; Van Vranken, N.E. *Jormungandr walhallaensis*: A New Mosasaurine (Squamata: Mosasauroidea) from the Pierre Shale Formation (Pembina Member: Middle Campanian) of North Dakota. *Bull. Am. Mus. Nat. Hist.* **2023**, *464*, 1–82. [CrossRef]
45. Longrich, N.R.; Polcyn, M.J.; Jalil, N.-E.; Pereda-Suberbiola, X.; Bardet, N. A Bizarre New Plioplatecarpine Mosasaurid from the Maastrichtian of Morocco. *Cretac. Res.* **2024**, *160*, 105870. [CrossRef]
46. Polcyn, M.J.; Bell, G.L., Jr.; Shimada, K.; Everhart, M.J. The Oldest North American Mosasaurs (Squamata: Mosasauridae) from the Turonian (Upper Cretaceous) of Kansas and Texas with Comments on the Radiations of Major Mosasaur Clades. In *Proceedings of the Second Mosasaur Meeting*; Everhart, M.J., Ed.; Sternberg Museum of Natural History, Fort Hays State University: Hays, KS, USA, 2008; Volume Special Volume; pp. 137–155, ISBN 978-0-615-23109-9.
47. Bell, G.L. A Phylogenetic Revision of North American and Adriatic Mosasauroidea. In *Ancient Marine Reptiles*; Elsevier: Amsterdam, The Netherlands, 1997; pp. 293–332, ISBN 978-0-12-155210-7.
48. Páramo-Fonseca, M.E. *Yaguarasaurus columbianus* (Reptilia, Mosasauridae), a Primitive Mosasaur from the Turonian (Upper Cretaceous) of Colombia. *Hist. Biol.* **2000**, *14*, 121–131. [CrossRef]
49. Bardet, N.; Suberbiola, X.P.; Jalil, N.-E. A New Mosasauroid (Squamata) from the Late Cretaceous (Turonian) of Morocco. *Comptes Rendus Palevol* **2003**, *2*, 607–616. [CrossRef]
50. Polcyn, M.J.; Bell, G.L. *Russellosaurus coheni* n. gen., n. sp., a 92 Million-Year-Old Mosasaur from Texas (USA), and the Definition of the Parafamily Russellosaurina. *Neth. J. Geosci.* **2005**, *84*, 321–333. [CrossRef]
51. Polcyn, M.J.; Everhart, M.J. Description and Phylogenetic Analysis of a New Species of Selmasaurus (Mosasauridae: Plioplatecarpinae) from the Niobrara Chalk of Western Kansas. In *Proceedings of the Second Mosasaur Meeting*; Everhart, M.J., Ed.; Sternberg Museum of Natural History, Fort Hays State University: Hays, KS, USA, 2008; Volume Special Volume; pp. 13–28, ISBN 978-0-615-23109-9.
52. Konishi, T.; Caldwell, M.W. Two New Plioplatecarpine (Squamata, Mosasauridae) Genera from the Upper Cretaceous of North America, and a Global Phylogenetic Analysis of Plioplatecarpines. *J. Vertebr. Paleontol.* **2011**, *31*, 754–783. [CrossRef]
53. Leblanc, A.R.H.; Caldwell, M.W.; Bardet, N. A New Mosasaurine from the Maastrichtian (Upper Cretaceous) Phosphates of Morocco and Its Implications for Mosasaurine Systematics. *J. Vertebr. Paleontol.* **2012**, *32*, 82–104. [CrossRef]
54. Palci, A.; Caldwell, M.W.; Papazzoni, C.A. A New Genus and Subfamily of Mosasaurs from the Upper Cretaceous of Northern Italy. *J. Vertebr. Paleontol.* **2013**, *33*, 599–612. [CrossRef]
55. Willman, A.J.; Konishi, T.; Caldwell, M.W. A New Species of *Ectenosaurus* (Mosasauridae: Plioplatecarpinae) from Western Kansas, USA, Reveals a Novel Suite of Osteological Characters for the Genus[1]. *Can. J. Earth Sci.* **2021**, *58*, 741–755. [CrossRef]
56. Kiernan, C.R.; Ebersole, J.A. Two New Plioplatecarpine Mosasaurs (Mosasauridae; Plioplatecarpinae) of the Genus *Ectenosaurus* from the Upper Cretaceous of North America. *PaleoBios* **2023**, *40*, 1–28. [CrossRef]
57. Bell, G.L.; Barnes, K.R.; Polcyn, M.J. Late Cretaceous Mosasauroids (Reptilia, Squamata) of the Big Bend Region in Texas, USA. *Earth Environ. Sci. Trans. R. Soc. Edinb.* **2012**, *103*, 571–581. [CrossRef]
58. Bell, G.L.; Polcyn, M.J. *Dallasaurus turneri*, a New Primitive Mosasauroid from the Middle Turonian of Texas and Comments on the Phylogeny of Mosasauridae (Squamata). *Neth. J. Geosci.* **2005**, *84*, 177–194. [CrossRef]
59. Lively, J.R. Taxonomy and Historical Inertia: *Clidastes* (Squamata: Mosasauridae) as a Case Study of Problematic Paleobiological Taxonomy. *Alcheringa Australas. J. Palaeontol.* **2018**, *42*, 516–527. [CrossRef]
60. Everhart, M.J. Revisions to the Biostratigraphy of the Mosasauridae (Squamata) in the Smoky Hill Chalk Member of the Niobrara Chalk (Late Cretaceous) of Kansas. *Trans. Kans. Acad. Sci.* **2001**, *104*, 59–78. [CrossRef]
61. Holwerda, F.M.; Mitchell, M.T.; Van De Kerk, M.; Schulp, A.S. Mosasaur Feeding Ecology from the Campanian Bearpaw Formation, Alberta, Canada: A Preliminary Multi-Proxy Approach. *Diversity* **2025**, *17*, 205. [CrossRef]
62. Russell, D.A. *A New Species of Globidens from South Dakota, and a Review of Globidentine Mosasaurs*; Fieldiana, Geology; Field Museum of Natural History: Chicago, IL, USA, 1975; Volume 33.
63. Schulp, A.S.; Polcyn, M.J.; Mateus, O.; Jacobs, L.L.; Morais, M.L. A New Species of *Prognathodon* (Squamata, Mosasauridae) from the Maastrichtian of Angola, and the Affinities of the Mosasaur Genus *Liodon*. In Proceedings of the Second Mosasaur Meeting; Fort Hays State University Fort Hays: Hays, KS, USA, 2008; Volume 3, pp. 1–12.
64. DeBraga, M.; Carroll, R.L. The Origin of Mosasaurs as a Model of Macroevolutionary Patterns and Processes. In *Evolutionary Biology*; Hecht, M.K., MacIntyre, R.J., Clegg, M.T., Eds.; Springer US: Boston, MA, USA, 1993; pp. 245–322, ISBN 978-1-4613-6248-7.

65. Polcyn, M.J.; Lindgren, J.; Bardet, N.; Cornelissen, D.; Verding, L.; Schulp, A.S. Description of New Specimens of *Halisaurus arambourgi* Bardet and Pereda Suberbiola, 2005 and the Relationships of Halisaurinae. *Bull. De La Société Géologique De Fr.* **2012**, *183*, 123–136. [CrossRef]
66. Polcyn, M.J.; Augusta, B.G.; Zaher, H. Reassessing the Morphological Foundations of the Pythonomorph Hypothesis. In *The Origin and Early Evolutionary History of Snakes*; Gower, D.J., Zaher, H., Eds.; Systematics Association; Cambridge University Press: Cambridge, UK, 2022; Volume Special Volume Series, pp. 125–156, ISBN 978-1-108-93889-1.
67. Koch, P.L.; Tuross, N.; Fogel, M.L. The Effects of Sample Treatment and Diagenesis on the Isotopic Integrity of Carbonate in Biogenic Hydroxylapatite. *J. Archaeol. Sci.* **1997**, *24*, 417–429. [CrossRef]
68. Passey, B.H.; Cerling, T.E.; Levin, N.E. Temperature Dependence of Oxygen Isotope Acid Fractionation for Modern and Fossil Tooth Enamels. *Rapid Comm Mass Spectrom.* **2007**, *21*, 2853–2859. [CrossRef]
69. Kusaka, S.; Nakano, T. Carbon and Oxygen Isotope Ratios and Their Temperature Dependence in Carbonate and Tooth Enamel Using a GasBench II Preparation Device: Letter to the Editor. *Rapid Commun. Mass Spectrom.* **2014**, *28*, 563–567. [CrossRef]
70. Swart, P.K.; Burns, S.J.; Leder, J.J. Fractionation of the Stable Isotopes of Oxygen and Carbon in Carbon Dioxide during the Reaction of Calcite with Phosphoric Acid as a Function of Temperature and Technique. *Chem. Geol. Isot. Geosci. Sect.* **1991**, *86*, 89–96. [CrossRef]
71. Chesson, L.A.; Kenyhercz, M.W.; Regan, L.A.; Berg, G.E. Addressing Data Comparability in the Creation of Combined Data Sets of Bioapatite Carbon and Oxygen Isotopic Compositions. *Archaeometry* **2019**, *61*, 1193–1206. [CrossRef]
72. Christiansen, P.; Bonde, N. A New Species of Gigantic Mosasaur from the Late Cretaceous of Israel. *J. Vertebr. Paleontol.* **2002**, *22*, 629–644. [CrossRef]
73. Polcyn, M.J.; Jacobs, L.L.; Schulp, A.S.; Mateus, O. The North African Mosasaur *Globidens phosphaticus* from the Maastrichtian of Angola. *Hist. Biol.* **2010**, *22*, 175–185. [CrossRef]
74. Mateus, O.; Polcyn, M.J.; Jacobs, L.L.; Araújo, R.; Schulp, A.S.; Marinheiro, J.; Pereira, B.; Vineyard, D. Cretaceous Amniotes from Angola: Dinosaurs, Pterosaurs, Mosasaurs, Plesiosaurs, and Turtles. In Proceedings of the V International Conference on Dinosaur Paleontology and their Environment, Salas de los Infantes, Burgos, 26 July 2012; pp. 71–105.
75. Lindgren, J. Dental and Vertebral Morphology of the Enigmatic Mosasaur *Dollosaurus* (Reptilia, Mosasauridae) from the Lower Campanian (Upper Cretaceous) of Southern Sweden. *Bull. Geol. Soc. Den.* **2005**, *52*, 17–25. [CrossRef]
76. Fernandez, M.; Martin, J.E. Description and Phylogenetic Relationships of *Taniwhasaurus antarcticus* (Mosasauridae, Tylosaurinae) from the Upper Campanian (Cretaceous) of Antarctica. *Cretac. Res.* **2009**, *30*, 717–726. [CrossRef]
77. Fischer, V.; Bennion, R.F.; Foffa, D.; MacLaren, J.A.; McCurry, M.R.; Melstrom, K.M.; Bardet, N. Ecological Signal in the Size and Shape of Marine Amniote Teeth. *Proc. R. Soc. B Biol. Sci.* **2022**, *289*, 20221214. [CrossRef] [PubMed]
78. Jagt, J.W.; Mulder, E.W.; Dortangs, R.W.; Kuypers, M.; Peeters, H.; Verding, L. Recent Additions to the Late Maastrichtian Mosasaur Faunas of Liège-Limburg (The Netherlands, Belgium). *Sargetia (Acta Musei Devensis Ser. Sci. Naturae)* **2002**, *19*, 13–26.
79. Dortangs, R.W.; Schulp, A.S.; Mulder, E.W.A.; Jagt, J.W.M.; Peeters, H.H.G.; De Graaf, D.T. A Large New Mosasaur from the Upper Cretaceous of The Netherlands. *Neth. J. Geosci.* **2002**, *81*, 1–8. [CrossRef]
80. Clementz, M.T. *Sea Cows, Seagrasses, and Stable Isotopes: Biogeochemical Evaluation of the Ecology and Evolution of the Sirenia and Desmostylia*; University of California: Santa Cruz, CA, USA, 2002.
81. Clementz, M.T.; Fox-Dobbs, K.; Wheatley, P.V.; Koch, P.L.; Doak, D.F. Revisiting Old Bones: Coupled Carbon Isotope Analysis of Bioapatite and Collagen as an Ecological and Palaeoecological Tool. *Geol. J.* **2009**, *44*, 605–620. [CrossRef]
82. Lee-Thorp, J.A.; Sealy, J.C.; Van Der Merwe, N.J. Stable Carbon Isotope Ratio Differences between Bone Collagen and Bone Apatite, and Their Relationship to Diet. *J. Archaeol. Sci.* **1989**, *16*, 585–599. [CrossRef]
83. Tieszen, L.L.; Fagre, T. Effect of Diet Quality and Composition on the Isotopic Composition of Respiratory CO_2, Bone Collagen, Bioapatite, and Soft Tissues. In *Prehistoric Human Bone*; Lambert, J.B., Grupe, G., Eds.; Springer: Berlin/Heidelberg, Germany, 1993; pp. 121–155, ISBN 978-3-662-02896-4.
84. Toperoff, A.K. *Examination of Diet of Harbor Porpoise (*Phocoena phocoena*) from Central California Using Stomach Content and Stable Isotope Analysis from Multiple Tissues*; San Jose State University: San Jose, CA, USA, 2002.
85. Walker, J.L.; Macko, S.A. Dietary Studies of Marine Mammals Using Stable Carbon and Nitrogen Isotopic Ratios of Teeth. *Mar. Mammal Sci.* **1999**, *15*, 314–334. [CrossRef]
86. Mendes, S.; Newton, J.; Reid, R.J.; Frantzis, A.; Pierce, G.J. Stable Isotope Profiles in Sperm Whale Teeth: Variations between Areas and Sexes. *J. Mar. Biol. Assoc. UK* **2007**, *87*, 621–627. [CrossRef]
87. Rey-Iglesia, A.; Wilson, T.; Routledge, J.; Skovrind, M.; Garde, E.; Heide-Jørgensen, M.P.; Szpak, P.; Lorenzen, E.D. Combining δ^{13}C and δ^{15}N from Bone and Dentine in Marine Mammal Palaeoecological Research: Insights from Toothed Whales. *Isot. Environ. Health Stud.* **2023**, *59*, 66–77. [CrossRef]
88. Benner, R.; Biddanda, B.; Black, B.; McCarthy, M. Abundance, Size Distribution, and Stable Carbon and Nitrogen Isotopic Compositions of Marine Organic Matter Isolated by Tangential-Flow Ultrafiltration. *Mar. Chem.* **1997**, *57*, 243–263. [CrossRef]

89. Boeuf, B.J.L.; Crocker, D.E.; Grayson, J.; Gedamke, J.; Webb, P.M.; Blackwell, S.B.; Costa, D.P. Respiration and Heart Rate at the Surface Between Dives in Northern Elephant Seals. *J. Exp. Biol.* **2000**, *203*, 3265–3274. [CrossRef]
90. Truchot, J.-P. *Comparative Aspects of Extracellular Acid-Base Balance*; Zoophysiology; Springer: Berlin/Heidelberg, Germany, 1987; Volume 20, ISBN 978-3-642-83132-4.
91. Randall, D.J.; Burggren, W.W.; French, K.; Eckert, R. *Eckert Animal Physiology: Mechanisms and Adaptations*, 5th ed.; W.H. Freeman and Co: New York, NY, USA, 2002; ISBN 978-0-7167-3863-3.
92. Lutz, P.L.; Storey, K.B. Adaptations to Variations in Oxygen Tension by Vertebrates and Invertebrates. In *Comprehensive Physiology*; Prakash, Y.S., Ed.; Wiley: Hoboken, NJ, USA, 1997; pp. 1479–1522, ISBN 978-0-470-65071-4.
93. Ackerman, R.; White, F. Cyclic Carbon Dioxide Exchange in the Turtle *Pseudemys scripta*. *Physiol. Zool.* **1979**, *52*, 378–389. [CrossRef]
94. Lutcavage, M.E.; Lutz, P.L. Diving Physiology. In *The Biology of Sea Turtles*; Lutz, P.L., Musick, J.A., Eds.; CRC Press: Boca Raton, FL, USA, 1997; Volume 1, p. 410, ISBN 978-0-203-73708-8.
95. McConnaughey, T.A.; Burdett, J.; Whelan, J.F.; Paull, C.K. Carbon Isotopes in Biological Carbonates: Respiration and Photosynthesis. *Geochim. Et Cosmochim. Acta* **1997**, *61*, 611–622. [CrossRef]
96. Biasatti, D.M. Stable Carbon Isotopic Profiles of Sea Turtle Humeri: Implications for Ecology and Physiology. *Palaeogeogr. Palaeoclimatol. Palaeoecol.* **2004**, *206*, 203–216. [CrossRef]
97. Roe, L.J.; Thewissen, J.G.M.; Quade, J.; O'Neil, J.R.; Bajpai, S.; Sahni, A.; Hussain, S.T. Isotopic Approaches to Understanding the Terrestrial-to-Marine Transition of the Earliest Cetaceans. In *The Emergence of Whales*; Thewissen, J.G.M., Ed.; Springer: Boston, MA, USA, 1998; pp. 399–422, ISBN 978-1-4899-0161-3.
98. Lindgren, J. Stratigraphical Distribution of Campanian and Maastrichtian Mosasaurs in Sweden—Evidence of an Intercontinental Marine Extinction Event? *GFF* **2004**, *126*, 221–229. [CrossRef]
99. Lindgren, J.; Siverson, M. The First Record of the Mosasaur *Clidastes* from Europe and Its Palaeogeographical Implications. *Acta Palaeontologica Polonica* **2004**, *49*, 219.
100. Jagt, J.W.M. Stratigraphic Ranges of Mosasaurs in Belgium and the Netherlands (Late Cretaceous) and Cephalopod-Based Correlations with North America. *Neth. J. Geosci.* **2005**, *84*, 283–301. [CrossRef]
101. Mulder, E.W.A.; Formanoy, P.; Gallagher, W.B.; Jagt, J.W.M.; Schulp, A.S. The first North American record of *Carinodens belgicus* (Squamata, Mosasauridae) and correlation with the youngest in situ examples from the Maastrichtian type area: Palaeoecological implications. *Neth. J. Geosci* **2013**, *92*, 145–152. [CrossRef]
102. Wendler, I. A Critical Evaluation of Carbon Isotope Stratigraphy and Biostratigraphic Implications for Late Cretaceous Global Correlation. *Earth Sci. Rev.* **2013**, *126*, 116–146. [CrossRef]
103. Cavin, L.; Tong, H.; Boudad, L.; Meister, C.; Piuz, A.; Tabouelle, J.; Aarab, M.; Amiot, R.; Buffetaut, E.; Dyke, G.; et al. Vertebrate Assemblages from the Early Late Cretaceous of Southeastern Morocco: An Overview. *J. Afr. Earth Sci.* **2010**, *57*, 391–412. [CrossRef]
104. Grigoriev, D. Redescription of *Prognathodon lutugini* (Squamata, Mosasauridae). *Proc. Zool. Inst. RAS* **2013**, *317*, 246–261. [CrossRef]
105. Gilmore, C.W. A New Mosasauroid Reptile from the Cretaceous of Alabama. *Proc. United States Natl. Mus.* **1912**, *41*, 479–484. [CrossRef]
106. Martin, J.E. A New Species of the Durophagous Mosasaur *Globidens* (Squamata: Mosasauridae) from the Late Cretaceous Pierre Shale Group of Central South Dakota, USA. In *The Geology and Paleontology of the Late Cretaceous Marine Deposits of the Dakotas*; Special Paper; Geological Society of America: Boulder, CO, USA, 2007; pp. 177–198, ISBN 978-0-8137-2427-0.
107. Dollo, L. Nouvelle Note Sur Les Vertébrés Fossiles Récemment Offerts Au Musée de Bruxelles Par M. Alfred Lemonnier. *Bull. Société Belg. Géologie PaléontologieEt D'hydrologie* **1889**, *3*, 214–215.
108. Schulp, A.S.; Polcyn, M.J.; Mateus, O.; Jacobs, L.L. Two Rare Mosasaurs from the Maastrichtian of Angola and the Netherlands. *Neth. J. Geosci.* **2013**, *92*, 3–10. [CrossRef]
109. Woodward, A.S. III.—Note on Tooth of an Extinct Alligator (*Bottosaurus belgicus*, sp. nov.) from the Lower Danian of Ciply, Belgium. *Geol. Mag.* **1891**, *8*, 114–115. [CrossRef]
110. Thurmond, J. New Name for the Mosasaur *Compressidens* Dollo, 1924. *J. Paleontol.* **1969**, *43*, 1298.
111. Bardet, N.; Suberbiola, X.P.; Iarochène, M.; Amalik, M.; Bouya, B. Durophagous Mosasauridae (Squamata) from the Upper Cretaceous Phosphates of Morocco, with Description of a New Species of *Globidens*. *Neth. J. Geosci.* **2005**, *84*, 167–175. [CrossRef]
112. Lingham-Soliar, T. Mosasaurs from the Upper Cretaceous of Niger. *Palaeontology* **1991**, *34*, 653–670.

MDPI AG
Grosspeteranlage 5
4052 Basel
Switzerland
Tel.: +41 61 683 77 34

Diversity Editorial Office
E-mail: diversity@mdpi.com
www.mdpi.com/journal/diversity

www.ingramcontent.com/pod-product-compliance
Lightning Source LLC
Chambersburg PA
CBHW061438210926
51850CB00008B/1542

* 9 7 8 3 7 2 5 8 4 7 7 1 6 *